LASER APPLICATIONS IN SURFACE SCIENCE AND TECHNOLOGY

LASER APPLICATIONS IN SURFACE SCIENCE AND TECHNOLOGY

H.-G. RUBAHN
MPI für Strömungsforschung, Bunsenstr.10, D-37073 Göttingen, Germany

JOHN WILEY & SONS
Chichester · New York · Weinheim · Brisbane · Singapore · Toronto

©B.G. Teubner Stuttgart 1996: Rubahn: *Laseranwendungen in der Oberflächenphysik und Materialbearbeitung*.Translation arranged with the approval of the publisher B.G. Teubner, Stuttgart, from the original German edition.
English translation copyright © 1999 by John Wiley & Sons Ltd,
Baffins Lane, Chichester,
West Sussex PO19 1UD, England

National   01243 779777
International (+44) 1243 779777
e-mail (for orders and customer service enquiries): cs-books@wiley.co.uk
Visit our Home Page on http://www.wiley.co.uk
or http://www.wiley.com

All rights reserved. No part of this publication may be reproduced, stored in a retrieval system, or transmitted, in any form or by any means, electronic, mechanical, photocopying, recording, scanning or otherwise, except under the terms of the Copyright, Designs and Patents Act 1988 or under the terms of a licence issued by the Copyright Licensing Agency, 90 Tottenham Court Road, London, W1P 9HE, UK, without the permission in writing of the Publisher.

Other Wiley Editorial Offices

John Wiley & Sons, Inc., 605 Third Avenue,
New York, NY 10158-0012, USA

WILEY-VCH Verlag GmbH, Pappelallee 3,
D-69469 Weinheim, Germany

Brisbane • Singapore • Toronto

Library of Congress Cataloging-in-Publication Data

Rubahn, H.-G. (Horst-Günter).
   [Laseranwendungen in der Oberflächenphysik und
Materialbearbeitung. English].
   Laser applications in surface science and technology / H.-G.
Rubahn.
      p.   cm.
   Includes bibliographical references and index.
   ISBN 0-471-98449-3 (cloth : alk. paper). — ISBN 0-471-98450-7
(pbk. : alk. paper)
    1. Laser beams.   2. Laser beams — Industrial applications.
3. Surfaces — Effect of radiation on.   4. Surface technology.
I. Title.
QC688.R8313   1999
620'.44 — dc21   98-51716
                                                                         CIP

British Library Cataloguing in Publication Data

A catalogue record for this book is available from the British Library.

ISBN 0 471 98449 3 (hardback)
ISBN 0 471 98450 7 (paperback)

Typeset by the author.
Printed and bound in Great Britain by Biddles Ltd, Guildford, Surrey.
This book is printed on acid-free paper responsibly manufactured from sustainable forestry, in which at least two trees are planted for each one used in paper production.

Contents

Preface vii

1 Light and Matter 1
  1.1 Coherent Light Sources 1
      1.1.1 Continuous lasers 13
      1.1.2 Pulsed lasers 15
  1.2 Surfaces 20
      1.2.1 Structure 21
      1.2.2 Dynamics 24
      1.2.3 Preparation 29
      1.2.4 Analysis 39

2 Adsorption, Desorption and Diffusion 53
  2.1 Laser-Induced Thermal Desorption 53
      2.1.1 MALDI 62
  2.2 Desorption Following Electronic Excitation 63
  2.3 Other Laser-Induced Desorption Processes 67
  2.4 Diffusion 73

3 Spectroscopy 81
  3.1 Spectroscopic Methods 81
      3.1.1 Absorption spectroscopy 85
      3.1.2 Laser-induced fluorescence 88
      3.1.3 Multiphoton ionization 92
      3.1.4 Spectroscopy with evanescent waves 94
      3.1.5 Two-photon photoemission 98
  3.2 Fluorescence Spectroscopy Near Interfaces 101
  3.3 Surface-Enhanced Raman Scattering 114
  3.4 Second Harmonic Generation 120
  3.5 Sum Frequency Generation 139
  3.6 Higher Order Wave Mixing 141

4 Dynamics and Ultrafast Studies 151
4.1 Electron Relaxation Dynamics 155
4.1.1 Surfaces and ultrathin films 155
4.1.2 Embedded nanoparticles 160
4.1.3 Supported metal clusters 161
4.2 Relaxation of Nuclear Motion 171
4.2.1 Vibrational state dynamics 171
4.2.2 Coherent-phonon oscillations 174
4.3 Ultrafast Phase Transitions 175
4.4 Photochemistry 177
4.5 Picosecond Electron Diffraction 181

5 Fundamental Laser Surface Treatment 185
5.1 Heating and Melting 185
5.2 Plasma Generation 211

6 Advanced Treatment 215
6.1 Laser Annealing 215
6.2 Laser Ablation 218
6.2.1 Polymers 219
6.2.2 Insulators 242
6.2.3 Metals 247
6.3 Laser Cleaning 250
6.4 Laser-Induced Periodic Structures 254
6.5 Laser–LIGA 259
6.6 Laser CVD 260
6.7 Pulsed Laser Deposition 263

7 Laser Medicine 267
7.1 Medical Laser Surface Treatment 267
7.2 Lasers in Ophthalmology 268
7.3 Basic Mechanisms 272

Bibliography 277

Index 331

Preface

A simple experiment, which uses nothing but a commercial laser pointer and two lenses, illustrates the peculiarities of the interaction of laser light with surfaces. The laser pointer is a source of coherent light, and the lenses serve to focus or to defocus it. Increasing the diameter of the laser beam by the use of lenses and irradiating a white sheet of paper results in a granular pattern with spots that appear sharply independent of spatial position. This so–called 'speckle pattern' (Dainty, 1975) is the result of scattering of coherent light from surface roughnesses. If one of the lenses is used to focus the laser beam into a tiny spot a few micrometers wide, and if this light hits a black area, then within the burn spot the strong concentration of energy results in rapid heating of the surface. Exchanging the continuous laser pointer with a pulsed laser leads to more dramatic phenomena, namely a melting of the surface and the explosive ablation of material. This kind of surface modification can be restricted to act on spatially and temporally very well defined spots. Consequently, the *materials treatment* of a solid was acknowledged as a potential area of application of laser light shortly after the experimental realisation of the laser principle in 1960 (Breech and Cross, 1962). Indeed, nowadays high–power continuous or pulsed lasers are widely used tools for drilling, marking, cutting or welding of materials. Low–power lasers are mainly used for spectroscopic or metrological applications.

In this monograph the physical aspects of laser–surface interactions are presented. These are at the heart not only of materials treatment, but also of laser spectroscopies on surfaces, which allow one to obtain important information about the electronic nature of the solid surface and of the rovibronic nature of its adsorbates. It is worth remembering that the specific surface properties (represented, for example, by the reconstruction of the atoms of the bulk lattice) are extended only a few atomic layers (of the order of a nanometer) into the bulk. On the other hand, laser light penetrates the solid much deeper. Typical values of penetration depth, given by the inverse absorption coefficient, are between 100 nanometers (metals), a few micrometers (strongly absorbing insulators) and several thousand meters (optical fibers). The laser–solid interaction, although often being accompanied by drastic modifications of the solid, therefore is not *per se* surface–specific. Under appropriate conditions (*e.g.*, small thermal diffusion length and ultrashort pulses), however, laser–induced effects such as heating might be localized at the surface. In the case of selective surface desorption or spectroscopy of adsorbates the surface sensitivity is also evident. Finally, the laser–induced *nonlinear* polarization, which adds reflected light waves at higher harmonics to the scattered light wave at the fundamental frequency, is in the case of centrosymmetric bulk materials also intrinsically surface-sensitive and its use has become an important surface analysis method.

A thorough understanding of the peculiarities of laser–surface interactions relies on a knowledge of the intrinsic properties of lasers and surfaces. Therefore, the first chapter of this monograph reminds the reader of some basics of laser and surface physics. The laser types most commonly used for the investigation of surfaces and some widespread surface–characterization methods are discussed. In addition, a few methods of surface preparation are mentioned, which have emerged only recently but have gained importance in the context of sophisticated new laser–surface treatments.

The use of laser techniques to study fundamental adsorption, desorption and diffusion processes on surfaces is discussed in Chapter 2. Important linear and nonlinear spectroscopic methods as well as the basic physical processes and some applications are explained in Chapter 3, since in the last two decades mainly *spectroscopy* with laser light has been introduced into many laboratories. Recently the development of ultrashort laser pulses has allowed one to observe directly the dynamics of electronic and vibronic surface excitation and relaxation processes. Thus Chapter 4 deals with the application of ultrashort laser pulses to surface structural, dynamic and photochemical studies. Since the methodology is fairly new, this chapter focuses on very recent experimental and theoretical findings. Possibly, some of these findings might have to be considered under a different point of view in the future, which obviously is a characteristic of active science.

The final three chapters are devoted to materials treatment, beginning in Chapter 5 with a discussion of fundamental *changes* of the plain surface induced by laser radiation, namely heating, melting and plasma generation. Here, the word 'plain' does not mean inevitably that the surface under discussion is free from adsorbates and that the discussed processes must proceed under vacuum conditions. It merely implies that the primary mechanism of interaction is not dominated by the coverage of the surface with, for example, a water film.

Structuring of the surface in terms of controlled crystallization processes, ablation of surface material, generation of periodic structures and generation of thin films is accounted for in Chapter 6.

Most of the basic mechanisms discussed in this monograph are used in laser medical applications. Here, a major field is the use of lasers in ophthalmology, which is highlighted as an example in more detail in Chapter 7.

Obviously, a rapidly expanding, innovative and important field such as the

Preface

interaction of photons with surfaces can by no means be discussed exhaustively in a single monograph. The purpose of the present book is to introduce possible laser-surface interaction mechanisms and processes and to make the reader curious to learn more about this fascinating scientific field. It is hoped that the book can provide some guidance about important new developments that one should have an eye on in the near future.

Details can be found in the literature that is extensively referenced throughout the text. In addition, recently published and more specific monographs might be consulted to fill the remaining gaps. For example, a discussion of various aspects of laser ablation processes (Miller, 1994); a compendium of articles on laser spectroscopy and photochemistry at metal surfaces (Dai and Ho, 1995); comprehensive discussions of the application of UV laser light to surfaces (Duley, 1996); a general description of recent advances in laser–tissue interactions (Niemz, 1996); an overview over laser-assisted processing on surfaces, including a most detailed discussion of ablation models (Bäuerle, 1996b); and industrial applications of lasers (Ready, 1997).

At this point I wish to thank my wife Katharina for her patience and encouragement and my colleague Frank Balzer for his indefatigable and inspired assistance as well as countless hours of extensive discussions. I am grateful also to all my other colleagues who have shared their knowledge with me. I extend special thanks to Professors E. Matthias and H.-L. Dai for constructive criticism and to Prof. J.P. Toennies for more than ten years of intellectual support at the MPI für Strömungsforschung.

Horst-Günter Rubahn

Göttingen, December 1998

1
Light and Matter

1.1 Coherent Light Sources

Lasers are sources of temporally and spatially coherent light. Compared with the light generated by a thermal lamp laser light is especially suited for applications in surface and materials science due to its coherence, intensity and due to the possibility to generate short and ultrashort pulses. In what follows we explain these properties in more detail.

Temporal and spatial coherence

Two partial waves of a light source are called *coherent* if their phase differences are constant, which leads, upon superposition, to interference phenomena. *Temporal coherence* is equivalent to the amplitudes of the emitted electromagnetic wave remaining constant over a considerably long time. This is demonstrated in Fig. 1 by comparison of the temporal evolution of the electric field amplitude emitted from a laser source and that emitted from a thermal light source.

The 'coherence length' L_c is defined as the difference in optical length that results in a phase difference between two partial waves of smaller than π; hence that difference that allows interferences to occur. The corresponding 'coherence time' $\tau_c = L_c/c$ is related to the coherence length via the speed of light c in the investigated medium.

The coherence length of thermal light behind an interference filter with spectral bandwidth of 1 Å at λ=500 nm is about 0.8 mm; that of a laser with a bandwidth of 10 MHz about 10 m. Hence the temporal coherence of a laser is several orders of magnitude higher compared with that of light generated by a thermal lamp and its light is well suited for applications which rely on interference phenomena such as optical holography.

High temporal coherence results in light that is strongly monochromatic. However, strongly monochromatic light is not necessarily highly temporally coherent since the photon statistics (and hence the amplitude fluctuations) of light from a lamp (Bose–Einstein statistics) and that from a laser (Poisson statistics) are fundamentally different. The degree of monochromasy can be

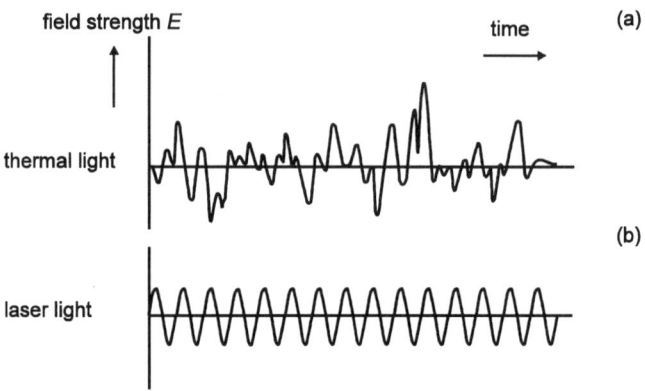

Figure 1 Temporal change of the electromagnetic field strength E for a thermal light source (a) and a laser (b).

defined via the ratio between the linewidth $\delta\nu$ and the carrier frequency ν, $\delta\nu/\nu$ ('spectral resolution') Using interference filters it is possible to obtain $\delta\nu/\nu \approx 10^{-3}$ without significant loss of intensity. The corresponding spectral range is shown in Fig. 2. An unstabilized Ar$^+$ laser allows us to define the wavelength and thus the energy to within 0.2 Å. A commercial, actively stabilized dye laser provides a spectral resolution of 10^{-5} Å or $\delta\nu/\nu \approx 10^{-9}$; more sophisticated systems used in basic research projects operate with linewidths in the Hz range, hence $\delta\nu/\nu \approx 10^{-15}$. This spectral resolution is comparable to that obtainable in Mößbauer spectroscopy and allows us, in principle, to determine the energies of substrate- or adsorbate-electronic states with a resolution of better than 1 μeV.

High *spatial coherence* means that laser light is parallel over a wide range. Analogously to temporal coherence light is called 'spatially coherent' if the phase difference between two partial waves is smaller than π. This condition is fulfilled for all partial waves within the Airy diffraction cone with radius a. The limiting angle for the diffraction cone of zeroth order is $\Theta_{max}=1.22\lambda/a$. It corresponds to the divergence angle of light. Laser light, which is emitted spatially coherent from an aperture with large diameter $2a$ (*e.g.*, being the outcouple mirror of the laser), has a corresponding small divergence and can be strongly focused down to the diffraction limit. Fig. 3 illustrates this behavior by showing the phase front of a light wave from a thermal source, which results from the superposition of the phase fronts of independent spherical waves (Huygen's principle). This corresponds to strong divergence. In the case of laser light the divergence is given by diffraction of the light wave at

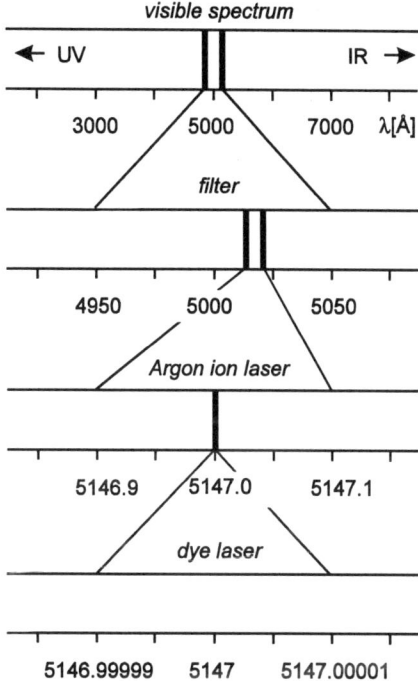

Figure 2 The monochromasy of laser light is demonstrated by comparing the spectral width (energetic uncertainty) of the emitted light from a single mode dye laser and an Ar$^+$ laser with that obtained by narrowing the visible spectrum with an interference filter.

the exit mirror of the laser cavity.

Due to the design of the laser resonator, laser light is emitted in most cases in the fundamental Gaussian mode TEM$_{0,0}$. Upon focusing the Gaussian beam at its waist with a thin lens of focal length f a beam waist with radius w_0^* is generated (Yariv and Yeh, 1984):

$$w_0^* = \frac{\lambda f}{\pi w_\mathrm{L} n} \cdot \left[1 + \left(\frac{\lambda f}{\pi w_\mathrm{L}^2 n}\right)^2\right]^{-0.5}. \qquad (1.1)$$

In this equation, n is the index of refraction of the material and w_L the radius of the laser beam that hits the lens. In the near field, which is dictated by Fresnel diffraction, the beam waist increases until the beam area is twice that inside the focus (Fig. 4, 'Rayleigh-range'). This distance $s_0 = \pi\, w_0^2\, /\lambda$ defines the parallelity of laser light ($2s_0$ is usually called the 'confocal parameter'). In the far field, for $s \gg s_0$, Fraunhofer diffraction results in geometrical optics

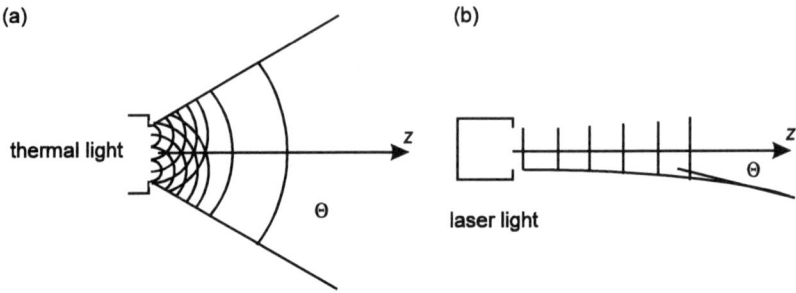

Figure 3 Divergence of light, emitted from a thermal (a) and a laser source (b). The divergence angle Θ is of the order of fractions of a mrad.

with a constant angle of divergence $\Theta \approx \lambda/\pi n w_0$.

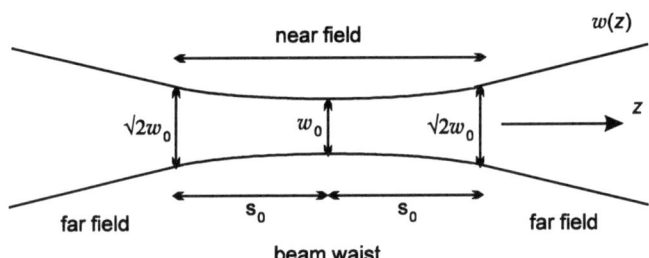

Figure 4 Focal range of a Gaussian beam (TEM$_{0,0}$–mode), propagating along the optical axis z. The Rayleigh range is denoted by s_0, the beam waist by w_0. See Eq. (1.11).

A helium–neon laser usually comes with a resonator that is constructed such that the beam waist is located at the position of the outcouple mirror. Given an aperture of 1 mm this results in a confocal parameter of $2s_0$=2.48 m and a divergence of Θ=0.7 mrad. Focusing this laser with a lens of focal length 3.5 mm would result in a focus diameter at the position of the beam waist of 1.8 µm.

Intensity
Lamps rarely achieve powers of more than kilowatts. Peak powers of

1.1. COHERENT LIGHT SOURCES

lasers, on the other hand, easily exceed 10^5 W during continuous operation ('cw') or 10^{11} W during pulsed operation. These high peak powers make a regime of laser–matter interaction readily accessible, where nonlinear optical interactions occur. The 'intensity' of a light source is defined as power per solid angle. In the context of interaction with surfaces the 'irradiance' (power per area) is a more useful quantity, since it affects directly the irradiated material.[1] It is important to distinguish average power per area (W/cm^2) from pulsed power per area or 'fluence' (J/cm^2), i.e., the energy density, if one wants to estimate the influence of the laser on the surface. Since laser light might be focused without significant losses of intensity close to the diffraction limit, average powers of 10^5 to 10^{11} W result in irradiances[2] between 10^{12} and 10^{18} or even 10^{20} W/cm^2 (Mourou, 1997). To visualize the magnitude of this irradiance one might compare it with the low irradiance of the sun (solar constant 0.137 W/cm^2), which nevertheless results on a clear day in a well detectable heating of the skin. An irradiance of 10^{12} W/cm^2 corresponds to a photon flux density of more than 10^{30} photons per square centimeter and second. Such a flux density corresponds to a black-body temperature of about one million degrees Kelvin — a temperature which is found only under extreme conditions, such as the center of a star. Obviously, such high irradiances open up a wide range of new physical phenomena including Gigagauss magnetic fields, terabar light pressures and relativistic plasmas (Perry and Mourou, 1994).

For a pulsed laser with an energy of 1 mJ/pulse the number of photons per pulse is $5.034 \cdot 10^{12} \cdot \lambda$ [nm] and the number of photons per second $4.72913 \cdot 10^{21} \cdot \lambda$ [nm]$/\tau$ [ns] with τ meaning the pulse length. The power P [kW] is given by $939.435/\tau$ [ns], and the average power can be calculated by multiplication with the repetition rate. Especially for the observation of nonlinear optical effects the induced electrical field strength is important, which is given for a laser with beam waist w_0, which irradiates a medium with index of refraction n, by $(245/w_0$ [cm]$) \cdot \sqrt{P[\mathrm{kW}]/n}$ [V/cm] (Birge, 1983).

Pulse generation

Light pulses might be generated from a thermal light source by a mechanical shutter or by electrical switching of the lamp voltage. The method is limited in the former case by the mechanical stability of the chopper wheel and results in pulses of a few microseconds (μs) length. In the latter case the minimum pulse length is limited by secondary processes during the discharge

---

[1] It is also instructive to compare the average spectral irradiances of thermal lamps and lasers. For example, at 633 nm a Xe arc lamp with 150 W electrical input power and an area of the radiation source of 0.5×2.2 mm^2 provides 0.24 Wcm^{-2}sr^{-1}nm^{-1}, whereas a He-Ne laser with 2 mW optical output power and a circular aperture of 0.63 mm diameter provides 10^8 Wcm^{-2}sr^{-1}nm^{-1}. The sun would have 2.2 Wcm^{-2}sr^{-1}nm^{-1}.

[2] Nonlinear interactions such as self-defocusing usually restrict the maximum obtainable irradiance.

and ultimately by the electronic lifetime of the spontaneous photon emission process (nanoseconds, ns). This problem can be overcome by using light that is dominated by a stimulated emission process, i.e., by exploiting the intrinsic properties that make laser light a very special kind of light. For example, by the use of mode coupling techniques pulses with lengths of the order of femtoseconds (1 fs = 10^{-15} s) can be generated. Since light travels within, for example, 100 fs just 30 µm, those ultrashort pulses allow one to map spatially the motion of particles with extremely high spatial resolution. For example, the fragments of a molecular photodissociation process depart with a relative velocity of typically a few thousand meters per second (corresponding to the nuclear motion correlated with a vibrational frequency of 10^{12} s^{-1}). Hence a temporal delay of 100 fs between a pump pulse (which excites the molecule into a dissociating state) and a probe pulse (which checks, for example, via laser-induced fluorescence the state of the initial molecule or its fragments) corresponds to a distance of a few tenths of a nanometer (Fig. 5). This is the characteristic length scale of a typical atom–molecule or molecule–surface potential energy curve (cf. Fig. 17). The use of two laser pulses, mutually delayed by a few tens of femtoseconds, thus allows one to observe the bond breaking of a molecule or the desorption from a surface in real time.

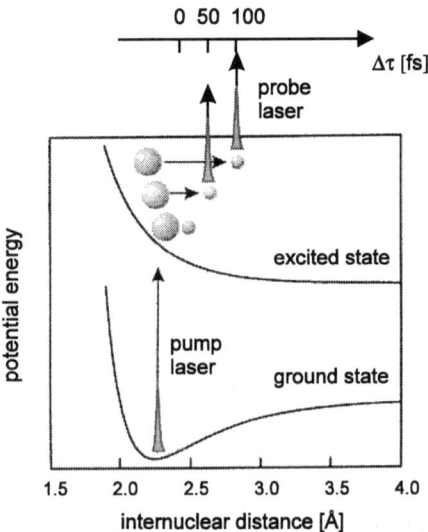

Figure 5 Laser-induced bond breaking of a molecule, which has been adsorbed on a surface. The departing fragments are probed by pulses from a second laser, which are temporally delayed by $\Delta \tau$.

1.1. COHERENT LIGHT SOURCES

Light emitted from a thermal source and laser light differ *in principle* by their *photon statistics*. Photons are bosons (even-numbered spins) and their statistics are thus described by a Bose–Einstein distribution with a mean number of photons per mode of

$$\bar{n} = \frac{\langle E \rangle}{h\nu} = \frac{1}{exp(h\nu/k_B T) - 1}. \qquad (1.2)$$

Hanbury-Brown and Twiss verified these statistics experimentally 1957 by counting the number of coincidences of detecting two photons from a light beam which has been split by a beam divider into two parts (Hanbury-Brown and Twiss, 1957a; Hanbury-Brown and Twiss, 1957b). The observed Bose–Einstein distribution reveals a 'bunching' of the photons, such as is demonstrated in Fig. 6. Here, each photon that is counted as a function of time is plotted as a horizontal line. The probability of finding a photon in the case that another one has already been observed is higher than the probability of finding a photon in the case that no photon has been observed before. The corresponding probability distribution for the observation of a certain number of photons, given an average number of photons per mode, then is very broad. The variation of numbers of photons is small for small $\bar{n} \ll 1$ and is proportional to $\bar{n}$: the photons behave like classical particles, which are independent of each other. For large $\bar{n}$ the wavefield character of thermal light results in a quadratic dependence of the variation on $\bar{n}$.

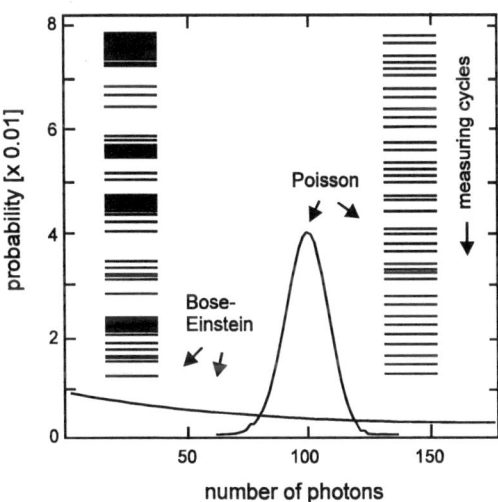

Figure 6 Bose–Einstein and Poisson distribution of observed photons as a function of time. Probability distributions are plotted as solid lines for an average number of photons per mode of $\bar{n} = 100$.

Even in the case of good spatial and temporal filtering the intensity amplitude of light is strongly fluctuating for large $\bar{n}$ (which is what one is interested in). In contrast, the light from a laser can be described by a Poisson distribution, which is the distribution found for a forced oscillator. The photons no longer 'bunch' (Fig. 6, right-hand side), and the probability distribution has a maximum at the average number of photons per mode. The mean square fluctuation is proportional to $\bar{n}$ even for large $\bar{n}$, meaning that the photons in that case too behave like independent particles ('coherent light'). Laser photons obey these statistics since they have been generated mainly by *stimulated emission* and not via spontaneous emission such as the photons of thermal light.

> LASER = Light Amplification via Stimulated Emission of Radiation.

A LASER is based on three main elements (Fig. 7)(Schawlow and Townes, 1958; Maiman, 1960): an *energy pump* (*e.g.*, a flash lamp or an electrical plasma discharge), which irradiates the *laser-active medium* (gas, solid, liquid or plasma) and that way excites particles from the ground into high lying electronic states, from which they relax by emitting photons. A *resonator* exerts a selective feedback to the system by restricting the number of allowed eigenfrequencies (modes) that can start oscillating and by coupling the emitted photons back to the emitting particles. If the total gain per round trip of photons in the resonator exceeds the total loss, then the condition for self-excited oscillation is fulfilled and the laser starts lasing.

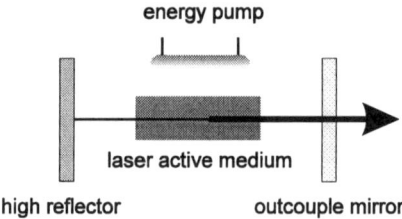

Figure 7 Fundamental constituents of a laser. High reflector and outcouple mirror build a resonator.

Let us assume for simplicity a two-level system. Since we are going to talk about large photon numbers and since we are mainly interested in processes that are stimulated by a classical radiation field it is well justified to describe the change of photon number n inside the resonator by a classical rate equation:

1.1. COHERENT LIGHT SOURCES

$$\frac{dn}{dt} = N_2 A_{21} + N_2 \rho(\nu) B_{21} - N_1 \rho(\nu) B_{12} - \frac{n}{t_1}, \quad (1.3)$$

where N_1 describes the population of the ground state and N_2 the population of the excited state (Fig. 8).

The population N_2, which has been excited by the energy pump, decays with the rate for spontaneous emission, A_{21}, which is inversely proportional to the spontaneous lifetime τ_2.[3] Simultaneously the radiation field with the spectral energy density $\rho(\nu) = n(\nu)\bar{n}h\nu$ induces an emission from N_2 ('stimulated emission') and an absorption from state $|1\rangle$ into state $|2\rangle$ ('stimulated absorption'). The rate for stimulated processes is determined by the energy density of the field and the Einstein B coefficient, which is connected to the A coefficient via $n(\nu)$:

$$B_{21} = \frac{A_{21}}{n(\nu)h\nu}. \quad (1.4)$$

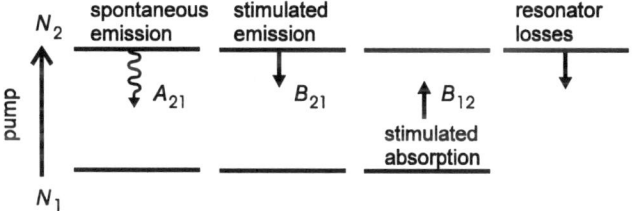

Figure 8 Two-level scheme with photon-induced pump processes and Einstein coefficients for spontaneous emission, A_{21}, stimulated emission, B_{21}, and stimulated absorption, B_{12}.

Emission and absorption might be different due to different statistical weights $g_j = 2J+1$, where J describes the angular momentum of a free atom in energy eigenstate E_j, $j=1,2$, $B_{12} = (g_2/g_1)B_{21}$. Stimulated emission processes increase the number of photons inside the resonator, and stimulated absorption processes decrease it. Another factor, which decreases the number of photons in the resonator, is the lifetime t_1 of the photons due to the reflectivities R_1 and R_2 of the resonator mirrors. If L denotes the length of the resonator and c is the speed of light inside it, then one finds

---

[3] In the case of equal statistical weights of ground and excited states the Einstein A coefficient is directly related to the oscillator strength f: $A_{21}[\text{s}^{-1}] = 6.67 \times 10^{13} \, f/(\lambda \, [\text{nm}])^2$.

$$t_1 = \frac{2L}{\ln(R_1 \cdot R_2)c}. \tag{1.5}$$

Here, diffraction losses due to the finite dimensions of the mirrors and their surface roughness have been neglected.

For photon amplification by the laser-active medium the number of photons must increase inside the resonator as a function of time:

$$\boxed{dn/dt > 0} \quad ,$$

which is equivalent to $\bar{n} > 1$. This latter condition is easily deduced: if one neglects resonator losses, then Eq. (1.3) can be rewritten in the case of equal statistical weights of ground and excited states ($B_{21} = B_{12}$):

$$dn/dt = N_2 A_{21} + \rho(\nu) B_{21} D, \tag{1.6}$$

introducing the 'inversion' $D = N_2 - N_1$. In thermal equilibrium we find that N_1 is always larger than N_2 and that the number of photons will decrease due to absorption by the radiation field. Photon amplification ($dn/dt > 0$) can be obtained solely far from thermal equilibrium, in the case that the probability for stimulated emission is higher than the probability for spontaneous emission processes, $B_{21}\rho(\nu) > A_{21}$, or $\frac{A_{21}}{n(\nu)h\nu}\rho(\nu) > A_{21}$, i.e.,

$$\frac{n(\nu)\bar{n}h\nu}{n(\nu)h\nu} = \bar{n} > 1. \tag{1.7}$$

Fig. 9 shows, that $\bar{n}$ is always much smaller than 1 for moderate radiation temperatures and especially in the case of visible light (photon energy of the order of eV, $k_B T$ of the order of tens of meV). The short wavelength of visible light results in very large numbers [4] of modes per volume element, over which the photons are distributed. In that case spontaneous emission processes dominate the stimulated processes by several orders of magnitude.

The laser resonator restricts the number of possible modes by external boundary conditions, resulting in a strong increase in the number of photons per mode. The tangential component of the electromagnetic field strength vanishes for transverse electromagnetic (TEM) waves on the mirrors of the resonator (Fig. 10). Hence the condition for standing waves inside the resonator is that even-numbered multiples of half of the wavelength, $q \cdot \frac{\lambda}{2}$, fit into given resonator length L.

Hence we find q longitudinal modes along the resonator axis (z-axis) with a mode spacing

$$\delta\nu_{q,q+1} = \frac{c}{2L} \tag{1.8}$$

---

[4] The number of modes per volume element and frequency interval $d\nu$ is given for transverse electromagnetic waves by $n(\nu)d\nu = \frac{8\pi\nu^2}{c^3}d\nu$. For $\lambda = 500$ nm one finds within a typical Doppler width of $d\nu = 1$ GHz a mode density of 3×10^{14} m^{-3}.

1.1. COHERENT LIGHT SOURCES

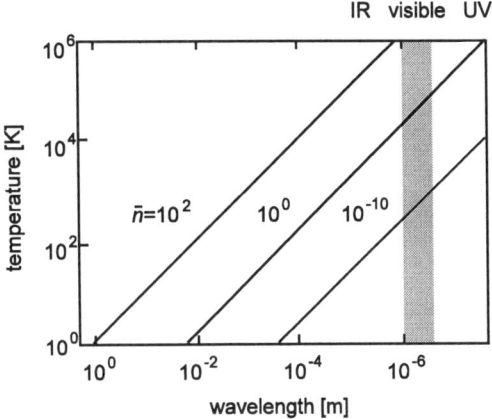

Figure 9 Calculated number of photons per mode $\bar{n}$ in thermal equilibrium as a function of radiation temperature and frequency.

(in the case of a resonator with plane mirrors ('Fabry–Perot')).

The field strength distribution for the fundamental mode of a beam of width d and wavelength λ in (x,y)-direction is a radial symmetric Gaussian distribution (cf., Fig.4)

$$E_{00}^2 \propto I_{00}(x,y) = I_0 \exp\left(-\frac{(x^2+y^2)}{w^2}\right) \tag{1.9}$$

with mode radius

$$w^2(z) = \frac{\lambda d}{2\pi}\left[1 + \left(\frac{2z}{d}\right)^2\right] \tag{1.10}$$

and beam waist

$$w_0 = \sqrt{\frac{\lambda d}{2\pi}} \tag{1.11}$$

instead of a focal point (cf., Fig. 4).

Higher order transverse modes are constructed by the superposition of spherical waves and can be classified by a radial mode number p and an azimuthal mode number l. According to Fig. 10, p describes the number of rings with minimum intensity and l the number of lines (azimuths) with minimum intensity. The number of transversal modes can be restricted without large effort to the fundamental mode TEM$_{00q}$ by inserting apertures inside the resonator or by using appropriately dimensioned resonators (long and small diameters). For longitudinal mode selection wavelength selective elements such as filters or etalons have to be inserted inside the resonator.

'Stable' resonators are characterized by the property that an ideal beam close to the axis never leaves the resonator if one uses perfectly reflecting

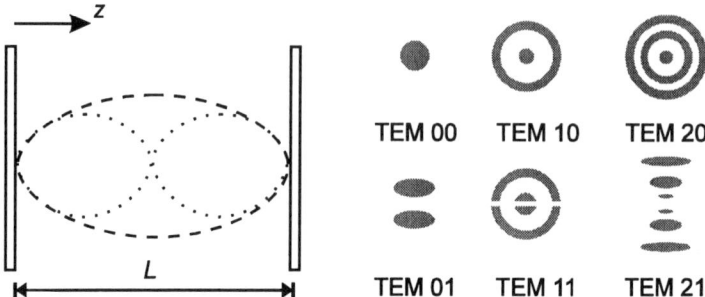

Figure 10 Longitudinal modes of a Fabry-Perot resonator (left-hand side) and transverse modes of a confocal resonator with spherical mirrors (right-hand side). For the transverse electromagnetic modes, TEM$_{pl}$, the areas of maximum field strength are plotted in dark color.

mirrors. Among this group of resonators one finds the plane (Fabry–Perot) and the confocal resonators (distance between mirrors = radius of curvature). Of course, the mode structure of those two most-often used resonators is different: confocal resonators possess a better transversal mode selection compared with plane resonators, but higher transversal modes are degenerate with longitudinal modes. The diffraction losses of confocal resonators are significantly smaller compared with those for plane resonators, and the alignment is much simpler since for multiple reflections of the photons in the resonator the parallelity of the mirrors is not as critical as in the case of Fabry–Perot resonators. The advantage of better use of the laser-active volume in the plane resonator is partially cancelled by the disadvantage of possible spatial holeburning effects. Hence, most lasers are equipped nowadays with confocal, long and narrow resonators. Plane resonators are used mainly for spectroscopic applications since in that case it is highly advantageous to have the possibility to vary the longitudinal mode spacing and hence the spectral resolution by simply modifying the length of the resonator, cf. Eq. (1.8).

General applications in surface science make it useful to discriminate between two types of lasers:
a) lasers with high spectral resolution but low peak power, and
b) lasers with high peak power but low spectral resolution.

Most of the continuous lasers belong to the first category, while most pulsed lasers provide only low temporal coherence. Since pulse length τ and bandwidth $\Delta\nu$ are inversely proportional, one finds that a pulse length of one nanosecond results in an energetic uncertainty or bandwidth of more than 100 MHz. Ultrashort laser pulses of a few femtoseconds thus correspond

1.1. COHERENT LIGHT SOURCES

to spectral uncertainties of a few ten nanometers. Fortunately, for the identification of electronic band structures in or on the surface of solids this is very often a sufficient spectral resolution.

1.1.1 Continuous lasers

For surface materials treatment in most cases CO_2 or Ar^+ lasers are applied. Spectroscopy of surfaces or of adsorbates on surfaces is mainly performed by the use of dye, Ti/sapphire or solid state lasers, the wavelength of which can be scanned over a more or less wide spectral range.

CO_2 laser

Fig. 11 shows term schemes of some lasers which are used for surface treatment. Obviously — and in contrast to the simplified two-level treatment that we have performed above — in most cases multi-level systems are used. In order to obtain inversion usually at least a three-level-system is necessary. The laser transition then occurs from the excited state, in which the population was accumulated, via stimulated emission into an energetically lower lying state, from which the population can be removed very quickly. Both processes (establishment of the population in the upper and depletion of the population in the lower laser state) have to occur fast compared with other photon-loss processes that might occur in the resonator.

Another important characteristic of a laser system is its efficiency η, which might be defined as the ratio between invested energy for pumping (mainly electrical) and obtained laser energy (Fig. 11). In the case of molecular lasers — especially for the CO_2 laser — this ratio is high (up to 0.3), since the energy for the pump photons is not much higher as compared with the energy of the laser photons.

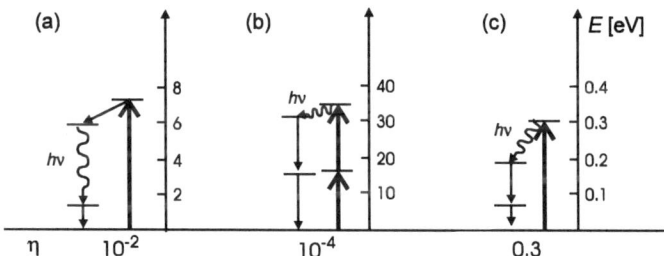

Figure 11 Efficiency η and term schemes of typical lasers that are important in surface treatment: (a) KrF excimer laser (pulsed), (b) Ar^+ laser, (c) CO_2 laser. Thick arrows denote pump transition, wavy arrows laser transition and thin arrows radiative and non-radiative relaxation channels.

A CO_2 laser consists of a glass tube, which is filled with a mixture of CO_2 (1 Torr), N_2 (1 Torr), He (6 Torr) and H_2O (0.1 Torr). A high-voltage discharge is ignited in this mixture, which excites 10% – 30% of the nitrogen molecules into the ($\nu=1$) first vibrationally excited state. This long-living state ($\tau > 0.1$ s) is quasi-resonant ($\Delta E=2$ meV) with the first vibrationally excited state ν_3 (asymmetrical stretch vibration) of CO_2. The latter state can thus easily be populated via resonant energy transfer. The relaxation of this state into an excited ν_2-state (symmetrical stretch vibration) of CO_2 leads to the emission of photons at a wavelength of 10.6 μm. An efficient resonator for this wavelength can be obtained by use of germanium-coated mirrors, which combine high reflectivity with high damage threshold in the infrared spectral range.

The continuous power of a CO_2 laser might be of the order of kilowatts. Due to the long lifetime of the excited level, Q-switches delivering pulses of $m$us length can be applied efficiently and thus peak powers of more than 10^5 W can be obtained. Pulsed CO_2 lasers (TEA, transverse excited atmospheric pressure laser) with pulse lengths of a few 100 ns deliver up to several 10 to 100 MW peak power.

Ar^+ laser

The laser-active medium is low-density argon gas in which a plasma discharge is ignited. The current density for the self-excitation of the discharge is 50 A/cm^2. In order to achieve such a high current density, beryllium-oxide discharge tubes with small diameters of 1 – 4 mm are used. Since the hot plasma would destroy the tubes, axial magnetic fields focus it in the middle zone of the tube.

Inside the discharge the Ar atoms are first ionized by collisions with electrons and then excited into the 4p-states. In the course of relaxation into the 4s-states, photons are emitted which are amplified inside the resonator (Fig. 12). The 4s-states relax very fast into the ionic ground state (350 ps), and thus a high inversion can conveniently be obtained. The low efficiency of the Ar^+ laser of $\eta \approx 10^{-4}$ is mainly due to the low ionization probability. The most intense lines of this laser (up to 10 W) are in the blue-green and ultraviolet spectral ranges, which — due to their high energy — make the laser especially interesting for pumping dye or Ti-sapphire lasers. If one uses discharges in other gases (*e.g.*, krypton) laser action in the red spectral range can be obtained.

The hot plasma results in a homogeneous linewidth of 800 MHz due to saturation and collisional broadening, which is significantly larger compared with the longitudinal mode spacing (150 MHz for resonators with 1 m length). As a consequence, one expects that different longitudinal modes can start oscillating statistically independent, which reduce mutually their inversion ('mode cannibalism'). While a transversal mode selection to the $TEM_{0,0}$ mode via an aperture in the case of clean tube windows is simple, especially due to the

1.1. COHERENT LIGHT SOURCES

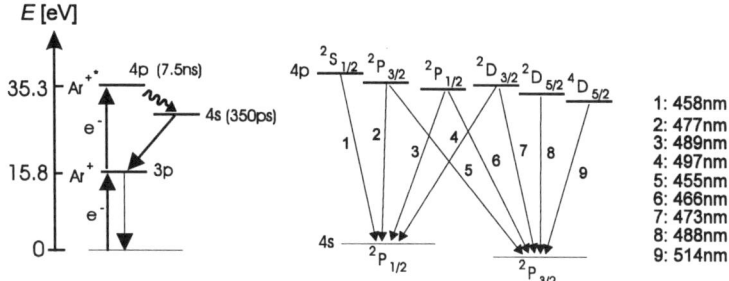

Figure 12 Simplified term scheme and strongest transitions of an Ar$^+$ laser. The radiative lifetimes of two electronically excited states are denoted in parentheses.

long, small resonator, a longitudinal mode selection is difficult due to instabilities of the plasma, thermal movements of the mirrors, etc.

1.1.2 Pulsed lasers

In materials treatment, where high energy densities are necessary, the most widely applied lasers in the infrared spectral range are pulsed CO_2 lasers, in the UV excimer lasers, and in the visible spectral range frequency doubled Nd:YAG lasers. For spectroscopic applications wavelength-tunable dye or solid state lasers (*e.g.*, Ti-sapphire lasers) are used, which are pumped via excimer or Nd:YAG lasers.

Short laser pulses might be generated either via *Q-switch* or via *mode coupling* of continuous lasers. In what follows some fundamental pulse generation mechanisms are discussed, but the wide range of coherent light sources that deliver nowadays ultrashort pulses (pico- to femtoseconds) is only briefly touched. More exhaustive overviews can be found in (Herrmann and Wilhelmi, 1987; Diels and Rudolph, 1996).

Q-switch

The most simple realization of a Q-switch consists of a light-switch inside the resonator, which is closed a long time for the build-up of high inversion and is then instantaneously opened. In that way the population is accumulated in the upper laser level far above the steady-state inversion level. In most cases in solid state lasers Pockels cells are used as optical switches. The birefringence in the corresponding crystals is changed via the linear electro-optical effect as a function of an external electrical field and hence they either transmit or block the polarized radiation inside the resonator. Switching times (nanoseconds are possible) have to be adjusted to the relaxation rates of the laser transitions so as to avoid detrimental effects such as 'spiking' (the occurrence of multiple peaks) or low output power. In the case of the ruby laser the lifetime of the excited state is 3 ms, meaning that the switch can be kept close for about 100 ns. During this period a high inversion is generated, which is depleted following the opening of the switch, resulting in a giant pulse of several nanoseconds length and MW peak power.

Nd:YAG laser

The Nd:YAG laser is a solid state laser. An yttrium–aluminum–garnet ($Y_3Al_5O_{12}$) crystal is doped with about one volume-percent triply positively charged neodymium-ions, which are electronically excited via irradiation by a flash lamp (Fig. 13). The radiation that is emitted during the decay into the ground state is amplified in a linear resonator with a Q-switch (Pockels cell). Due to the interaction of the excited ions with the crystal lattice the initially sharp electronic states are broadened into electronic bands. This spectrally broad gain profile ($\Delta \nu \approx$120 GHz) results in mode cannibalism — similar to the case of the Ar^+ laser — and makes longitudinal mode selection a difficult task. On the other hand, the broad gain profile allows one to obtain extremely short pulses via mode coupling. If one 'seeds' the high power laser with a low power laser that has a narrow spectral profile it is possible to narrow the spectral profile of the Nd:YAG laser by more than an order of magnitude ('injection seeding'). The spectrally narrow seed laser effectively lowers the threshold for a special mode to start oscillating inside the resonator and thus suppresses the mode cannibalism.

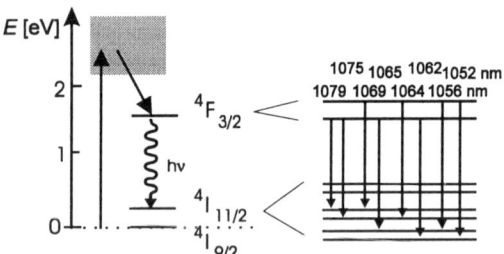

Figure 13 Level scheme and most important transitions of a Nd:YAG laser.

In the case of gas lasers (*e.g.*, Ar^+ or CO_2 lasers) one might use an acousto-optical modulator (AOM) as a Q-switch instead of a Pockels cell. This modulator serves simultaneously as an outcouple mirror ('cavity dumping').

1.1. COHERENT LIGHT SOURCES

The AOM consists of a birefringent crystal (, $e.g.$, TeO_2), inside which a standing acoustic wave with a modulation frequency of a few tens of MHz is generated via a piezoelectric transducer. This acoustic wave with wavelength λ_{sound} diffracts an incoming light wave of wavelength λ according to the Bragg condition

$$\lambda = 2\lambda_{sound} sin\Theta. \qquad (1.12)$$

The amplitude of this standing wave (and thus the diffracted intensity) can be modulated by an AC voltage which is applied to the transducer. In that way a sequence of light pulses of up to several MHz can be obtained via a sequence of ultrasound pulses, whereas the pulse width ranges from 10 to 100 ns. Compared with cw operation a power enhancement of not more than a factor of 10 – 100 can be achieved since no additional gain is generated in the laser-active medium. Short pulses that are generated that way, however, can be easily pulse-amplified by applying subsequent Nd:YAG or excimer lasers.

Excimer laser
The artificial word 'excimer' is an acronym for 'excited dimer' and characterizes molecules, which fulfill in an ideal manner the conditions for the generation of high inversion: a thermally instable ground state (well depth of the order of $k_B T$) and excited states, which decay in short time (nanoseconds) (Fig. 14). Such molecules are, for example, rare gas–halogen combinations RgX (Rg=Ar, Kr, Xe ; X=F, Cl), which are generated in self-contained high pressure gas discharges of, for example, 98% buffer gas (He, Ne), 1.5% Rg and 0.5% X in an ion–recombination reaction

$$Rg^+ + X^- + M \rightarrow RgX^+ + M; \; M=He,Ne \qquad (1.13)$$

or a harpooning-reaction

$$Rg^* + XH \rightarrow RgX^* + H. \qquad (1.14)$$

Since one needs a high pump power of MW/cm^2 to achieve a high inversion density and since the discharge ends after a few tens of nanoseconds due to instabilities, one has to switch high currents in short time. This becomes possible by introducing fast high-voltage switches ('thyratrons'). However, although significant improvements in this technology have been obtained during the last few years (for example 'magnetic switch control', MSC), the switches still are one of the most prominent sources of fault for excimer lasers. Also, one has to note that a short discharge time means that the photons can oscillate only a few times inside the resonator. This is detrimental for obtaining a high degree of mode selection and results in low spatial and temporal coherence of the emitted light; the divergence is about 1×3 mrad and line widths amount to a few hundred wavenumbers. On the plus side, the efficiency is about $\eta \approx 0.01$,

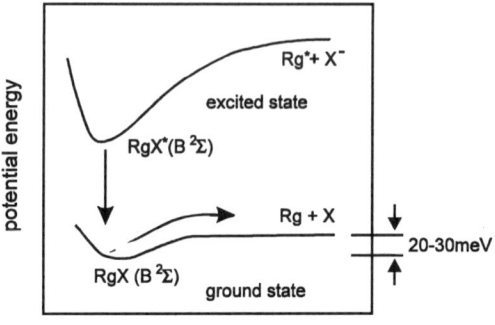

Figure 14 Ground state and a possible excited state of a rare gas (Rg)–halogen (X) dimer. The ground state is unstable at room temperature ($T=300$ K) since the potential well depth (32 meV for XeCl*) is of the order of $k_B T$ (26 meV).

and high laser powers of a few Joules in the UV spectral range[5] are possible. If one focuses this energy on a spot of a few tens of micrometers diameter, one ends up with irradiances of more than 10^{14} W/cm². Ultrashort pulse excimer lasers allow one to obtain irradiances that are even a few orders of magnitude higher.

Mode-coupling

The most elegant and 'generic' way of generating short pulses by a laser is the active or passive coupling of the phases ϕ_q of the longitudinal resonator modes q, which have the spacing $\Delta\omega_{q,q+1} = c/2L$, where L is the resonator length. In the case of statistically independent, fluctuating phases the resulting field strength is given by

$$E(t) = \sum_{q=-n}^{n} E_{0q} \exp(i(\omega_0 + q \cdot \Delta\omega_{q,q+1})t + i\phi_q), \quad (1.15)$$

i.e., the total intensity equals the sum of the intensities of individual modes,

$$I \propto \sum_q |E_{0q}|^2. \quad (1.16)$$

On a time scale this is equivalent to Gaussian distributed noise. If one achieves a constant phase relation between individual modes,

---

[5] Commercially available wavelengths are: 157 nm (F_2), 193 nm (ArF), 248 nm (KrF), 308 nm (XeCl), 351 nm (XeF).

1.1. COHERENT LIGHT SOURCES

$$\phi_{q+1} - \phi_q = \phi = \text{const.}, \tag{1.17}$$

then the resulting field strength is given, just as in the case of diffraction, by a grating[6] by

$$E(t) = A(t)\exp(i\omega_0 t), \tag{1.18}$$

where the carrier wave oscillates fast with the light frequency ω_0 and the amplitude of the field varies as

$$A(t) = E_0 \frac{sin((2n+1)/2)(\Delta\omega_{q,q+1}t + \phi)}{sin(1/2)(\Delta\omega_{q,q+1}t + \phi)}. \tag{1.19}$$

If one plots the intensity distribution on a temporal scale, then this results in maxima whenever the denominator approaches zero (i.e., whenever $\Delta\omega_{q,q+1}t + \phi = m \cdot \pi$). Hence, the pulse maxima are separated from each other by $\tau_p = 2L/c$, and they are more intense if more modes interact phase-coupled (the intensity scales with the number of modes squared). Instead of obtaining a number of photons in statistically independent modes one essentially works with a single pulse oscillating in the resonator, which contains all modes oscillating with fixed phase difference. The pulse length (the distance between maxima and minima) is $\Delta\tau_p \approx 1/\delta\nu$ if the pumping is strong enough to allow all modes within the gain profile $\delta\nu$ to oscillate. It becomes shorter if the gain profile becomes wider. For an Ar$^+$ laser with $\delta\nu \approx 5$ GHz one achieves pulses of 200 ps length, for a Nd:YAG laser even 5 ps and for dye-lasers with $\delta\nu$ of the order of 10 nm several tens of femtoseconds.

Active mode coupling is often achieved by implementing an AOM into the resonator, the frequency Ω of which is a multiple of the longitudinal mode spacing $c/2L$, $\Omega = 2\pi c/2L$. Due to the nonlinear rate equation the amplitude modulation of the initially oscillating mode at ν_0 results in side bands at $\nu_0 + \delta\nu$, where further modes can start oscillating synchronously. A technical problem in the context of this method are the strong restrictions concerning the stability of the resonator length in order to conserve the resonance of the mode frequency at the AOM frequency.

This problem can be overcome by introducing a saturable absorber as a passive mode locker instead of the AOM into the resonator. For high light intensities this material (usually a dye) becomes transparent, resulting in a loss-modulation of the light pulses that oscillate inside the resonator. The most intense mode experiences the lowest absorption and hence the highest gain. In that way all modes that start oscillating are passively locked to the strongest mode and one obtains again very short laser pulses. Since the saturable absorber degrades the initial part of the pulse (it is transparent

---

[6] Instead of an angle variable, one has to calculate with time as the variable.

only above a certain intensity) and since the gain medium degrades the final part of the pulse (the inversion is quickly reduced following the start of the pulse oscillation) one obtains an even shorter pulse compared with active mode coupling (the whole gain profile is used). Of course, this only works if the oscillation frequency of the pulse in the resonator is larger than the relaxation time of the laser-active medium, which is fulfilled usually in the case of dyes. An obvious disadvantage of the saturable absorber is its small spectral working range (usually centered around 600±20 nm).

During the last decade the development of a new laser-active medium with a high gain and a high damage threshold (titanium–sapphire, Ti:Al_2O_3) has allowed one to develop spectrally widely tunable, mode-coupled lasers with 'intrinsic' saturable absorbers (Kapteyn and Murnane, 1994). In these KLM ('Kerr lens mode locking' or 'self mode locking') lasers, Ti:Al_2O_3 is the laser-active medium, which is excited with an Ar^+ laser. If the light intensities inside the Ti:Al_2O_3 crystal are sufficiently high, then the optical Kerr effect results in self-focusing of the laser beam. The index of refraction n of the medium then possesses, besides the intensity-independent, linear component $n_0(\omega)$ an intensity-dependent, nonlinear term $n_2 I(t)$.

The phase $\phi = \omega t - \omega n z/c$ depends on $I(t)$ and the same is the case for the frequency $\omega = d\phi/dt$. The frequency of the initial part of the pulse is decreased, $dI/dt > 0$, and the frequency of the final part is increased, $dI/dt < 0$. Thus the pulse is spectrally broadened by this kind of self-phase modulation. At the same time the linear part of the index of refraction, $n_0(\omega)$ for normal dispersion $(dn_0/d\lambda < 0)$, smears out the pulse in time. If one now compresses the pulse again by a grating or a prism combination inside the resonator, then a bandwidth-limited pulse is generated which is shorter than the initial pulse since its spectral width is larger. In that way pulse lengths of 17 fs have been generated (Huang et al., 1992). Again, an obvious problem with these kinds of lasers, besides the small pulse-power of nJ, is their sensitivity to mechanical shocks, which easily disturb the self-mode-lock operation.

1.2 Surfaces

We call that region of the solid where the symmetry and the bondings between particles are intrinsically different from that in the bulk, a *surface*. As a consequence, in many cases surfaces are characterized by a high reactivity (Somorjai, 1994). The diameter of a laser spot on a surface amounts to at least a few micrometers and the penetration depth of visible light is at least a few ten nanometers. Hence many phenomena occurring in the course of laser–surface interactions can be described in macroscopic terms. However, if one is interested in more subtle physical phenomena, such as symmetries in the nonlinear optical response of the surface, diffusion processes, reactivities or adsorption and desorption kinetics (Masel, 1996), then the crystallographic, *microscopic* structure of the surface and its elementary electronic and lattice

1.2. SURFACES

dynamics become important.

1.2.1 Structure

Crystalline surfaces can be described as the result of a cut through the bulk of the solid and the following separation of both halves. This picture is the basis for characterizing the orientation of crystal surfaces by *Miller's indices*.[7] The indices (h, k, l) define atomic lattice planes and directions within the crystal (Fig. 15). They are the smallest even-ordered triples of numbers, which have the same ratio as the inverse of the axes that are represented as multiples of the lattice vectors (a, b, c). Hence they describe a vector, which is pointing perpendicularly to the multitude of atomic lattice planes. It is called the 'reciprocal lattice vector' $\vec{g}_{hkl}$. The length of the vector represents the distance between the planes. The total of all reciprocal lattice vectors is called the *reciprocal space*.

The introduction of Miller's indices is especially useful for the discussion of diffraction methods, which provide an image of the surface in reciprocal space and are used for the determination of surface structures. For example, the Laue condition for constructive interference of scattered waves $\vec{k}_f$ as a result of incoming waves $\vec{k}_i$ on a crystal is

$$\vec{k}_f - \vec{k}_i = \vec{g}_{hkl}. \tag{1.20}$$

The density of packing of a surface depends on the crystal type and orientation. For a fcc crystal (face centered cubic) the (111) surface has the most dense packing of spheres, whereas for a bcc crystal (body centered cubic) the (110) surface has the most dense packing (Fig. 15). The asymmetry of the surroundings of the surface atoms changes the structure as compared with the bulk. For example, the strongly localized bondings of a semiconductor are cut at a surface ('dangling bonds'). In order to minimize the free surface energy the surface atoms rearrange within the surface plane ('reconstruction'). The lack of atoms above the surface means that the forces that are exerted on the surface atoms by the atoms below it are not compensated. This results in a contraction or dilatation of the uppermost surface layer distances (relaxation').

At the surface, due to the elimination of translational symmetry, new electronic states, 'surface states', are generated inside the electronic band gap (Davison and Steslicka, 1992). In contrast to bulk electronic states, which show an oscillating amplitude in the (x,y,z)-direction and are described by real wavevectors, the amplitude of the surface states decays exponentially in the direction of the surface normal (z). These states do not possess dispersion in the z-direction (Fig. 16) and are described by complex wavevectors $\underline{k}$ in

---

[7] Indices of planes are denoted usually by parentheses, directions by curly brackets.

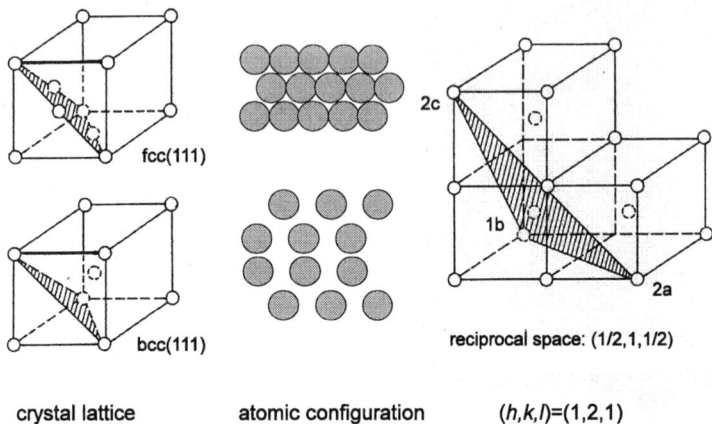

Figure 15 Crystal lattice, reciprocal lattice vectors and Miller's indices (h,k,l). On the left-hand side the (111)-planes are shown for a fcc and a bcc crystal and in the middle of the figure the corresponding atomic packings. Obviously, for a fcc crystal the (111) surface contains the most dense packing, which is not the case for a bcc crystal. On the right-hand side the derivation of the nomenclature for the shaded higher-index plane (121) is exemplified. Here, (a,b,c) are the basis vectors of the translational lattice, which means that the (121) plane is described by $(2a, 1b, 2c)$, giving the reciprocal triple $(1/2,1,1/2)$ and the smallest integral-order triple $(1,2,1)$.

the form of Bloch waves $\psi_k = e^{ikr} \cdot u_k(r)$ with $u_k(r)$ being lattice-periodic functions. Their localization at the surface means that they are especially sensitive to adsorbate coverages of the surface.

Surface states were postulated for the first time in basic theoretical work by Tamm (Tamm, 1932) and later by Shockley (Shockley, 1939). Usually, a 'Shockley state' is distinguished from a 'Tamm state' by means of the interaction strength between the electrons and the involved surface atoms. Shockley states result from the sudden change in periodicity of the potential at the surface. They are generated by the interaction of the *free* electron gas with the surface and can extend several layers deep into the solid. Hence the energetic position of the Shockley states is independent of the true shape of the potential. Tamm states, on the other hand, are localized by the bonding potential of the electrons to the surface atoms, which differs from the bonding potential inside the solid due to the asymmetrical termination of the potential. Both Tamm and Shockley states can be well described by Bloch waves with amplitudes that are exponentially decaying into the crystal and into the

1.2. SURFACES

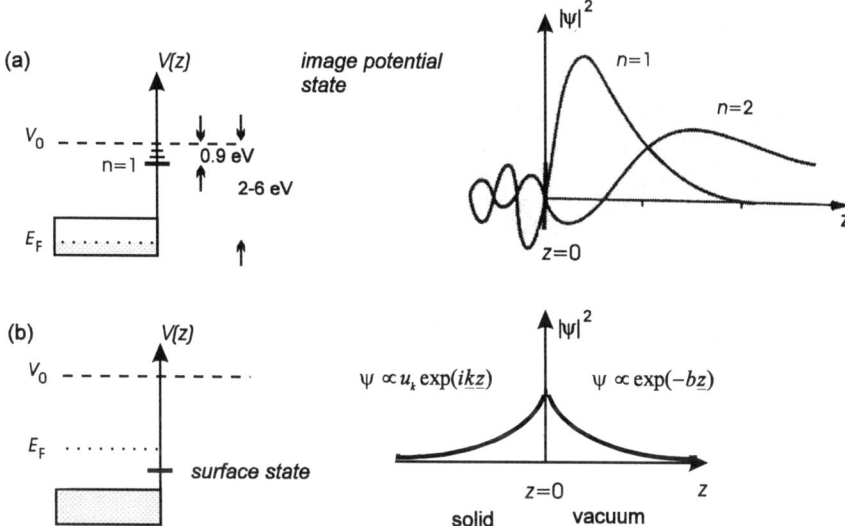

Figure 16 (a) Energetic position (left-hand side) and probability density (right-hand side) of image potential states ($n=1,2$) at a metal surface. (b) Possible surface state of a semiconductor. E_F represents the Fermi energy and V_0 the vacuum energy (determined by the work function of the solid, which is between 2 and 6 eV for a metal). The parameter b in the argument of the exponentially decaying wavefunction outside the semiconductor is given by $\sqrt{\frac{2m}{\hbar^2}(V_0 - E)}$. Note that on the left-hand side of $|\psi|^2$ only the envelope of the function is plotted.

vacuum (Fig. 16).

Another class of surface states, which differ principally from the above discussed states, are the image potential states. If one removes an electron from a metal surface, attractive Coulomb forces are generated due to the image potential. The potential $V(z) = -q^2/4z$ supports a Rydberg series of bound states (Fig. 16b). In order to determine the energetic position of these unoccupied states, one might irradiate the surface with electrons, which in the case of resonance with a surface state results in the emission of photons of defined energy ('inverse photoemission'). By the use of conventional photoelectron spectroscopy (irradiation with a high-energy photon and observation of the emitted electrons; see Section 1.2.4) only occupied states can be detected since electrons can be excited only from those states. However, one might excite electrons from an occupied bulk state into an unoccupied surface state by a laser pulse. A second laser pulse then detects the laser-populated states (two-photon photoemission; see Chapter 3, Section 3.1.5).

This is an alternative method to inverse photoemission spectroscopy.

In semiconductors surface states play an important role for the electronic properties of the surface. In metals their contribution to the total electronic density of states at the Fermi level is less important. Nevertheless, in combination with defects or adsorbates they can change the optical response of the surface. An example is the resonance enhancement of the nonlinear surface susceptibility of second order, $\chi_S^{(2)}$ (Chapter 3, Section 3.4), in the case that the frequency of the exciting laser corresponds to a dipole-allowed transition between the occupied bulk or surface state and an unoccupied surface state.

1.2.2 Dynamics

Dissociative adsorption

Consider a molecule A_2 which is adsorbed on a surface with a small binding energy of less than 100 meV ('physisorbed'). The attractive force between A_2 and the surface results from the dispersion interaction, i.e., the coupling of electronic motion in the surface and in the molecule ('van der Waals force'). Molecules with permanent electric moments experience an additional attraction due to induction and electrostatical forces. Those forces too are a function of the distance between A_2 and the surface. In Fig. 17 the *potential* is plotted, the derivation of which with respect to distance z results in the exerted force. The attractive forces result, with decreasing distance to the surface, in an increase in the negative potential energy: A_2 becomes bound. If the molecule is close enough to the surface that its electronic cloud overlaps partially with the electronic cloud of the surface, then the Pauli exclusion principle (two electrons in the same eigenstate is not allowed) results in a repulsive interaction. Thus the molecule cannot penetrate the surface. The equilibrium between attractive and repulsive forces results in a potential well, the deepest point of which marks the equilibrium distance of the particle from the surface. For physisorbed molecules this distance is about 0.4 nm. The depth of the potential well corresponds to the binding energy of the adsorbed particle, and its shape determines the energy levels of possible discrete vibrations of the adsorbate with respect to the surface.

If electrons are exchanged between adsorbate and substrate in the vicinity of the surface, then this is usually referred to as 'chemisorption'. In this case the binding energy is of the order of a few electronvolts. The equilibrium distance is between 1 and 3 Å. In Fig. 17 the process of *dissociative adsorption* is sketched, which transforms a physisorbed two-atomic molecule into two chemisorbed atoms (Zangwill, 1988). Since in this example, which corresponds to H_2/Cu(100), the dissociation energy of the molecule A_2 is smaller as compared with the energy required for chemisorption, the molecule will chemisorb dissociatively only if one adds an activation energy E_A. This might be achieved by laser excitation or via adsorption of the molecules with high translational energy, *e.g.*, from a molecular beam onto the surface. In the lat-

1.2. SURFACES

ter case the transferred energy from the normal component of the collision en

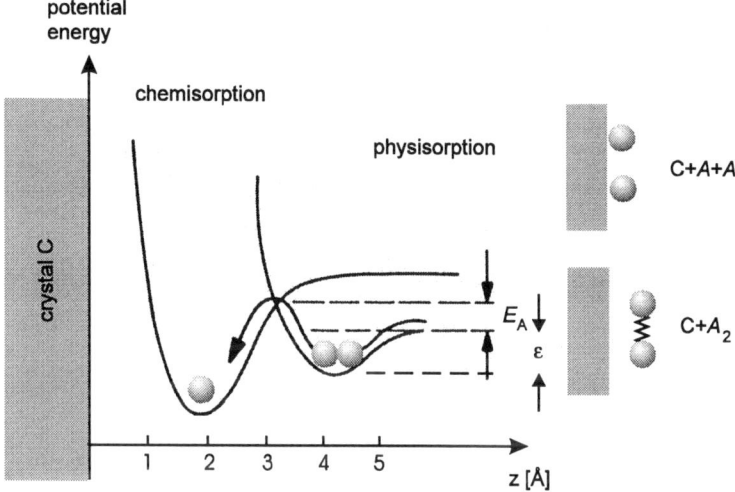

Figure 17 Dissociative adsorption of a molecule A_2 on a surface with activation energy E_A. A physisorbed molecule with well depth ϵ is transformed into two chemisorbed atoms.

Surface phonons

The energy quanta of the collective vibrations of the lattice are called *phonons*. They can be described as waves in the crystal which are polarized longitudinally (parallel to the direction of evolution) or transversally (perpendicular to the direction of evolution). Usually the vibrational energies of longitudinally polarized phonons are larger compared with those of transversally polarized phonons. The vibrational energy depends on the wavevector; hence the waves show *dispersion*.[8] The collective translational motion of the lattice atoms results in one longitudinal and two transversal 'acoustic' phonon branches (parallel or 'in phase' motion of neighboring lattice atoms). If the number of atoms per unit cell, n, is larger than 1, then in addition $3(n-1)$ 'optical' branches exist. In the latter case neighboring atoms move counterparallel or 'out of phase'. In the case that they carry different electric charges a temporally varying dipole moment is created, allowing the interaction with electromagnetic radiation. This is the case, for example, for alkalihalides such as NaCl and explains the historical name of these modes.

---

[8] An exception are the dispersionless Einstein modes, which are due to adsorbate atoms that vibrate perpendicularly to the surface plane without mutual interaction.

Among the optical modes (n—1) are polarized longitudinally and $2(n$—1) transversally.

Surface phonons are characterized by their amplitude being decaying exponentially normal to the surface and changing periodically parallel to the surface. They result as projections of bulk phonons or are generated as unique new modes by the existence of the surface. Thus, in addition to the bulk-phonon bands in (x,y,z)-coordinates surface-phonon bands in (x,y)-coordinates are generated, which have different wavelengths compared to the bulk phonons according to their projection to the surface. The energetically lowest lying surface phonon, which is related to a transversal acoustical phonon of the bulk, is called a 'Rayleigh phonon'.[9]

Surface plasmons
A useful approximation for the description of the electronic properties of a metal relevant to conduction is to describe all of the electrons collectively as a plasma of density N_e, which oscillates longitudinally with respect to the crystal lattice with the bulk-plasma frequency $\omega_p^2 = N_e e^2/(m_e \epsilon_0)$. Following the initial excitation by an external field the electric field itself that is generated by the expansion acts as a restoring force. The quanta of the electronic oscillation have energies $h\nu$ of the order of a few electronvolts.

At the border between a metal with the complex dielectric function $\epsilon_1 = \epsilon_1' + i \cdot \epsilon_1''$ and the vacuum with the constant ϵ_2 the electromagnetic continuity conditions for p-polarized light allow the existence of a new class of solutions of Maxwell's equations, the 'surface plasmons' (sometimes called 'surface plasmon polaritons'). They move along the surface with a wavevector $k_x = 2\pi/\lambda_x$ which depends on frequency according to the dispersion relation

$$k_x = \frac{\omega}{c}\sqrt{\frac{\epsilon_1 \epsilon_2}{\epsilon_1 + \epsilon_2}}. \qquad (1.21)$$

Along a smooth surface the surface plasmons decay via damping in the adsorbing metal (imaginary part of the dielectric function). The decay length is, for example, 22 μm for a silver surface excited at 514 nm. The initial energy of the plasmons is transferred into Joule-heating of the metal. In the case of a rough surface the plasmons can also decay radiatively or by loss of coherence of the initial collective excitation due to electron electron scattering. This might reduce the decay length drastically (Perner et al., 1997; Klein-Wiele et al., 1998).

In the direction of the surface normal (z-direction) the field amplitude decays exponentially with the characteristic length (1/e—depth) (Raether, 1988)

---

[9] For example, the long-wave components of earthquakes are Rayleigh waves.

1.2. SURFACES

$$z_i = \frac{\lambda}{2\pi}\sqrt{\frac{\epsilon_1' + \epsilon_2'}{\epsilon_i' \epsilon_i'}}, \quad i = 1, 2. \tag{1.22}$$

For silver at a wavelength of 600 nm the decay depth in the metal (index 1) is 24 nm, the decay depth in the vacuum (index 2) is 390 nm. Thus the plasmon oscillation is localized at the surface of the metal and close to the initial place of excitation.

In Fig. 18 the dispersion relations of surface plasmons (Eq. (1.21)) are plotted. For large wave vectors k_x (short wavelengths) the surface plasma frequency is a factor of $\sqrt{2}$ smaller[10] as compared to the corresponding bulk plasma frequency: $\omega_s = \omega_p/\sqrt{2}$. If one is to cover the metal surface by a dielectric film with dielectric function ϵ_2, the plasma frequency is reduced by a factor $\sqrt{1 + \epsilon_2}$.

The straight line in Fig. 18 is the 'light line', representing the dispersion relation for light in a vacuum: $k_x = \omega/c$. Since the dispersion curve for surface plasmons is restricted to the right-hand side of the light line and no crossing point exists, on an ideally flat surface, surface plasmons cannot decay radiatively into photons and they cannot be excited by light.[11] In order to couple light to surface plasmons an increase in the wavevector by Δk_x is necessary for a given photon energy (corresponding to a rotation of the light line to the right in Fig. 18). This might be achieved by modification of the surface with a lattice structure or more generally by invoking surface roughness, which adds additional reciprocal lattice vectors to the initial wavevector of the light.

Another useful method for coupling light to surface plasmons, which was introduced as early as 1968, is to exploit the total internal reflection inside a prism which has been attached to the metal film and which has the dielectric function ϵ_p (attenuated total reflection, ATR). In the so-called Otto configuration (Fig. 19a) a gap of the order of the incoming wavelength (generated by using thin spacer layers) is applied between the glass prism and the metal film that is to be investigated (Otto, 1968). In the Kretschmann–Raether configuration (Fig. 19b) the metal film of thickness 10–50 nm (Kretschmann and Raether, 1968; Kretschmann, 1971) is adsorbed directly on the prism. The prism defines a second light line (inside the prism) with the dispersion relation $k_x = \sqrt{\epsilon_p}\omega/c$, which is rotated by $\sqrt{\epsilon_p}$ to the right in Fig. 18 and thus has a crossing point with the dispersion curve for the surface plasmons. Between the two light lines (within the region of

---

[10] For a sphere the dipole plasma frequency is within the quasi-static limit even a factor $\sqrt{3}$ smaller compared with the bulk plasma frequency. The quasi-static limit is valid for particles with radii which are significantly smaller compared with the wavelength of the exciting light; in other words, for a spatially constant phase of the electromagnetic field.
[11] An excitation is possible via irradiation with electrons since the momentum transfer is varied inside the solid by variation of the scattering angle.

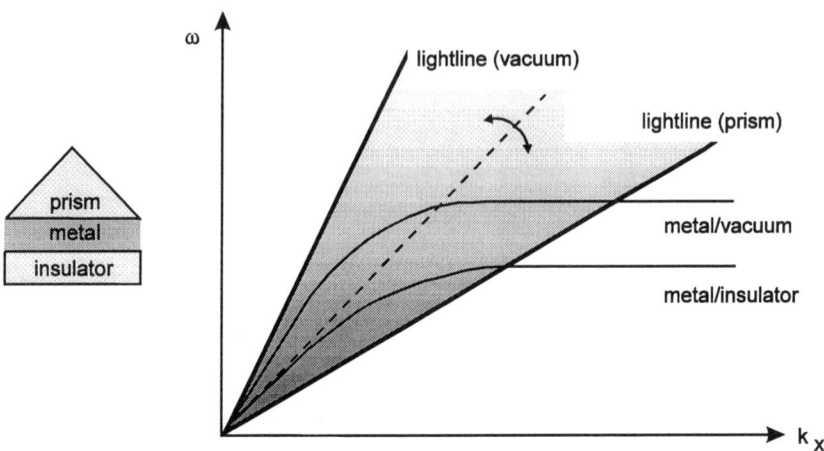

Figure 18 Dispersion relations for surface plasmons at the metal/vacuum and metal/dielectric interfaces. Within the shaded area surface plasmons can be excited by the ATR method at both interfaces. The dashed line characterizes the variation in the effective photon wavevector as a function of angle of incidence of the light with respect to the normal on the prism hypotenuse.

total internal reflection at the prism hypotenuse) plasmons can be excited at the metal/vacuum interface.[12] The exact crossing point is given by the angle of incidence Θ_i on the film, which defines the projection $sin\Theta_i$ of the momentum onto the surface. The advantage of the Otto method is that the surface at which plasmons are generated is not disturbed by the prism. The Kretschmann–Raether method is experimentally simpler since an ultrathin spacer layer is not necessary.

If one changes the angle of incidence for a given metal film, a minimum in the reflected intensity can be observed (Fig. 19), the position of which being characteristic for the involved dielectric function (Eq. (3.7)). The depth of the minimum, i.e., the efficiency with which the energy is coupled into the film, is determined by the film thickness since the thickness defines the degree of damping of the incident and reflected waves and also the phase relation between the partial waves that are generated at the prism/metal and metal/vacuum interfaces. For a silver film at an excitation wavelength of

---

[12] The total reflection results in an evanescent wave at the prism surface, which can be phasematched to the plasmon wave.

1.2. SURFACES

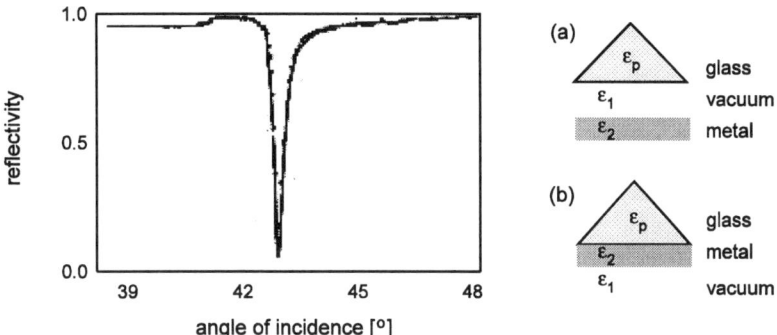

Figure 19 Left-hand side: ATR minimum at the silver/air interface following irradiation by an HeNe laser with 1.5 mW. Reprinted from *Surf. Sci.*, (Rothenhäusler et al., 1984), Copyright 1984, with permission from Elsevier Science. Right-hand side: Two possible configurations for the excitation of surface plasmons in metal films. (a) Otto configuration; (b) Kretschmann–Raether configuration.

500 nm minimum reflectivity is observed for a thickness of 55 nm.

The ATR minimum corresponds to an enhancement V_{max} of the electromagnetic field at the metal/vacuum interface:

$$V_{max} = \frac{1}{\epsilon_2} \frac{2|\epsilon_1'|^2}{\epsilon_1''} \frac{\sqrt{|\epsilon_1'|-1}}{1+|\epsilon_1'|}. \quad (1.23)$$

For a rough silver film irradiated at 600 nm one obtains a field-enhancement by a factor of 200, which can effectively be used for nonlinear optical spectroscopy (cf., Fig. 61 and Chapter 3, Section 3.3) or to reflect thermal atoms by generating a 'surface-plasmon' mirror (Esslinger et al., 1993), which is based on the radiation force that atoms experience in the evanescent wave (Cook and Hill, 1982).

1.2.3 Preparation

In air and at room temperature adsorbates cover most surfaces within a short time (especially water molecules with their high binding energies of, for example, 80 meV on stainless steel). The initial adsorption rate can be defined via the *sticking coefficient*, which is the ratio between the number of collisions that lead to adsorption to the total number of collisions between the background gas and the surface. The sticking coefficient depends on surface temperature and increases with decreasing temperature.[13] If one evacuates the surroundings of the crystal, then the adsorption rate is decreasing since the number of available adsorbate molecules decreases. A common rule-of-thumb is that in a vacuum of 10^{-6} mbar and for a sticking coefficient of unity, in one second one monolayer of background gas adsorbs.[14]

Especially for intrinsic surface sensitive laser spectroscopic methods such as SHG (second harmonic generation) preparation of the surface before the laser experiments is of eminent importance for reproducibility of the results. But even less specific processes such as laser-induced desorption might be strongly affected by the crystallographic nature of the irradiated surface, especially due to the site specificity of the bond-breaking efficiency (Tanimura and Kanasaki, 1998).

Depending on the type of solid, different methods of preparation are applicable. For *conducting and semi-conducting surfaces* sputtering by irradiation with high-energetic ions (1–2 kV) can be used. 1.8 kV He ions, for example, transfer on a tungsten surface at normal incidence a recoil energy of about 150 eV. This energy leads to explosive desorption of the adsorbates and topmost surface layers and results in a clean but microscopically very rough surface. In order to regain the initially smaller surface roughness[15] the surface is reconstructed by annealing it to several hundred degrees Celsius.

In selected cases well-defined *insulator* surfaces can be obtained by the cleavage of single crystals in an ultrahigh vacuum. Irradiation with intense pulses of UV laserlight from an excimer laser also results in clean surfaces. In the latter case the fast heating of the upper surface regions or the explosive evaporation of the uppermost layer, which has been penetrated by the laser light, causes the cleaning (see Chapter 6, Section 6.3).

Another way of preparation is to grow the material of interest via evaporation or sputtering on an appropriate substrate made of different material (heteroepitaxy). In that way single crystalline surfaces can be formed from materials, from which bulk crystals are difficult to grow. Depending on the binding energies between adsorbate atoms and between the adsorbate and substrate and depending on growth conditions (evaporation rate, surface temperature) different growth modes might be observed: layer-by-layer growth ('Frank–Van der Merwe') if the adsorbate–substrate binding (adhesion energy) dominates; cluster growth ('Volmer–Weber') if the binding energy between the adsorbate and substrate is much smaller compared with the binding energy between the adsorbate and adsorbate (cohesion energy); or mixed layer-

---

[13] In the case of dissociative chemisorption, as shown in Fig. 17, the sticking coefficient might increase with increasing temperature since the higher surface temperature helps to overcome the barrier to dissociative adsorption.

[14] The rate of 10^{-6} Torr· s is called "1 Langmuir". A monolayer corresponds to about $10^{14} - -10^{15}$ atoms per cm^2, depending on the geometry of the adsorbate binding sites.

[15] The topic 'surface roughness' is discussed in more detail in the context of the SERS effect (Chapter 3, Section 3.3).

1.2. SURFACES

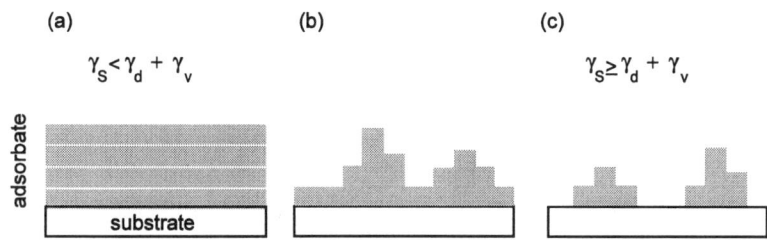

Figure 20 Growth modes: (a) Frank–Van der Merwe, (b) Stranski–Krastanoff, (c) Volmer–Weber. The limiting conditions of surface tensions are marked as discussed in the text.

and-cluster growth ('Stranski–Krastanov') if the adhesion energy initially dominates, but then lattice distortion between the adsorbate and substrate enforces island growth (Venables, 1994). If one assume the islands on the surface to be droplets, then a more quantitative discrimination between the different growth modes can be obtained by introducing the contact angle ϕ of the border of the droplet with the surface plane. It is

$$\gamma_d \cdot \cos\phi = \gamma_s - \gamma_v, \tag{1.24}$$

where γ_d is the surface tension between droplet and vacuum, γ_s that between substrate vacuum and γ_v that of the interface between droplet and substrate. Obviously, if $\phi \to 0$, then a layer is grown on the surface, while for $\phi > 0$ droplets or islands are grown. The Stranski–Krastanoff mode results if initially the elastic energy is used to relax the adsorbate lattice in order to fit with the substrate lattice. After a monolayer has been formed, this energy is no longer necessary for lattice relaxation and can be used for droplet formation. A more extended discussion, including supersaturation phenomena, can be found in (Reichelt, 1988). The application of scanning tunneling microscopy and extensive numerical simulations has recently allowed one to understand the early stages of epitaxial metal film growth (Brune, 1998) and also of three-dimensional island growth (Jensen et al., 1998) in much detail.

Figure 21 shows a possible scenario for the island growth mode, namely the growth of metal clusters on dielectric surfaces following the thermal evaporation of metal atoms. Those systems are of special interest for optical applications since they consist of an array of strongly polarizable particulates, the optical properties of which depend on morphology and electronic structure (cf. Chapter 4, Section 4.1.3). Due to the small binding energy between the metallic atoms and the insulator surface (about 0.1 eV) the adsorbed atoms are very mobile. Hence they will migrate over the surface and stick at defects of the surface (dislocations, borders, edges, etc.) or at already adsorbed atoms

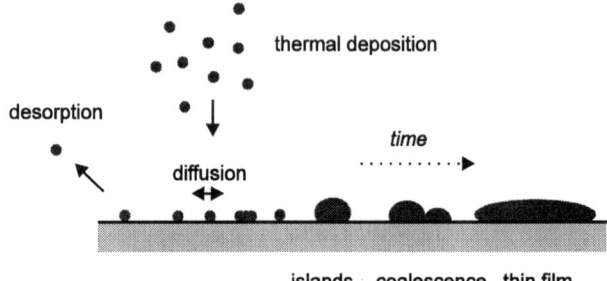

Figure 21 Thermal adsorption of metal atoms on an insulator with defects. Depending on the sticking coefficient, some of the atoms will be reflected from the surface and some will be adsorbed. The adsorbed atoms have an initial high mobility even at surface temperatures well below room temperature and form aggregates (islands or 'clusters') at sites with high binding energy. With increasing adsorption time the islands coagulate and form rough metallic films.

(binding energy between metal atoms of the order of eV), where they start forming islands. The islands coagulate and form ultrathin metallic films, the roughness of which depends on the exact conditions of the growth (surface temperature, surface defect density, flux of the adsorbing atoms, kinetic energy, etc.). This kind of metal film growth has been investigated over many years for alkali metals (Rasigni and Rasigni, 1973), and especially for ultrathin gold film growth, *e.g.*, (Schmeisser and Harsdorff, 1970; Golan et al., 1992; Levlin et al., 1997).

Of course, thermal-evaporation-induced growth is by its nature a statistical process and lacks a strong regularity of the generated structures. Thus, for example, the islands possess a very broad distribution of sizes (Andersson and Granqvist, 1977) (Chapter 4, Section 4.1.3, Fig. 101). In the case of mixed semiconductor islands, the process can be largely improved by more sophisticated methods of molecular beam epitaxy, such as growth interruption (Ma et al., 1998). If stored in ambient air, the average radius a_0 of the islands increases by diffusion limited mass transfer (thermally driven Ostwald ripening (Zinke-Allmang et al., 1992)) with characteristic time t_c as

$$a_0(t) = a_0(t=0)\left(1 + \frac{t}{t_c}\right)^{0.25}, \qquad (1.25)$$

and this results in an ensemble of well separated 'self-organized dots' (SODs). Repulsive interactions between growing semiconductor islands on metals lead to gratings with atomic dimensions (Pascal et al., 1997) and for the preparation of metal islands on metals one might take advantage of strain relief phenomena during the growth process (Roeder et al., 1997),

1.2. SURFACES

which again result in self-organized nanostructures (Brune et al., 1998). A variety of other, partially laser-based methods for the generation of periodic structures on a nanometer scale has recently emerged. Among those are nanosphere lithography (Hulteen and Duyne, 1995), surface electromagnetic wave etching (Kumagai et al., 1992), electron-beam lithography (Gotschy et al., 1996a) scanning force methods using electromagnetic near-field enhancement (Jersch and Dickmann, 1996; Jersch et al., 1998) or tip-induced metal deposition (Engelmann et al., 1998), atomic beam lithography using ultrathin organic films (Lison et al., 1997), and laterally self-limited laser-induced thermochemistry or biological templating (Gorbunov et al., 1997). It has been demonstrated that such manufacture of particle arrays even in the case of metals allows one to enhance the Raman scattering probabilities (Kahl et al., 1998) but also to tailor the optical properties such as extinction spectra (Gotschy et al., 1996b) or the surface plasmon enhanced near-field (Krenn et al., 1997). The latter one is a very interesting feature with respect to possible applications in nanooptics. Of course, procedures of that kind are also very well known for semiconductor nanoparticle arrays (see, for example, (Wang and Herron, 1991; Alivisatos, 1996; Shah, 1996; Fendler, 1998)).

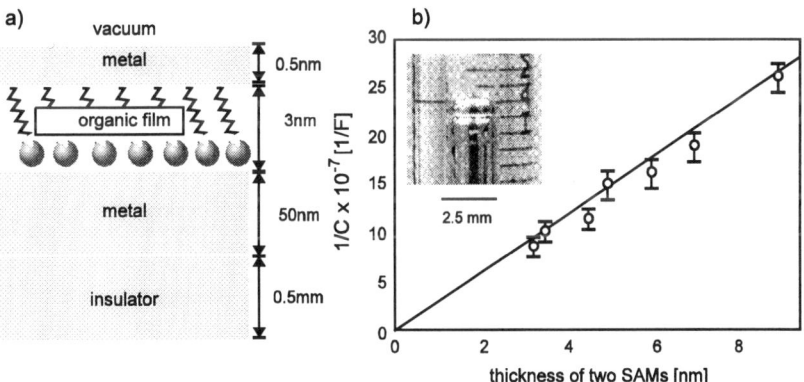

Figure 22 (a) Example of a nanoscaled layered system, which separates an ultrathin conducting layer from another layer via a dielectric spacer layer. This system could be used as a nano-capacitor, the capacity of which is changeable via change in the chain lengths of the organic films. (b) Chain-length dependence of the inverse of the capacitance of a Hg-SAM/SAM-Hg capacitor. Reprinted with permission from (Rampi et al., 1998). Copyright 1998 American Institute of Physics. The inset shows a photograph of the Hg-SAM drop system inside a syringe.

If one is interested in enlarging the structures, which have been designed on the subnanometer scale, to macroscopic dimensions or to manufacture them as an industrial series, then one has to rely even more on the *self-organizing* processes in the materials (Tredgold, 1994). Besides the above noted self-organized growth mechanisms in semiconductor, metal or glass systems, also supramolecular nanostructures including organic materials, 'nanocrystal superlattices' (Laitenberger et al., 1997; Collier et al., 1998) have found interest. Here, especially monomolecular organic films (Grunze, 1993) have gained importance during the last few years. Goals span a wide range from the fabrication of ultrafast optical switches via optical biosensors (Ding et al., 1997) and lithographic applications (Nowak et al., 1996) to microstructured frequency doublers on a waveguide-basis (Neuschäfer et al., 1994).

An example of an artifically constructed layered and nanoscaled structure is shown in Fig. 22a. The building blocks for this kind of molecular architecture of layered systems with new linear and nonlinear optical properties are ultrathin hydrocarbon films, labelled 'organic films' in the figure.

It has long been known that fatty acid molecules, spread on water, form a monomolecular film of high quality if they are compressed with a floating Teflon barrier ('Langmuir–Blodgett technique', named for K. Blodgett and I. Langmuir (Blodgett and Langmuir, 1937; Ulman, 1991)). The hydrophilic end groups of the films stick at the water surface, while the hydrophobic end groups (*e.g.*, CH_3) point into the air. By moving the floating barrier over the surface, the pressure is measured. A sudden, strong increase signals that an ordered, nearly incompressible film has been formed (Fig. 23a).

If one transfers a substrate into the water trough and removes it by sliding it through the fatty acid film on the water surface, then the hydrophilic end groups will be removed from the water and stick on the substrate. A second cycle of dipping and removing results in two more adsorbed monolayers on the substrate (Fig. 23b). The thickness of the individual films depends solely on the number of CH_2 groups in the organic molecule and the tilt angle with respect to the substrate normal. For cadmium-arachidic acids ($[CH_3-(CH_2)_{18}-COO^-]_2 Cd^{2+}$) one finds 26.4 ± 0.1 Å (Steiger, 1971). This method allows one to prepare multiple layers up to several thousand layers thickness (several tenths of a μm) (Möbius and Bücher, 1972).

A disadvantage of the LB method is that the chain ends are not chemisorbed at the surface (i.e., their binding is weak) and that the uncovered part of the surface due to growth defects is relatively large (about 30%). An alternative method is provided by the use of self-organizing molecules (SAMs), *e.g.*, docosane thiol, $CH_3-(CH_2)_{21}-SH$, the sulfuric end groups of which are chemisorbed (binding energy $E_b \approx 2$ eV (Evans and Ulman, 1990)), while the CH_3 tail groups again point away from the surface (Fig. 24b and c) (Dubois, 1992). In the past 15 years following the observation that alkane thiols form well-organized monolayers on gold surfaces (Allara and Nuzzo, 1983; Porter et al., 1987) the range of possible applications has been greatly extended. This

1.2. SURFACES

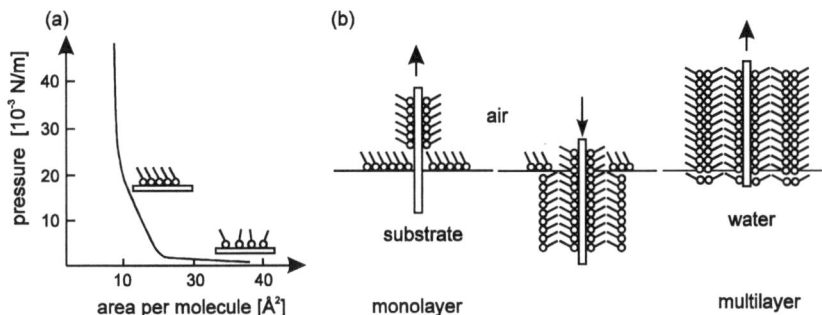

Figure 23 The Langmuir–Blodgett technique for the generation of monomolecular films on surfaces. The numerical values are for Cd arachidic acid. (a) Pressure increase signals the formation of an incompressible film at the water surface. (b) Multilayer formation by dipping the film-covered substrate several times into the trough.

is mainly due to the simplicity of preparation of the ultrathin organic films by dipping the metallic substrate into ethanolic solutions. However, the inherent disadvantage of the simplicity of this method is that it is difficult (although possible (Tillman et al., 1989)) to adsorb more than one monolayer on the sample. Convenient methods to change the thickness of the layer are to vary the length of the hydrocarbon chain or to use chains with functionalized end groups that allow the growth of additional layers of organic films (Ulman and Tillman, 1989).

The quality of such SAM films on metal surfaces has been investigated with numerous methods such as ellipsometry (thickness measurement) (Ulman, 1991), X-ray diffraction (crystallographic bulk structure, tilt angle) (Fenter et al., 1994; Fenter et al., 1997; Fenter et al., 1998), infrared and Raman measurements (vibrational spectroscopy, tilt angle) (Delamarche et al., 1996), second-harmonic generation (dynamics) (Buck et al., 1995), electron microscopy (overall homogeneity) (Strong and Whitesides, 1988), metastable induced electron spectroscopy (orientation) (Heinz and Morgner, 1997), atomic force microscopy (microscopic order in the nanometer range, phase transitions) (Flörsheimer et al., 1993), scanning tunneling microscopy (real space structure) (Delamarche and Michel, 1996), and electron (Gerlach et al., 1997) and atomic beam diffraction (Camillone et al., 1997) (crystallographic surface structure). As shown in Fig. 25, the different methods result in complementary information on different parts of the films (head group, chain, tail group).

Using low-energy electron diffraction (LEED, Fig.27) one observes for short

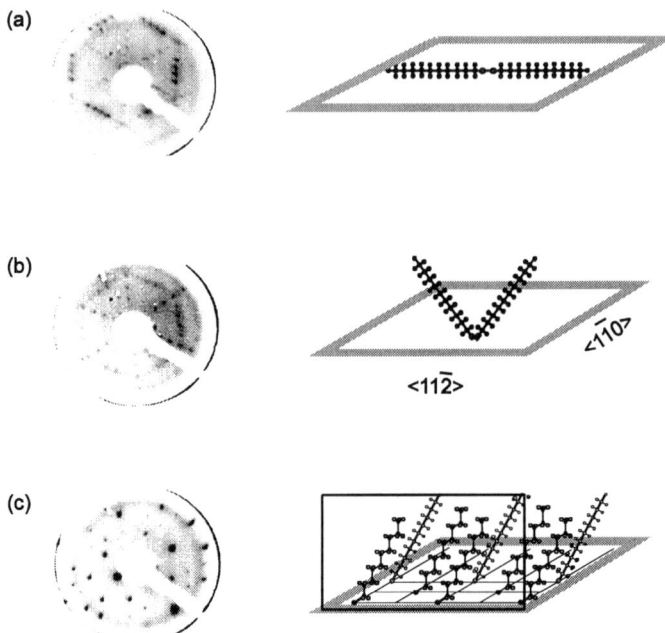

Figure 24 Structure and orientation of alkane thiols on gold surfaces. The three rows characterize — with increasing coverage from the top to the bottom — possible structures (right-hand side) for decane thiol molecules, as deduced from measured LEED pictures (left-hand side). Reprinted with permission from (Gerlach et al., 1997). Copyright 1997 Springer-Verlag.

chains (up to 12 CH_2 groups) adsorbed on room temperature gold surfaces sharp diffraction spots (Gerlach et al., 1997). Apparently the electrons, which have a penetration depth of several nanometers, are scattered by the whole carbon backbone of the chain. If one increases the chain length, then the overall statistical movement of the chain increases too and the Debye–Waller factor (which depends on the Debye temperature, i.e., the 'softness' of the films) smears out the diffraction peaks. The Debye–Waller factor also prohibits obtaining a sharp diffraction pattern using atomic beam diffraction, a method that is especially sensitive to the chain tail groups. In that case one has to cool down the surface to less than 90 K and then one obtains, even for long-chain films (more than 16 hydrocarbon groups), sharp peaks (Camillone et al., 1993). LEED and HAS studies prove that the long chain alkane thiols form films with crystallographic structure. More sophisticated LEED and atomic beam investigations reveal a variety of possible stable film structures, including

1.2. SURFACES

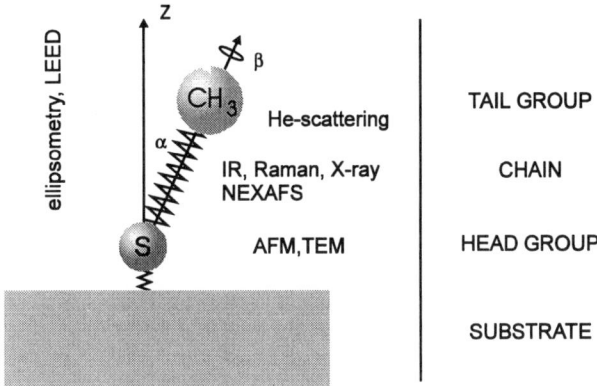

Figure 25 Side view of long chain alkane thiol molecules with typical tilt angle with respect to surface normal, $\alpha=30°$, and azimuthal angle with respect to the plane of the paper, $\beta=50°$. The assessment range of different analytical methods is also shown.

molecules that are oriented parallel, perpendicular or tilted with respect to the surface normal (Fig. 24) (Gerlach et al., 1997).

An application of organized organic mono and multilayers are inert dielectric spacer layers for thin metallic films (Jung and Czanderna, 1994), as shown schematically in Fig. 22a. If one adsorbs a metal film on top of the organic film one has to avoid diffusion of the metal into the film in order to build up a well-defined nanoscaled structure. For the nano-capacitor element shown in Fig. 22b this problem has been solved by bringing two liquid mercury (Hg) drops, covered with SAMs of different chain length, in direct contact within a syringe. Since the SAMs adsorb with tilt angel zero on a Hg surface[16], they do not form domains and the spacer layer thickness d increases in proportion to the chain length. The capacitance is given by

$$C = \epsilon_0 \epsilon \frac{A}{d}, \tag{1.26}$$

where A is the area of the capacitor. Hence one expects a linear increase in $1/C$ with chain length if the dielectric constant of the SAMs does not depend on the chain length. This is obviously the case, and one obtains $\epsilon=2.7 \pm 0.3$, very similar to the value for alkane thiols on Au ($\epsilon=2.6$). The measured value

---

[16] The tilt angle is determined by the interplay between distance between possible adsorption sites on the surface and the space requirement given by the van der Waals radii of the organic molecule chains.

of capacitance is, *e.g.*, for CH_3—$(CH_2)_{17}$—SH, C=8.6 nF.

Of course, the range of possible applications would be much larger if one were to design such nanostructured elements on solid surfaces. Here, one has to overcome the problem of defects in the organic films, which are leaks for the applied currents between the electrodes, as well as the above mentioned penetration of the metals into the organic film, the probability of which being inversely proportional to the reactivity of the metals at the SAM-surface (Jung and Czanderna, 1994). This means that silver atoms have the lowest probability to form a well-organized metallic overlayer, whereas copper, nickel, potassium, sodium, aluminum, chromium and titanium are, in this order, more appropriate. Indeed, a recent experiment on electric rectification in a nanoscaled system very similar to the one shown in Fig. 22 but using a lithographically prefabricated, small-scale semiconductor heterostructure, has incorporated a titanium film as the upper metallic electrode (Zhou et al., 1997).

Besides their possibility to form spacer layers with variable thickness in the nanometer regime, organized organic films are of interest also as optoelectronic elements. For example, functionalized side groups can be attached to them, which add optical anisotropy and make them birefringent. One obtains a linear optical filter with properties that can be changed by applying an electric field that orients the side groups.

More important even than the linear are the nonlinear optical properties of the organic films. These are mainly due to the highly polarizable conjugated Π-electron systems of the hydrocarbons, which connect an electron-acceptor tail and an electron-donor head group. Values of the nonlinear optical polarizability of second order as large as 10^{-13} esu (Marowsky et al., 1988) can be obtained for special dye films (cf. Table 1, Section 3.4). In combination with thin film waveguide technology (Chapter 3, Section 3.6) frequency doubling elements with atomic dimensions and high conversion efficiencies can be manufactured that way. In contrast to conventional semiconductor elements, which have (resonant) switching times of the order of nano- or picoseconds, in organic films transient switches can be generated on the femtosecond time scale (Yariv, 1985). A possible realization is a switching pulse which induces a change in the index of refraction in the organic film, which is transferred as a phase modulation to the subsequent pulse. Or a diffraction grating is written into the organic material, which switches the following pulse spatially between two subsequent wave guides (nonlinear optical directional coupler). Finally, the high nonlinearity of third order also allows one to exploit holographic diffraction phenomena ('four-wave mixing', Chapter 3, Section 3.6) on a time scale of femtoseconds. A possible application is the generation of a phase conjugated mirror inside an ultrathin film, which might be used for the correction of dispersion effects in optical data lines (Fuchs et al., 1991).

1.2. SURFACES

1.2.4 Analysis

Nowadays a variety of surface-investigation techniques are commercially available, which provide partially complementary information on the structure and dynamics of pure and adsorbate-covered surfaces. Recent, more technically oriented overviews can be found in (Walls and Smith, 1994; Yates, Jr, 1998). In principle one might divide the methods into *diffraction methods*, *direct-imaging methods* and *emission methods* (Fig. 26). In what follows the three groups are introduced by selected examples.

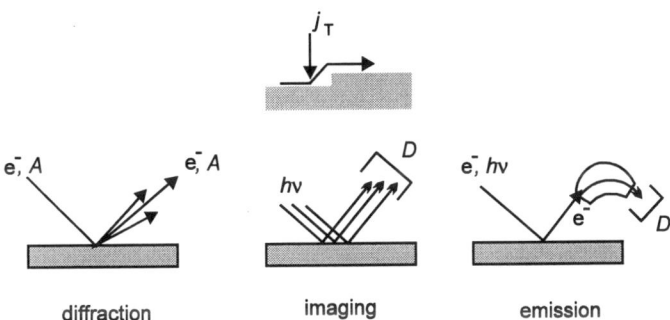

Figure 26 Schematic of different types of surface investigation methods. The symbols e^- and A denote electrons and neutral (H, He) or metastable (He*) atoms as well as molecules (H_2). Photons are denoted by $h\nu$ and I_T is the tunneling current. Photon or electron detectors are denoted by D.

Diffraction methods

Similar to the case of X-ray or neutron diffraction for the investigation of bulk structure one might deduce the structure of a surface from the angular pattern of a diffracted particle beam. A necessary prerequisite is that the de Broglie wavelength $\lambda_{\mathrm{dB}} = h/|\vec{p}| = h/\sqrt{2mE}$ of the particle beam is of the order of the lattice constant of the investigated surface; otherwise only specular or diffuse reflection occurs. For electrons one finds

$$\lambda_{\mathrm{dB}}[\text{Å}] = \sqrt{\frac{150}{E[\text{eV}]}}, \quad (1.27)$$

resulting in $\lambda_{\mathrm{dB}} = 1.2$ Å for 100 eV electrons. Comparison with typical lattice constants (Cu(100): 3.61 Å, LiF(001): 2.84 Å) reveals that electrons are valid for obtaining structural information from diffraction experiments.

The surface sensitivity of the detection method is given by the penetration depth of the particle beam. This depth is 0.5 – 1 nm for low energy electrons (a few tens of eV), meaning a sensitivity to the first two or three surface layers and possibly adsorbates on the surface (LEED, 'low-energy electron diffraction' (Hove et al., 1986)). A typical LEED set-up is shown in Fig. 27: the low-energy electron beam, which is generated by an electron gun, is diffracted in an ultrahigh vacuum (UHV) chamber from the investigated crystal surface. The diffracted beams are sampled on a fluorescent screen, where they generate a picture of the reciprocal surface lattice. Intensity, position and width of the fluorescent points are detected with a video camera. In order to avoid inelastically diffracted electrons, a retarding electric field is applied between the crystal and fluorescent screen.

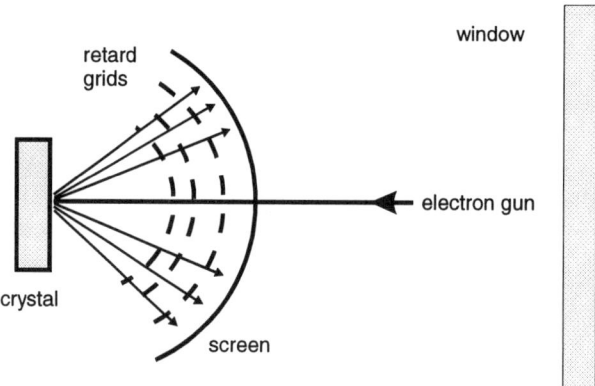

Figure 27 LEED set-up for the surface diffraction of low-energy electrons.

From the number and position of the diffraction maxima one can deduce within the 'transfer width' of the instrument (meaning the spatial range of coherent scattering) the period of the surface lattice or its possible disturbance by adsorbate superlattices. Typical transfer widths are from 10 nm for a conventional LEED up to 100 nm for a SPALEED, 'spot profile analysis LEED'. The intensity of the diffraction maxima contains information about the structure of the unit cell and the dynamics of the surface (*e.g.*, the Debye–Waller factor $\exp(-\vec{K}^2 \langle u^2 \rangle)$ with scattering vector $\vec{K}$ and mean deflection $\langle u^2 \rangle$ of the thermally excited vibrations of the surface atoms).

The dispersion curves of surface phonons can be determined via electron energy loss spectroscopy (EELS) (Ibach and Mills, 1982). The sample is

1.2. SURFACES

irradiated by monoenergetic electrons with energies up to a few hundred electronvolts and the elastically and inelastically scattered electrons are analyzed with an energy resolution down to a few meV. The interpretation of the data is difficult since the penetration depth of the electrons results in contributions to the signal intensity not only from the topmost but also from lower lying atomic layers. Since the cross section for photon excitation by electron collisions depends sensitively on the incidence energy and scattering geometry (meaning the surface projection of the involved momenta), a variation of those parameters in some cases allows one to discriminate between neighboring phonon dispersion curves in spite of the relatively low energy resolution of the method. Applications include investigations of metals and chemisorbed adsorbates. However, in the case of insulators severe problems arise due to surface charging effects.

Scattering of low energy, nearly monochromatic helium beams at a surface also results in a diffraction pattern since a helium beam has a de Broglie wavelength

$$\lambda_{dB}[\text{Å}] = \sqrt{\frac{20.4}{E[\text{meV}]}} \tag{1.28}$$

of about 1 Å at a velocity of 920 m/s (helium atomic beam scattering, HAS) (Kress and Wette, 1991; Hulpke, 1992)(Fig. 28). The reflected intensity distribution of an intense, monochromatic beam of ^{4}He atoms is detected by a magnetic mass spectrometer as a function of scattering angle with respect to the surface normal. One obtains information on the electronic density distribution about 0.4 nm *above* the position of the ionic cores of the lattice. At this position the interaction between the electrons of the He atoms and that of the substrate result in a classical point of inversion (Fig. 29). Hence HAS is a truly surface-sensitive method, which is well suited for the investigation of growth and change of adsorbate coverages.

Similar to LEED the small de Broglie wavelength results in a high structural resolution. The low kinetic energy (meV) and the monochromasy of the beam enable one to determine precisely the energy distribution of the scattered atoms via time-of-flight (TOF) measurements. The TOF spectra show distinct maxima for a given energy loss or gain, which are due to the excitation or annihilation of phonons (Benedek and Toennies, 1994). Dispersion curves can be determined as a function of scattering angle. Due to the closed electronic shell of the ^{4}He atoms the attractive van der Waals forces are weak; hence the atoms are — except under special angles of incidence which lead to 'selective adsorption' in the potential well which is a few meV deep — not trapped at the surface and do not result in contamination of or change in the initial surface. The coupling of the incident atoms to the phonons occurs via the repulsive part of the interaction potential.

HAS is a valid method also for the investigation of the concentration of

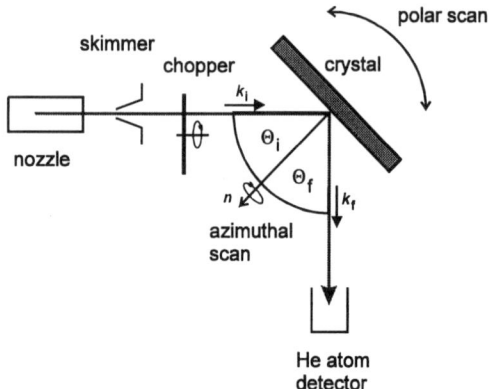

Figure 28 HAS set-up for surface diffraction of low-energy atoms. The nearly monochromatic helium atomic beam is generated by expansion of helium gas under high pressure through a nozzle with small orifice (cf. Fig. 30).

defects at a surface (steps, edges, dislocations, kinks). Statistically distributed defects result in diffuse scattering, in contrast to scattering from the periodically ordered lattice atoms. The cross sections σ for such processes are of the order of the gas phase interaction cross sections between helium and the defect atoms (100 Å^2). Thus the method is nondestructive, but has a sensitivity that allows one to detect defect concentrations of less than 1% of a monolayer.

Using the same argument adsorbate coverages of the surface can be investigated with unprecedented precision and absolutely destruction-free. For example, upon statistical adsorption of adsorbates (Volmer–Weber growth, Fig. 20) the initial specularly scattered He signal intensity from the ordered substrate surface will decrease due to an increasing contribution of diffuse scattering from the adsorbates, which represent randomly distributed defect sites. Assuming that the adsorbates are perfectly diffuse scatterer the surface analog of Beer's law, Eq.3.4, namely the lattice gas formula, leads to (Poelsema and Comsa, 1989)

$$\frac{I}{I_0} = (1 - \Theta)^{\sigma n_S}, \qquad (1.29)$$

where n_S is the number of substrate atoms per unit area and σ again is the scattering cross section of He atoms from the adsorbate atoms. Hence Eq. (1.29) allows one to deduce quantitatively the coverage Θ of the surface

1.2. SURFACES

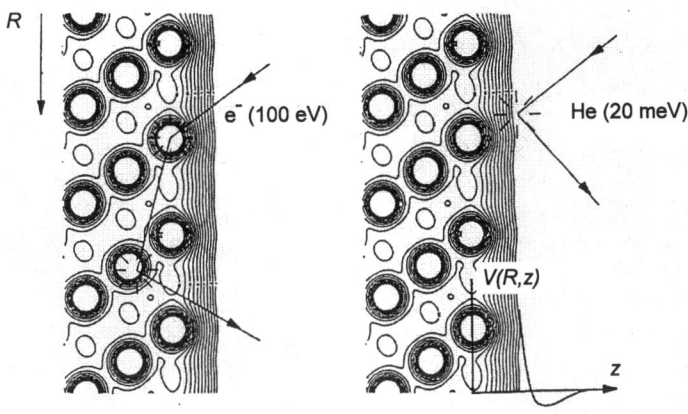

Figure 29 Comparison of surface sensitivities of electron and low-energy atom diffraction from surfaces. The open circles symbolize the atomic cores, the lines are cuts through the planes of equal electron density. The interaction potential of the helium atoms with the surface $V(R,z)$ for fixed coordinates along the surface, R, and as a function of a coordinate normal to it, z, is shown too. Reprinted with permission from (Lahee and Toennies, 1993). Copyright 1993 Physics World.

with adsorbate atoms.

A necessary prerequisite for a successful application of the helium diffraction method to the investgation of surface structure and dynamics is the generation of an intense helium beam with small velocity smearing $\Delta v/v \leq 0.05$ and small de Broglie wavelength. This can be achieved by expansion of the helium gas under high pressure (≈ 100 bar) through a nozzle of small diameter (≈ 10 μm) at low nozzle temperatures (≈ 100 K) into a vacuum chamber (Toennies and Winkelmann, 1977) (Fig. 30). In the course of the expansion into the vacuum chamber the particles interact and cool down adiabatically by collisions. The velocity components perpendicular to the beam axis and also the smear-out in the direction of expansion decrease, whereas the mean velocity increases. The undirected thermal energy (in the form of translational and (for molecules) rotational and vibrational motion) is transformed into directed translational energy: a supersonic beam evolves. This is the case if enough collisions occur in order to perform the transformation. That is, the mean free path λ_f of the particles in the nozzle orifice has to be smaller than the nozzle diameter d:

$$\frac{\lambda_f}{d} = \frac{kT_0}{\sqrt{2} \cdot p_0 \sigma d} \ll 1. \tag{1.30}$$

Here, σ denotes the total collision cross section between the expanding particles ($\sigma \approx 100$ Å^2 for helium (Lambert, 1977)), T_0 and p_0 are the mean temperature und pressure inside the nozzle. The ratio between the mean free path and nozzle diameter is usually called the 'Knudsen number'. A small Knudsen number is equivalent to a small velocity smearing.

Downstream the nozzle the area close to the beam axis is separated from the background gas by a 'skimmer', which is a conically formed aperture (Fig. 30) that allows maximum flux of particles with parallel velocity vectors to reach the subsequent vacuum chamber. In this chamber a chopper generates particle pulses and thus enables one to perform an energy analysis of the scattered particles via a TOF measurement.

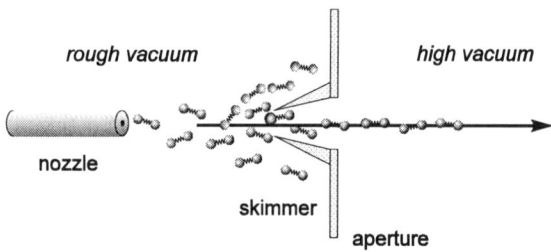

Figure 30 Generation of a beam of directed particles by expanding a gas at high stagnation pressure and low temperature through a nozzle with small orifice into a vacuum chamber. The conically shaped aperture ('skimmer'), placed about 12 mm downstream the nozzle, separates the free supersonic nozzle beam of noninteracting particles from the less directed part of the beam and the background gas.

If one investigates a surface with a pulsed laser it might be advantageous to use a *pulsed* nozzle for the generation of the primary beam (Fig. 31). The average gas load is significantly smaller compared with a continuous nozzle, which results in a gain of beam intensity by the factor $T/\Delta t$, where T means the pulse period and Δt the pulse length. An electromagnetically pulsed nozzle such as the one shown in Fig. 31 has a minimum pulse period of about 4 ms (corresponding to a repetition rate of 250 Hz) and an optimum pulse length of about 200 μs, which results in a gain with respect to the continuous beam

1.2. SURFACES

by a factor of 20. As long as the characteristic ratio of the pumping system, $\tau = V/S$, is small compared with the pulse period, this gain factor can be enhanced by using shorter pulses. The pumping speed S of the high vacuum pump (typically 1500 l/s) has to be large compared with the volume V of the chamber (typically 5 l). The intensity of a pulsed beam, $j_{\Delta t}$, is deduced from the intensity of a continuous beam, j_{cw}, if one takes into account the pumping speed of the pump (Gentry, 1988)

$$j_{\Delta t} = j_{cw} \frac{T}{\Delta t} \cdot \left(\frac{T}{\tau} \cdot (1 - \exp\left(-\frac{T}{\tau}\right))\right)^{-1}. \quad (1.31)$$

Hence the finite pumping speed results in about 15% intensity loss compared with the roughly estimated value. The He particle flux at a pressure of 6 bar and a nozzle diameter of 0.8 mm is $3 \cdot 10^{21}$ s^{-1} with an average velocity of 1800 m/s. In order to retain a background gas of less than 10^{-3} mbar in the main chamber and thus avoid attenuation of the beam by background gas scattering, additional roughening pumps with high pumping power have to be added to the high vacuum pump. Due to the high particle density within the pulse and in order to avoid skimmer interference the distance nozzle skimmer has to be about a factor of three larger compared to a continuous expansion, i.e., about 35 mm.

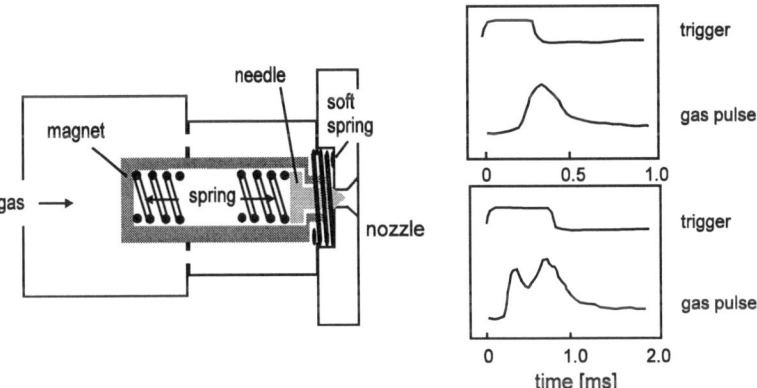

Figure 31 Schematic drawing of a pulsed electromagnetic nozzle. The inner chamber, which is filled with gas, is closed by a needle with a Teflon fitting. It can be opened against the strong spring by applying a voltage to the magnetic stamp. On the right-hand side the temporal evolution of typical gas pulses from a piezoelectric driven nozzle is shown. At the top $\Delta t = 300$ μs, and at the bottom $\Delta t = 800$ μs. Reprinted with permission from (Andresen et al., 1985). Copyright 1985 American Institute of Physics.

The generation of a cooled or heated beam becomes difficult in the case of the electromagnetically pulsed nozzle since the fittings and the movable parts are not temperature-resistant. Shorter pulses can be obtained with nozzles that work by the current-loop principle (Gentry and Giese, 1978) or piezo-electrically. In Fig. 31 helium gas pulses are shown that have been generated by a piezoelectrically driven nozzle. For longer pulses (lower part of the figure) the intensity drops due to skimmer interference. It is important before using such pulses in surface experiments to determine their characteristics and to use an additional chopper in order to select the smoothly shaped part of the pulse.

Direct-imaging methods

The diffraction of particles at a surface corresponds — similar to the diffraction of photons in an optical element — to a Fourier transformation of the incoming beam and — due to the finite diameter of the particle beam — to integration over real space. Information about the structure of the surface becomes available in reciprocal space as a diffraction pattern. Thus information about the local structure, *e.g.*, in the neighborhood of a step, is only obtainable if the structure has periodicity. Direct-imaging methods rely on either back-transforming the Fourier-transformed picture by a second transformation (light and electron microscopy), or the surface structure is directly scanned. The resolution in the former case is limited by the wavelength of the applied radiation, hence for light the resolution is of the order of a few hundred nanometers whereas for 60 kV electrons it is about a few percent of a nanometer. Lens errors (especially chromatic aberration due to the non-monochromaticity of the electron beam) lead in the case of a scanning electron microscope (SEM) to a resolution limit of about 1 nm.

Using low-temperature field ion microscopes a resolution of about a tenth of a nanometer — albeit along a very limited range of the surface — can be achieved (Fig. 32). In such microscopes (Müller, 1951) a high voltage of a few tens of kilovolts is applied between the sample (which is a sharp tip) and a fluorescent screen (Fig. 32a). He atoms from a low-pressure buffer gas (partial pressure 10^{-4} Torr) are polarized by the field and attracted by the tip, where they face field strengths of up to 10^{11} V/m. The atoms are positively ionized about half a nanometer in front of the tip and consequently repelled and accelerated in the direction of the fluorescent screen where they generate photons (Fig. 32b). Usually a multi-channel plate (MCP) is mounted in front of the fluorescent screen, which serves to amplify the electron flux. Due to the high acceleration voltage the He ions follow nearly straight trajectories, and thus the tip with radius of the order of 10 nm is magnified onto the screen at a distance of about 10 cm by a factor of approximately 10 cm/10 nm = 10^7. This is sufficient for a resolution in the subnanometer range and discrimination between individual atoms. Although the actually investigated range on the surface is only of the order of 50 – 200 nm diameter, the diffusional motions of individual atoms along the surface could be studied intensively in the past

1.2. SURFACES

that way.

Scanning tunneling microscopes (STM) (Fig. 33) allow one to obtain resolutions even in the sub-nanometer regime over wide spatial ranges on the surface (Chen, 1993). With the STM (Binnig et al., 1982) one images the surface structure by monitoring changes in the tunneling current intensity j_T between the conducting tip, which is mounted on piezo-elements and has a radius in the nanometer range, and the (conducting) surface. The tunneling current for a given voltage V between the tip and surface is

$$j_T = \rho(r, e_F) \frac{V}{z} \exp(-const. \cdot z \cdot \sqrt{W_B}) \quad (1.32)$$

and thus depends exponentially on the distance z to the surface (the barrier width) and the square root of the difference in work functions W_B between the surface and tip (the barrier height). The prefactor $\rho(r, e_F)$ is the spatial electron density distribution close to the Fermi level and is what one essentially mimics. Depending on whether the tip is biased positively or negatively with respect to the surface one detects the highest occupied molecular orbitals (HOMO) or the lowest unoccupied molecular orbitals (LUMO). Obviously, due to the exponential distance dependence, a small change in distance results in a strong variation in the tunneling current.

In order to measure the tunneling current one needs free carriers in the sample, which are not available in the case of insulating surfaces. Here, the atomic force microscope (AFM) can be applied, which uses the fact that the (vertical and torsional) deflections of a cantilever tip, induced mainly by the van der Waals force between the tip and surface, depend to a higher order power on the distance between the tip and surface (Magonov and Whangbo, 1996).

Even with light microscopes one might obtain a resolution below the diffraction limit if the sensor is removed from the far field into the near field (a distance closer than a wavelength) of the light source. This can be achieved by scanning an aperture with a diameter of a few tens of nanometers at a distance of a few nanometers above the surface and by measuring the transmitted light. The detection sensitivity of such a SNOM ('scanning near field optical microscope') can be increased by the use of a light fiber as the aperture, which is in most cases covered by an aluminum film. This tip is inserted into the near field of the investigated object. A small part of the light enters the fiber and is guided to the photon detector. By mounting the fiber tip on piezo-crystals and measuring lateral forces, in the course of scanning nanometer-resolved structural and spectroscopic information can be obtained just as in the case of an AFM. The limit is about 12 nm resolution: if the diameter of the aperture is smaller, then light is coupled into the aluminum coating and does not reach the detector.

It should be noted finally that all scanning imaging methods rely on a vibrationless mounting (sketched in Fig. 33 by springs) and (in the case of

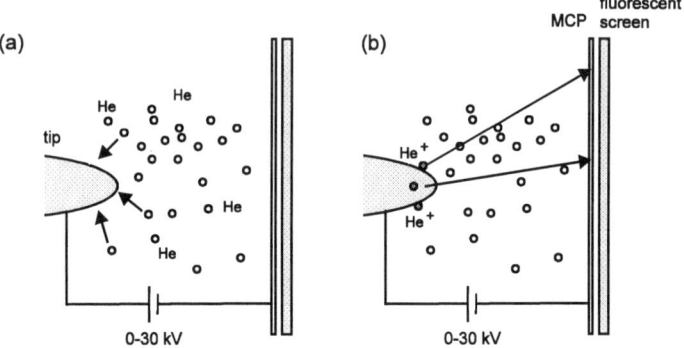

Figure 32 Field ion microscope (FIM). (a) Attractive force between the tip and He atoms due to polarization of the atoms by the field. (b) Repulsion and acceleration of the positively charged He ions onto the fluorescent screen behind a multi-channel plate (MCP) amplifier.

the SNOM) sensitive photodetectors and gradient fibers, which allow efficient coupling and guiding of the probe light.

Emission methods

Information on the chemical structure of the surface can be obtained by methods which result in the emission of electrons having characteristic energies after irradiation with high energetic photons or particles (Fig. 34). In photoelectron spectroscopy with deep ultraviolet or X-ray photons (UPS or XPS, respectively) electrons of the atoms A of the solid are excited from the valence bands (UPS) or from the K- and L-shells (XPS) into the continuum,

$$h\nu + A \rightarrow A^+ + e^-. \tag{1.33}$$

The subsequently emitted electrons are energetically analyzed by a spectrometer, *e.g.*, a cylindrical mirror analyzer (CMA). In the most simple arrangement this analyzer consists of two cylinders surrounding the target and the electron gun. Emitted electrons from the target enter the region between the cylinders through a small aperture and are repelled by the outer cylinder to exit the inner cylinder through a second aperture and to hit an electron multiplier, which is mounted in line with the target. The position where they hit the multiplier is given by their kinetic energy and the voltage between the two cylinders. The entrance and exit aperture of the inner cylinder determine

1.2. SURFACES

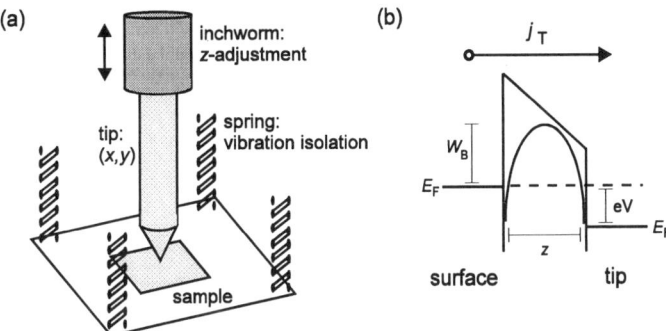

Figure 33 Scanning tunneling microscope (STM). (a) Principal set-up. A conducting tip is mounted on an (x, y, z)-adjustable piezo-stack. The distance from the surface, z, is adjusted by applying a voltage V between the tip and surface and measuring the tunneling current j_T. Another voltage is used to scan the tip in the (x, y)-direction over the surface. (b) Energy level scheme for a negatively biased tip, including Fermi energies E_F, effective work function of the barrier, W_B, and distance between tip and surface, z. The tunneling current is directed from the surface to the tip by applying a voltage V.

the transmission and thus the energy resolution of the CMA. An energy resolution of better than half a percent is easily obtained with transmissivities of the order of 10%.

From the energetic position and the intensity of the photoelectron maxima the abundance of specific elements at the surface can be deduced. From the shift with respect to the photoelectron maxima of free elements one obtains information about the binding strengths and positions of the elements at the surface. The line shapes in UPS allow one to determine the population density of the energy bands, and the angular dependence of the UPS intensities provides information about the electronic band structure.

Information about the short-range order around the surface atoms of selected elements can be obtained also via irradiation with X-rays with energies above the absorption edge. If one measures by use of an energy analyzer the emission rate of photoelectrons as a function of the photon energy in discrete, element-specific energy windows, then one observes oscillations in the case of short-range order (SEXAFS, 'surface extended X-ray absorption fine structure'). These are due to scattering of the photoelectrons, which are

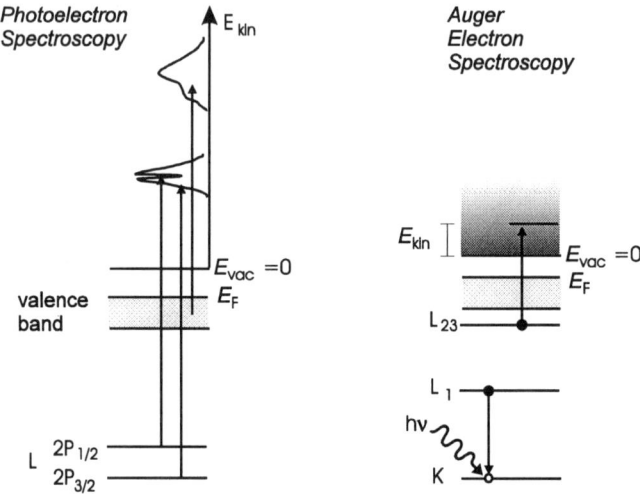

Figure 34 Energy level schemes for a photoelectron (UPS or XPS) and an Auger spectrometer (AES). As sketched, the emitted electrons possess a kinetic energy distribution that reflects the properties of the previous excitation process.

emitted from the initially excited surface atom, at neighboring atoms. Since this is a coherent scattering process, characteristic interference phenomena appear. The same kind of interferences are observed as oscillations in the emission cross section as a function of electron and thus photon energy. From the amplitude of the oscillations, average atomic distances and from their spacing the average number of interacting neighboring atoms can be deduced. Hence the method serves mainly for structural investigations. Close to the absorption edge (NEXAFS, 'near edge EXAFS'), one might obtain additional information about the electronic properties of the atoms, since the emitted photons can subsequently induce additional resonant excitations.

A sensitive chemical analysis of surface adsorbates that also provides a convenient manner of monitoring surface cleanness and calibrating surface coverage (cf. Fig. 81) is made possible by Auger electron spectroscopy (AES). The ionization of the inner electronic states (*e.g.*, of the K-shell) of the surface atom via irradiation with high-energy electrons (several keV) results in a hole in the K-shell, which is filled by the transition of an electron from, for example, the L_1-shell (Fig. 34). The difference in energy is used to release an electron

1.2. SURFACES

from a higher shell (L_{23}) from the solid. The kinetic energies of these Auger electrons, which are determined again with a CMA, are given by the differences between the relevant atomic binding energies E_b:

$$E_{kin}(KL_1L_{23}) = E_b(K) - E_b(L_1) - E_b^{eff}(L_{23}), \qquad (1.34)$$

where the superscript "eff" means that the ionization potential takes into account that the atom has already been ionized by the loss of an inner-shell electron. Since the kinetic energies depend only on the energy differences between electronic states in the atom, they are element-specific. Moreover, their absolute values are between 100 eV and 1000 eV and hence their penetration depth is small (0.5 – 1 nm) and the method is truly surface-sensitive.

The intensity of the Auger lines is given by

$$I_{KL_1L_{23}} = I_0 \cdot \sigma_K(E_0) \cdot P_{KL_1L_{23}} \cdot T(E_{kin}) \cdot D(E_{kin}), \qquad (1.35)$$

with $\sigma_K(E_0)$ the ionization cross section of the K-shell, $T(E_{kin})$ the transmission function of the CMA, $D(E_{kin})$ the detection sensitivity of the electron multiplier and $P_{KL_1L_{23}}$ the Auger decay probability. Especially for elements with larger masses radiative relaxation (i.e., fluorescence) competes with the Auger yield and makes an analysis of absolute intensities difficult. But even if one can neglect radiative decay and measures the energy spectrum of the generated electrons, broad valence band and core level emission tends to overlay the discrete line spectrum. Therefore one usually modulates the outer cylinder of the CMA with a small voltage and detects phase-sensitively the first derivative of the Auger electron signal. Groups of sharp peaks at predefined energies then provide a 'fingerprint' of the elements forming the surface under investigation.

2

Adsorption, Desorption and Diffusion

2.1 Laser-Induced Thermal Desorption

Irradiation of an adsorbate covered surface with a laser in most cases results in desorption of the adsorbate. It is useful to distinguish between direct desorption mechanisms via electronic excitation of the adsorbate or the substrate and indirect desorption mechanisms. The most commonly employed and already in the seventies investigated (Ertl and Neumann, 1972) form of laser desorption via an indirect mechanism is laser–induced thermal desorption (LITD). At the microscopic level it can serve to

- study the kinetics of surface desorption and adsorption with high temporal resolution
- study surface diffusion via determination of coverage gradients
- investigate in real time reaction kinetics on surfaces
- desorb fragile molecules such as organic polymers, biological relevant molecules, C_{60}, products of a surface reaction, etc. into the gas phase and make them available for mass spectrometric analysis ('MALDI')
- clean surfaces from adsorbates without destroying the surface structure of the substrate.

Even nanosecond pulsed lasers abruptly heat the substrate surface with heating rates of more than 10^{12} K/s. For comparison, with electron beam irradiation heating rates less than 10^3 K/s are obtained. In most cases the thermalization of the adsorbed laser energy proceeds even faster than the desorption process, meaning that the substrate is heated instantaneously. A valuable hint for the primarily thermal character of the desorption process are the state distributions of the desorbed molecules (rotation and vibration), which under certain conditions should be in thermal equilibrium with the surface temperature. These kinds of distributions can be monitored selectively using laser spectroscopy.

If the heating time is short compared with the time that is necessary for the heated molecules to react, then the desorption rate might be greater than the fragmentation rate by orders of magnitude. An example is LITD of methanol on Ni(100) (Hall, 1987). Methanol (CH_3OH) chemisorbs intact at 100 K surface temperature on nickel. If one heats the surface with conventional methods at a rate of 15 K/s the following surface reaction chain occurs as a function of surface temperature:

$$CH_3OH \text{ (s)} \to CH_3O \text{ (s)} + H \text{ (s)} \text{ at } T_s = 200 \text{ K}$$
$$CH_3O \text{ (s)} \to CO \text{ (s)} + 3H \text{ (s)} \text{ at } T_s = 250 \text{ K}$$
$$2H \text{ (s)} \to H_2 \text{ (gas)} \text{ at } T_s = 350 \text{ K}$$
$$CO \text{ (s)} \to CO \text{ (gas)} \text{ at } T_s = 470 \text{ K} \quad .$$

After about 25 s heating time the CO and H_2 molecules desorb from the surface. In the case of laser heating the surface temperature reaches 1300 K after 20 ns, and the methanol molecules desorb without fragmentation. The desorption rate

$$k_{des} = \nu_{des} \exp\left(-\frac{E_a^{des}}{k_B T}\right) \qquad (2.1)$$

has a pre-exponential frequency factor $\nu_{des} = 10^{14}$ s^{-1} and an activation energy of 14 kcal/mol. At 1300 K the internal energy of the methanol molecule is $E = (3n-6)kT = 12kT = 26$ kcal/mol, i.e., high enough that the major part of the molecules can desorb. The rate for the thermal fragmentation reaction at the surface

$$k_{frag} = \nu_{frag} exp\left(-\frac{E_{frag}}{k_B T}\right) \qquad (2.2)$$

possesses a significantly smaller activation energy, namely 9 kcal/mol. Hence 'continuous' heating results in a higher probability for fragmentation compared with desorption. On the other hand, the frequency factor $\nu_{frag} = 2 \times 10^9$ is more than four orders of magnitude smaller for the fragmentation reaction. This factor describes the number of trials that the molecule undertakes in order to desorb or to decompose. A higher value of ν means a higher entropy difference between the gas and adsorbate states. If the temperature is elevated fast enough (as is the case for laser heating), then the number of trials for fragmentation is too small and desorption takes over.

Fig. 35 shows the relative contributions of fragmentation and desorption as a function of heating rate. In the case of laser heating, fragmentation is negligible.

Desorption kinetics

Temperature programmed thermal desorption spectroscopy (TDS) allows one to obtain information about adsorption states and the nature of the

2.1. LASER-INDUCED THERMAL DESORPTION

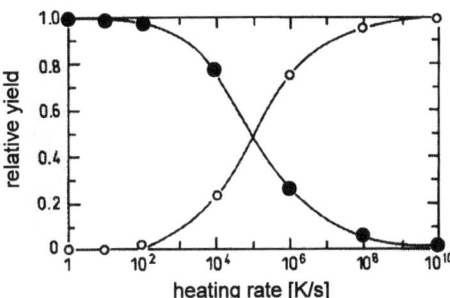

Figure 35 Relative fragmentation (•) and desorption yields (○) as a function of heating rate for $CH_3OH/Ni(100)$. Reprinted with permission from (Hall, 1987). Copyright 1987 American Chemical Society.

adsorbate–substrate or adsorbate–adsorbate interactions. A typical set-up is shown in Fig. 36. In an ultrahigh vacuum (UHV) chamber the substrate is mounted on a manipulator. It is heated from the rear by a current that flows through a tungsten coil. In order to increase the kinetic energy of the thermally emitted electrons (and thus the final temperature of the substrate), an acceleration voltage is applied between substrate and coil ('electron impact heating'). The temperature of the substrate is measured by a thermocouple. The desorbed particles are detected using a quadrupole mass spectrometer.

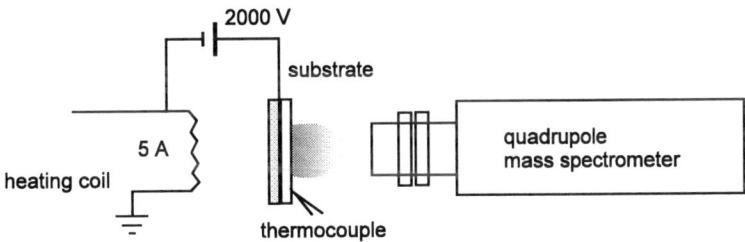

Figure 36 Set-up for a TDS measurement. Typical values of current and voltage are noted.

The desorption rate, i.e., the temporal change in surface coverage Θ via heating (King, 1975),

$$-\frac{d\Theta}{dt} = \nu_n \Theta^n \exp\left(-\frac{E_a}{k_B T}\right) \qquad (2.3)$$

is defined by the activation energy E_a, which usually is a function of Θ and the vibrational frequency ν_n in the surface potential,

$$\nu_n \approx \frac{k_B T}{h}, \qquad (2.4)$$

which is $\approx 10^{13}$ s^{-1} if the atoms can move nearly free on the surface. Also, the order n of the desorption process is important. This order characterizes the fraction of particles that participate in the critical desorption step. It depends on the adsorbate–substrate interaction and the coverage of the surface (except for $n=0$, which is desorption of the surface material and might also be called 'ablation').

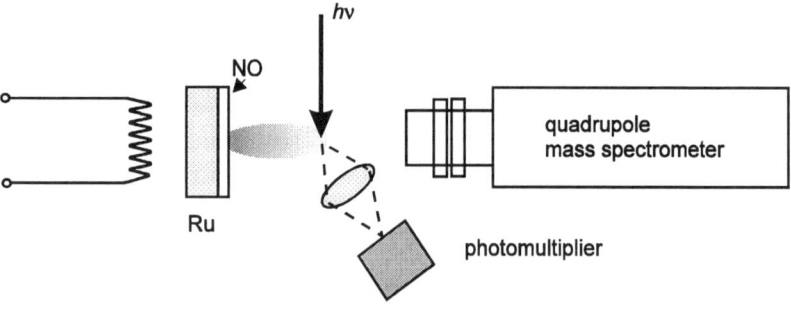

Figure 37 Experimental set-up for the determination of the rotational temperature of NO molecules, which desorb thermally from a Ru(001) surface.

The change of coverage is determined by the measurement of gain at selected masses for the constant heating rate dT/dt, hence

$$\frac{d\Theta}{dT} = \frac{d\Theta/dt}{dT/dt}. \qquad (2.5)$$

In the case of laser heating the heating rate is no longer constant but depends strongly on the time during the laser pulse. In order to determine the total mass gain, the temporal change in temperature $T(t)$ has to be calculated and Eq. (2.3) has to be integrated. For desorption kinetics of first order ($n=1$),

2.1. LASER-INDUCED THERMAL DESORPTION

i.e., mass loss with desorption of independent particles, one obtains (Zhu et al., 1989)

$$\Theta(t) = \exp\left(-\int_{T_0}^{T(t)} \frac{\nu_1}{dT/dt} \exp\left(-\frac{E_a}{k_B T}\right) dT\right). \quad (2.6)$$

An example is the thermal desorption of NO from Ru(001) (Cavanagh and King, 1981). Here, Ru(001) was chosen as the metal single crystal since it is highly reactive to NO and since the vibrational modes of adsorbed NO as well as the coverage–dependent binding sites are well known. Fig. 37 shows the set-up of the experiment. NO molecules are desorbed via resistance heating (heating rate 12 K/s) and are detected by a pulsed dye laser via laser-induced fluorescence (LIF) on the $X^2\Pi_{1/2} \to A^2\Sigma^+$-transition ($\nu \approx 44141$ cm^{-1}).

The observed LIF intensity is proportional to the population N' in the laser-excited state of the NO molecules (see Eq. (3.5)). In the case of saturation of the transition the particle density of the ground state is $N'' = N'$ and thus can be determined directly from the LIF intensity after performing a reference measurement in a cell with known particle density.

In the case of thermal equilibrium between the rotational states one finds

$$N'' \propto (2J'' + 1)\exp\left(-\frac{E_{J''}}{k_B T}\right), \quad (2.7)$$

i.e., $\ln I_{\rm LIF} \propto -E_{J''}$, with $E_{J''}$ the internal energy of the rotational states. Fig. 38 shows the LIF that has been measured that way. From the slope of the straight line one obtains a rotational temperature of 235±35 K.

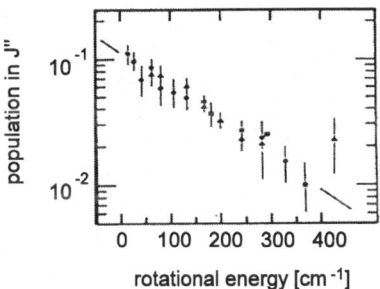

Figure 38 Measured rotational state distribution as a function of rotational energy for thermal desorption of NO/Ru(001). The line corresponds to a rotational temperature of 235 K. Reprinted with permission from (Cavanagh and King, 1981). Copyright 1981 American Physical Society.

The surface temperature for this experiment was set to 455±20 K, meaning that the NO molecules did *not* desorb in thermal equilibrium (free rotating) from the surface, but have most probably traversed a chemisorption surface state. Thus from the measured rotational and vibrational state distributions for given surface temperatures information about the existence of barriers perpendicular to the surface and thus on the topology of the molecule–surface potential can be derived.

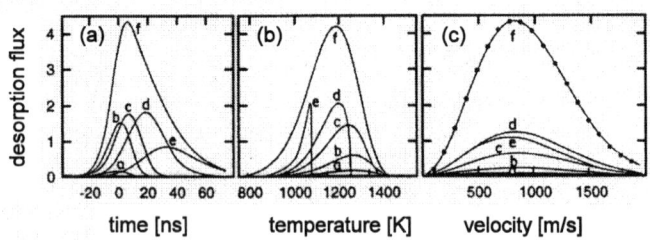

Figure 39 Calculated flux of desorbing model adsorbates (*e.g.*, CO) from a model surface (*e.g.*, ruthenium) for different positions from the center of the heating laser beam as a function of time (a), temperature (b) and velocity (c). The desorption is assumed to obey first order kinetics. The abbreviations denote: a: $r=0$ (center); b: $r=0.1\ r_p$ with r_p meaning the spatial FWHM of the laser pulse; c: $r=0.25\ r_p$; d: $r=0.3\ r_p$; e: $r=0.4\ r_p$ (that position where half of the adsorbate is desorbed); f: total flux. Reprinted from *Surf. Sci.*, (Brand and George, 1986), Copyright 1986, with permission from Elsevier Science.

In order to achieve desorption kinetics of second order ($n=2$), two adsorbate atoms have to migrate over the surface and have to desorb recombinatively. This is the case, for example, for the desorption of H_2 from ruthenium (Guthrie et al., 1982). The pre-exponential factor here is 10^{-3} cm^2/adsorbate·s .

From measured time-of-flight (TOF) velocity distributions of thermally desorbed particles that have experienced low heating rates one might derive directly surface temperatures. This becomes more difficult in the case of LITD. The main reason for possible problems is that the flux of desorbing adsorbates (which is the value that is typically measured in the case of laser desorption experiments) shows maxima at different times depending on the distance r from the center of the approximately Gaussian-shaped heating laser beam (curves a – e in Fig. 39a). The particles at the center of the laser beam are desorbed first.

As a consequence particles leave the surface distributed over a wide range

2.1. LASER-INDUCED THERMAL DESORPTION

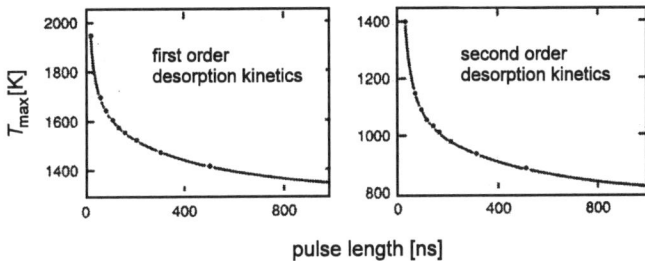

Figure 40 Calculated maximum temperatures as a function of pulse length for first order (left-hand side) and second order (right-hand side) model desorption kinetics Reprinted from *Surf. Sci.*, (Brand and George, 1986), Copyright 1986, with permission from Elsevier Science.

of temperatures (Fig. 39b) with particles desorbed at the center having the highest temperature. Whereas the maximum of surface temperature is at 1600 K, the most probable temperature of the total flux of desorbing particles is at 1250 K. In the observed TOF distribution of the desorption flux one has to attribute a different Maxwell–Boltzmann distribution to each determined temperature according to Eq. (6.16). However, the sum of these distributions, which results from particles desorbing from different spatial positions within the heating laser spot, leads to another Maxwell–Boltzmann distribution at a similar 'temperature' of 1190 K (Fig. 39c, curve f, dots). Hence the observation of Maxwell–Boltzmann TOF distributions alone is not unambiguous proof that the particles have been desorbed at thermal equilibrium with a given surface temperature. The observed distribution might also result from averaging over the sum of distributions, resulting from very different surface temperatures.

An appropriate choice of the laser pulse length enables one to obtain high desorption rates without experiencing the usual high surface temperatures, which always include the possibility of surface melting. In Fig. 40 the achieved maximum temperature is plotted as a function of laser pulse length assuming that half the adsorbate coverage has been desorbed at $0.4\,r_p$. The left-hand side of the figure has been calculated for desorption kinetics of first order, the right-hand side for kinetics of second order.

Obviously with long pulses one obtains the same desorption yields for significantly smaller temperatures as compared with short pulses, especially if the desorption kinetics is of second order. This is mainly because for long pulses the atoms have a higher probability of recombining at the surface prior to desorption.

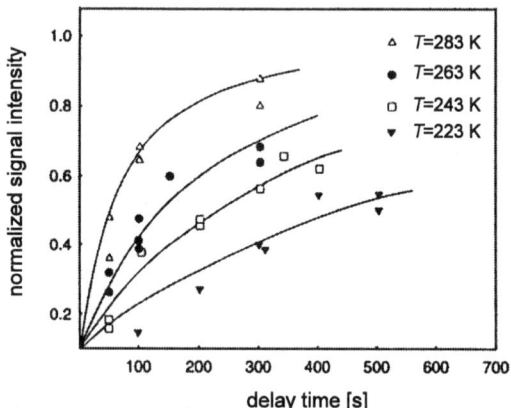

Figure 41 Normalized desorption signal intensity as a function of temporal delay between a desorption and a detection laser pulse for the system H/Ni(100) and different surface temperatures. Reprinted from *Surf. Sci.*, (George et al., 1985), Copyright 1985, with permission from Elsevier Science.

Measurements of diffusion

Laser-induced thermal desorption allows one not only to obtain information about the interaction potential *normal* to the surface, but also about the potential *along* the surface. This potential dominates the mobility of the adsorbates. Vice versa measurements of the diffusion of the adsorbates allow one to obtain information about the forces acting along the surface.

In order to measure the diffusion of the particles along the surface, the system is brought initially out of equilibrium by a desorption laser pulse, which removes the adsorbate within the laser focus diameter (Viswanathan et al., 1982). The following laser pulse, applied onto the same spot a few seconds later, desorbs those adsorbate particles that have refilled the initial hole in the coverage. The desorbed particles are detected with a mass spectrometer, which thus determines the particle rate that has moved along the surface.

If one normalizes the mass spectrometer signal intensity onto the initial signal intensity before application of the desorption laser, then one obtains as a function of desorption time τ between the desorption and detection pulses for the given surface temperature T_s a signal intensity $S(\tau)$, which increases slowly and approaches finally the initial signal intensity. This behavior is

2.1. LASER-INDUCED THERMAL DESORPTION

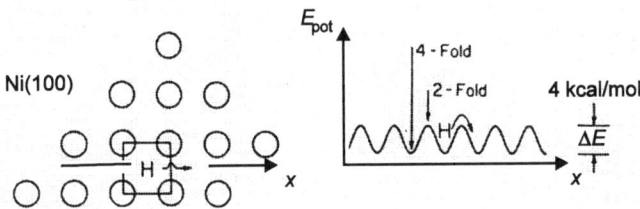

Figure 42 Surface diffusion of hydrogen atoms from a fourfold via a twofold into a neighboring fourfold-symmetry surface spot on a Ni(100) surface. The barrier between the two spots (4 kcal/mol = 174 meV) corresponds to the measured activation energy for diffusion. The circles on the left-hand side represent Ni atoms. On the right-hand side the potential energy of the hydrogen atoms is plotted in the x-direction along the surface. Reprinted from *Surf. Sci.*, (George et al., 1985), Copyright 1985, with permission from Elsevier Science.

plotted in Fig. 41 for the adsorption of hydrogen on a Ni(100) surface. The initial coverage with hydrogen was 12% of the saturation coverage and the desorption laser was a Nd:YAG laser. The diffusional motion into an initially depopulated spot of radius a can be obtained, assuming macroscopic mass transport, via the solution of Ficks second law (George et al., 1985)

$$S(\tau) = \frac{2}{a}\sqrt{\frac{D(T_s)t}{2\pi}}. \qquad (2.8)$$

The corresponding fit curves in Fig. 41 result in values for the diffusion constant $D(T_s)$ as a function of surface temperature. As seen, the diffusion rate increases strongly with increasing temperature. An exponential fit of the diffusion constant vs temperature results in an activation energy for the diffusion of 174 meV and a hopping frequency of 3.4×10^{-13} s^{-1}.

The value of the activation energy is about 17% of the value for the recombinative desorption of hydrogen on Ni(100). An activation energy E_a of about 5%–20% of the value of the binding energy has been found experimentally for many adsorbates (Zangwill, 1988). Apparently the value for E_a has to amount to at least the difference between the heats of adsorption for adsorbates on strongly and weakly bound places.

In the case H/Ni(100) the measured activation energy probably corresponds to the barrier for diffusional motion between a surface adsorption place in the middle of four Ni atoms ('fourfold hollow site') to the next place with equal symmetry (Fig. 42). The fourfold symmetry place is energetically more favorable compared with the twofold one. Therefore it is preferentially populated in thermal equilibrium.

The laser desorption method for the measurement of diffusional motion on surfaces is applicable in principle to all possible adsorbate–substrate systems. However, the evaluation of the measurements is only simple and unique if (i) the diffusion rate is independent of coverage; (ii) the laser desorption affects only a well-defined surface spot with known diameter; (iii) the diffusion rates are larger than about 10^{-8} cm^{2}s^{-1} and (iv) the desorption rate is negligible during the measuring period (about 100 s). Since the diffusion and desorption coefficients usually are not independent of each other (the ratio is in most cases about 1:6) this method of measurement is applicable only within a limited range of temperatures. Finally, (v) an important condition is that after each laser shot the same clean surface with identical morphology is obtained. In this context surface reactions within the laser spot are critical parameters that might affect the outcome of the measurements. If one takes all these restrictions into account it becomes clear why the method has been applied only to a few selected systems so far.

2.1.1 MALDI

Laser-induced thermal desorption processes might be used to transfer large organic molecules with negligible fragmentation rate into the gas phase. This is performed by diluting the biologically important molecules (*e.g.*, peptides or proteins with masses up to 300 000 amu) in an acid such as nicotinic acid (NiAc). Subsequently the mixture is applied to a metallic sample probe and is dried. Then an UV laser (*e.g.*, 266 nm) irradiates the sample, the energy of which is adsorbed mainly in the matrix and leads to its 'explosive' desorption. In the course of this process the biomolecules M are also desorbed and protonated by a reaction of the form (Karas et al., 1987)

$$\text{NiAc}^* + \text{M} \rightarrow (\text{M+H})^+ + (\text{NiAc-H})^-. \tag{2.9}$$

This protonation process allows one to monitor the ionic molecules directly in a mass spectrometer for large masses, such as a reflectron time-of-flight (Karatev et al., 1973) or a Fourier transform mass spectrometer (Marshall and Schweikhardt, 1992). The method is called MALDI (matrix-assisted laser-induced desorption and ionization) (Hillenkamp et al., 1991; Kirpekar et al., 1998).

Special care has to be taken that the matrix is dilute enough (at least a molar ratio of 100:1 of matrix to analyte) in order to ensure isolation of the biomolecules from each other. Also, the power density of the laser should be about 1 MW/cm^2 to achieve a sufficient ion yield but to also minimize the fragmentation processes. On the other hand, the same matrix, optimized for the laser desorption and ionization process, can be used for a variety of different biomolecules, and so the method has a high degree of universality.

2.2. DESORPTION FOLLOWING ELECTRONIC EXCITATION

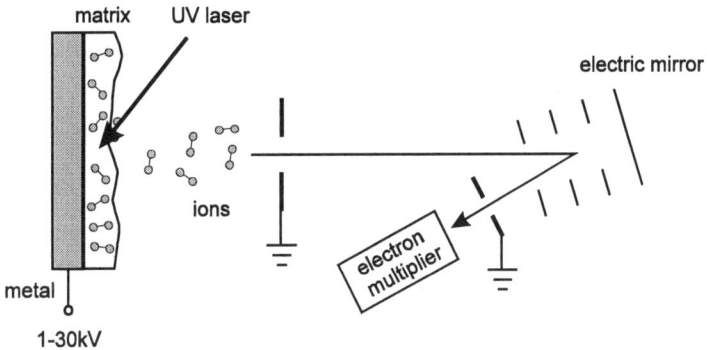

Figure 43 Schematic drawing of a MALDI apparatus for the detection of biologically important molecules, including a reflectron time-of-flight mass spectrometer. The protonated ions are accelerated by a high voltage into the reflectron.

An account of alternative laser-based mass-spectrometric techniques, such as laser microprobe mass spectrometry, post-ionization detection or secondary plasma source mass spectrometry, can be found in (Lubman, 1990). A review with special emphasis on the laser desorption of biomolecules is given by (Levis, 1994).

2.2 Desorption Following Electronic Excitation

The observed desorption processes are not necessarily dictated by heating of the substrate, but might also be the result of the direct electronic excitation of an antibonding state of the adsorbate or the excitation of hot photoelectrons in the substrate.[1] A recent review note of nonthermal laser-induced desorption processes is given in (Al-Shamery and Freund, 1996). See also Chapter 4, Section 4.4.

The electronic states of the adsorbates differ significantly from that of the free molecule, as exemplified in Fig. 44 for NO adsorbed on a platinum surface (Harris et al., 1995). For the equilibrium distance of the adsorbed NO molecules, z_{eq}=1.1 Å, at t_1 the resonance level is located 2.38 eV above the

---

[1] One usually characterizes as 'hot' those electrons that are not in thermodynamic equilibrium with the lattice heat bath. In the case of laser excitation with ultrashort pulses (femtoseconds) hot electrons are produced with high probability since the electron–phonon relaxation time constant (several picoseconds) is usually larger than the laser pulse length.

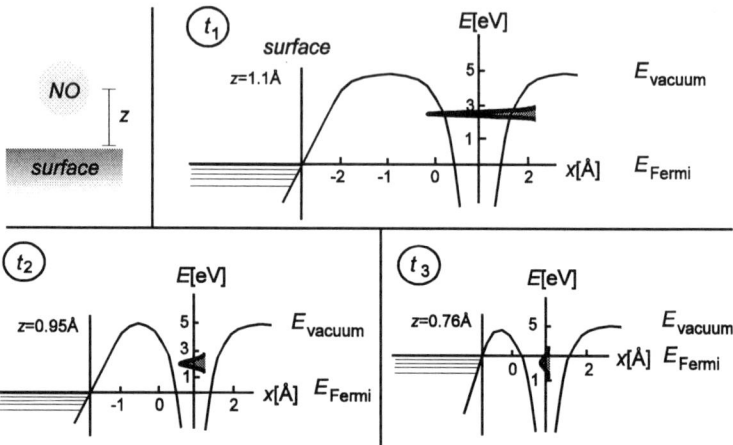

Figure 44 Plot of total one-electron potential energy E of a NO molecule (neglecting its internal degrees of freedom), adsorbed on a platinum surface as a function of distance x between the laser-excited electron and surface for constant molecule–surface distances z ($t_1 - t_3$). Also plotted is the associated resonance level. According to time-dependent quantum wave packet calculations. Reprinted with permission from (Harris et al., 1995). Copyright 1995 American Institute of Physics.

Fermi level with a lifetime of 14.5 fs (0.045 eV half width). It corresponds to the lowest unoccupied molecular orbital (LUMO) $2\pi^*$ of the NO–Pt complex. With decreasing distance z of the molecule to the surface the barrier to the substrate becomes narrower, the probability for tunneling of the excited electrons increases and the resonance broadens (0.16 eV at $z=0.95$ Å, 1 eV at $z=0.76$ Å). Its energy is lowered compared with the Fermi level. The total energy of the resonance, which is given by the exciting laser, of course stays constant. The decreasing distance between electron and surface corresponds to the movement of the excited wave packet on the excited potential energy curve between t_1 and t_3 following the Antoniewicz model (Fig. 45). The quantum mechanical rate for tunneling P_T depends on width and height of the barrier, i.e., on the equilibrium position of the adsorbates electrons with respect to the substrate,

$$P_T \propto \exp\left(-\int_{z_1}^{z_2} \sqrt{2m(V(x) - E)}\,dx\right). \tag{2.10}$$

2.2. DESORPTION FOLLOWING ELECTRONIC EXCITATION

Here, E and m denote the electron's energy and mass and $V(x)$ is the form of the barrier (to first approximation a parabola).

The electrons of the adsorbate might be excited by the laser into states above the Fermi level of the substrate, from which they tunnel into unoccupied states of the substrate (adsorbate-mediated desorption; the reverse, substrate-mediated process is sketched in Fig. 110). The lifetimes of the excited adsorbate states are inversely proportional to the tunneling rate, and in the case of metallic substrates are of the order of femtoseconds. Hence the laser-induced excitation is quenched within a few femtoseconds by substrate excitations. If one takes into account that photodissociation processes in the gas phase usually need tenths of picoseconds,[2] the question arises as to why a direct desorption from metallic surfaces can take place.

Obviously the 'lifetime' of the state at the surface is only a measure of the exponential decay of the population in the excited state and not a fixed value. Hence in the tail of the exponential decay curve one will always find particles that stay long enough in the excited state to be able to leave the surface.

A somewhat more quantitative model of this process in the framework of effective one-electron potentials might be obtained via the 'Menzel–Gomer–Redhead (MGR)' model (Menzel and Gomer, 1964; Redhead, 1964; Menzel, 1995) (Fig. 45). Following laser excitation into an antibonding state (repulsive potential) the molecule moves adiabatically along the excited state potential curve. Correspondingly, the distance to the substrate increases. Depending on the electronic structure of the adsorbate–substrate system the excited particle will lose energy during its movement on the excited state potential curve via quantum mechanical tunneling (nonirradiative) or via radiation of a photon. This means that it is de-excited to the ground state. Since the distance to the substrate has been increased, the molecule is electronically de-excited after recurrence to the electronic ground state, but it has gained kinetic energy E_{kin} by the movement on the repulsive potential energy curve. If this energy gain is larger than the remaining barrier to desorption, then the molecule will desorb with an energy E_{kin}^f.[3] If the energy gain is too small to allow desorption, the highly vibrationally excited molecule will lose its laser-induced extra energy by coupling to lattice oscillations (phonons) of the solid. Eventually the substrate is thus heated indirectly via adsorbate excitations.

The MGR model might be modified easily for physisorbed systems if one assumes a stronger (ionic) bound state to be the laser-excited state (Antoniewicz, 1980). As seen in Fig. 45 for desorption to occur again a gain of kinetic energy via traversing a potential curve is necessary. In contrast to the

---

[2] This time constant is given by the slope of the repulsive potential from which the dissociation occurs, in most cases a few to a few tens of eV/Å.

[3] By appropriate choice of the reaction coordinate laser-induced *dissociative* desorption, i.e., the intramolecular bond breaking in the course of the desorption, might be described analogously.

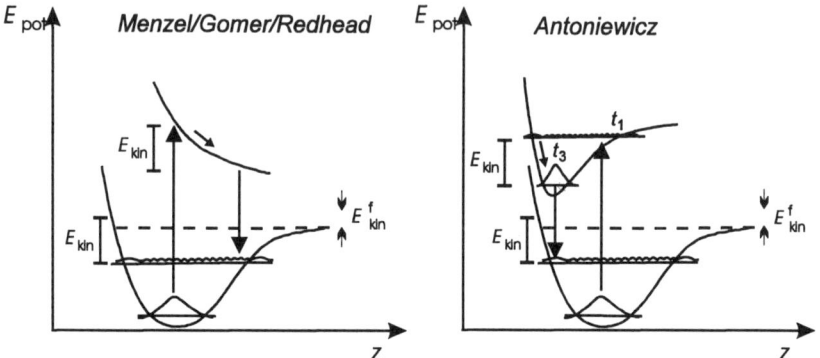

Figure 45 Models for laser-induced desorption of adsorbates. Left-hand side: Menzel–Gomer–Redhead; right-hand side: Antoniewicz (for ionic, excited states). The times t_1 and t_3 are the same as in Fig. 44. E_{kin} are the kinetic energies of the bound adsorbates, E_{kin}^f that of the desorbing adsorbates.

MGR model here the ion is attracted by its image charge. Therefore in the excited state the equilibrium distance is closer to the substrate. If the kinetic energy, which is gained following recurrence into the electronic ground state, is higher than the binding energy, then the adsorbate will desorb.

The desorption process from an insulator and that from a metal surface differ significantly by the importance of the electronic relaxation ('quench') processes. This is reflected in the resulting velocity distributions. For insulator surfaces with large band gap no significant electronic quenching takes place and the average velocities of the desorbing particles are dictated by energy conservation, i.e., photon vs binding energy. For metals and semiconductors, and especially for chemisorbed adsorbates, the electronic quenching times are of the order of 10^{-16} s to 10^{-14} s, i.e., short compared with the nuclear movements. It can be shown (Zimmermann and Ho, 1994) that in this case the resulting velocity distributions have the form of (flux–weighted) Maxwell–Boltzmann distributions although the underlying processes are clearly non-thermal. The measured densities are given by

$$n(t) \propto t^{-4} \exp\left(-\frac{m\Delta x^2}{2k_B T t^2}\right). \tag{2.11}$$

Here, T means an effective Boltzmann temperature which has no direct relation to the surface temperature and is merely a fit constant together with the mass m and the effective distance Δx between the detector and surface.

2.3. OTHER LASER-INDUCED DESORPTION PROCESSES

The Boltzmann distribution results mainly from the exponential decay of the adsorbate as a function of its residence time in the excited state. The exact form of the distribution (hence the temperature T) is given by the slope of the repulsive potential (in the case of NO/Pt(111), *e.g.*, 40 eV/Å) and the electronic relaxation time at the critical point.

The above discussed desorption processes are usually categorized under the term *DIET* ('desorption or dissociation induced by electronic transitions') and are the topic of a voluminous series of compendia of articles (Tolk et al., 1983; Brenig and Menzel, 1985; Stulen and Knotek, 1986; Betz and Varga, 1990; Burns et al., 1993; Szymonski and Postawa, 1995) which might be consulted for further discussion. Since laser-induced desorption from surfaces obviously is one of the most simple photochemical processes, one commonly subsumes many processes of surface *photochemistry* (Chapter 4, Section 4.4) also under the heading DIET. In this monograph we discuss photochemistry in more detail in the context of ultrafast processes on surfaces (Chapter 4).

2.3 Other Laser-Induced Desorption Processes

Desorption, induced by irradiating (adsorbate-covered) surfaces with intense laser light, is in most cases well described by invoking either thermal (LITD) or electronic processes (DIET). However, there are many variants and modifications, depending on adsorbate or substrate boundary conditions. The desorption must not lead to neutrals, but can also result in desorbed ions, which might result from a DIET (Helvajian and Welle, 1989) or a thermionic process (Campbell et al., 1992). The ionic (Shea and Compton, 1993) and the neutral yield (Lee et al., 1993) as well as the emission of electrons is much enhanced if the desorption process includes electromagnetic field enhancement effects via surface plasmon excitation. Substrate-mediated desorption mechanisms via hot carrier excitation have been identified not only for molecular desorption, but — in spite of the strong electronic coupling — also for desorption of metallic atoms from metallic layers (Hellsing et al., 1997). Of course there is also the possibility to resonantly excite adsorbate eigenenergies by the laser, which in the course of 'resonant heating' (Gortel et al., 1983a; Gortel et al., 1983b) or 'vibrational predesorption' (Heidberg et al., 1987) results in the desorption event.

In what follows as an example of a laser-induced desorption process that does not fit unambiguously into either the LITD or DIET categories, we discuss in some detail electromagnetic field enhanced desorption from metallic islands on insulator surfaces.

Electromagnetic field enhanced desorption
Field enhancement on a rough surface largely increases the probability for ablation processes as compared with a flat surface (Ertl and Küppers, 1985). Such a rough surface might consist of nanometer-sized islands (cf. Chapter 4,

Section 4.1.3) or 'clusters'. If an incoming transverse electromagnetic field hits the clusters, then a size- and wavelength-dependent longitudinal collective oscillation of the electron density might be excited in the islands, a *surface plasmon excitation* or *giant resonance* (Brechignac and Connerade, 1994). This is in contrast to the case of a continuous metallic film, where such a process is not possible due to momentum conservation. For the case of rough alkali cluster films a resonance-like behavior in the desorption probability for atoms (Hoheisel et al., 1988) and molecules (Viereck et al., 1997b) has been realized experimentally, in agreement with theoretical predictions (Monreal and Apell, 1990) which suggest the direct creation of antibonding pairs at defect sites in the course of a surface plasmon excitation.

The absorption coupling has been measured to be cluster size and laser wavelength dependent (Balzer et al., 1993b), resulting in a variation in desorption yield as a function of size, but not in a change in kinetic energy distributions (Balzer et al., 1997b). This indicates that the initial photon absorption proceeds via surface plasmon excitation, which is also the rate-limiting step and is largely decoupled from the final particle desorption process.

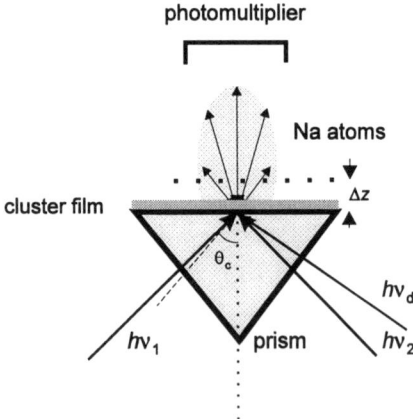

Figure 46 Set-up for an evanescent wave time-of-flight detector. Desorption laser $h\nu_d$ as well as detection lasers $h\nu_1$ and $h\nu_2$ hit the prism at an angle larger than the critical angle $\Theta_c = 41.3°$; i.e., they are totally internally reflected. The evanescent wave of all the lasers is coupled to a rough metal film on top of the prism hypotenuse and leads to the desorption of atoms, which are instantaneously detected by the detection lasers, using the two photon laser-induced fluorescence scheme of Fig. 59b.

Additional information for a more detailed desorption model has

2.3. OTHER LASER-INDUCED DESORPTION PROCESSES

become available by measurements of the characteristic time constants for photodesorption. This has been performed by applying evanescent-wave time-of-flight (TOF) spectroscopy (Gerlach et al., 1996a). There, advantage is taken of the spatial confinement of the evanescent wave above a prism hypotenuse by growing alkali island films on the prism surface, exciting them with a pulsed desorption laser at total internal reflection and detecting the desorbing atoms via two-photon laser-induced fluorescence also under the angle of total internal reflection (Fig. 46). Thus essentially a TOF detector with extremely small dimension normal to the surface is used, given by the decay length (about 250 nm) of the evanescent wave

$$\beta = \frac{\lambda}{2\pi} \frac{1}{\sqrt{n_1^2 \sin^2 \Theta_i - 1}}, \qquad (2.12)$$

with λ being either the wavelength of desorption or detection laser, $\Theta_i = 45°$ the angle of incidence and n_1 the index of refraction of the prism. Due to the fact that the detection lasers are counterpropagating, this detector is also velocity-selective along one direction on the surface.

The measurements result in velocity distributions (differential reflection coefficients) with velocity components in x-, v_x, and y-direction, v_y,

$$\frac{d^2R}{dt dv_x} \propto \int_{z_0}^{\infty} dz\, e^{-2k_L z} \frac{1}{z} \int_{-\infty}^{+\infty} dv_y\, \sqrt{v_x^2 + v_y^2 + \left(\frac{z}{t}\right)^2} \frac{d^3R}{dE_f\, d\Omega_f}, \qquad (2.13)$$

with E_f denoting the final energy of the scattered particles and Ω_f the final solid angle in which they are scattered. The velocity distributions have to be convoluted by the Gaussian laser pulse width, by the total decay time of the atoms which are resonantly excited into the $5S_{1/2}$ state (Fig. 58, right-hand side) and finally by the decay time τ_{des} of the adsorbed photons into the desorbed atoms. Here, $k_L = 1/\beta$ with β given by Eq. (2.12) and $z_0 \approx 10$ nm being that point above the surface where the energy levels of the alkali atoms are no longer strongly (with respect to the two-photon transition linewidth) perturbed by the van der Waals interaction with the prism surface.

In order to deconvolute τ_{des} from the measured distribution, the kinetic energy distribution of the desorbing atoms with final velocity v_f has to be known accurately. From a variety of systematic TOF measurements, as discussed in (Balzer et al., 1997b), it is concluded that this is given by the parameter-free multiphonon differential reflection coefficient

$$\frac{d^3R}{dE_f\, d\Omega_f} \propto \frac{\cos^2\theta_f}{v_f} \exp\left(-\frac{mv_f^2}{2k_B T_s}\right), \qquad (2.14)$$

which describes the fraction of particles of mass m that is scattered at a surface temperature T_s into the energy interval dE_f and the solid angle $d\Omega_f$. Here, θ_f

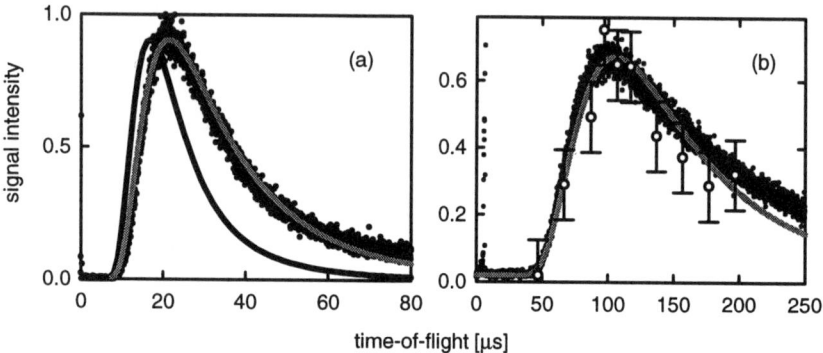

Figure 47 Time-of-flight distributions of Na atoms, desorbed upon laser excitation (λ=500 nm) of surface-bound Na clusters (mean radius 50 nm, surface temperature 300 K). Reprinted with permission from (Balzer et al., 1997b). Copyright 1997 American Institute of Physics.

is the scattering angle, measured from the surface normal. This distribution then proposes a $\cos^2$ angular distribution of the desorbed atoms, which is what has been observed experimentally (Balzer et al., 1993b). Note that in the case of a Maxwell–Boltzmann distribution the measured TOF distribution (for a density detector) would be given by

$$n(t) \propto t^{-4}\exp\left(-\frac{m\Delta x^2}{2t^2 k_B T}\right), \qquad (2.15)$$

where m is the particle's mass, k_B Boltzmann's constant and T a nominal equilibrium 'temperature'. This corresponds, as a function of the particle's velocity v_f, to the differential reflection coefficient

$$\frac{\mathrm{d}^3 R^{M-B}}{\mathrm{d}E_f \mathrm{d}\Omega_f} \propto \cos\theta_f\, v_f\, \exp\left(-\frac{mv_f^2}{2k_B T_s}\right). \qquad (2.16)$$

A comparison of measured TOF curves with the theoretical predictions of both Eq. (2.16) and Eq. (2.14) is shown in Fig. 47. Theoretical curves according to Eq. (2.14) are shown as solid grey lines. The measurements have been performed with the experimental arrangements shown schematically in Fig. 48. Fig. 47a shows two-photon laser-induced fluorescence detection of Na atoms following irradiation with a desorption laser of fluence 0.26 mJ/cm^2. Having traversed a flight distance Δx=11 mm, the desorbed atoms encounter

2.3. OTHER LASER-INDUCED DESORPTION PROCESSES

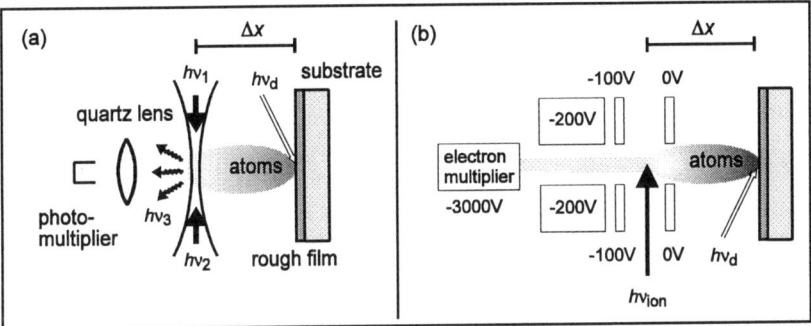

Figure 48 Schematics for the time-of-flight fluorescence detection of laser-desorbed atoms from a rough cluster film on an insulating surface. The desorption laser is denoted by $h\nu_d$.

two counterpropagating detection lasers $h\nu_1$ and $h\nu_2$, which induce blue shifted two-photon fluorescence at $h\nu_3$ according to the term scheme in Fig. 59b (at 330 nm for detected Na atoms). The fluorescence is a measure for the density of desorbed particles. The black line is a Maxwell–Boltzmann distribution at the surface temperature, according to Eq. (2.16).

In Fig. 47b resonance-enhanced two-photon ionization detection of Na atoms is plotted. Here, the desorbed atoms are ionized by a pulsed UV laser at 248 nm and the ions are detected by a time-of-flight spectrometer. The desorption laser fluence was 1.2 mJ/cm^2 and Δx=58 mm. Finally, instead of the pulsed UV laser a cw UV Ar$^+$ laser in combination with $h\nu_1$ has been used (open circles), which then is a resonance-enhanced two-photon ionization (RETPI) scheme (Chapter 3, Section 3.1.3).

Obviously, the Maxwell–Boltzmann distribution at the surface temperature does not reproduce the experimental finding, whereas Eq. (2.14) provides a satisfactory agreement. This is a convincing hint that the plasmon-induced desorption mechanism is more complicated as compared with a simple thermal desorption.

From the TOF measurements the relevant time constant is determined to be of the order of nanoseconds. It is thus suggested that the electronic excitation is lost by vibronic coupling and desorption occurs by a substrate mediated process. Following thermalization, a distribution of neutral alkali atoms has been generated, which are weakly bound to the defect states. Subsequently, these atoms gain energy from the clusters via multiphonon scattering processes. The probability for those processes is high (Manson,

1991; Manson, 1994), since the surface temperature is above the Debye temperature of the alkali cluster film, the particle mass is high and the masses of scattered and scattering atoms are equal. The kinetic energy distribution of the desorbed atoms then can be described by expressions from semiclassical multiphonon scattering theory (Eq. (2.14)) (Balzer et al., 1997b; Manson et al., 1996).

This model is not that unusual for desorption processes. In the course of photolysis of chemisorbed trimethylaluminum (Higashi, 1989), where the desorbed species is also strongly vibrationally coupled to the substrate, a linear increase in desorption yield with increasing power and subthermal photoproduct distributions have been observed, too. The velocity distributions corresponded to Boltzmann temperatures *below* the surface temperature due to vibrational energy loss into the substrate. Similar subthermal velocity distributions have been observed in the course of the thermal desorption of Ar from Pt(111), a system that has a shallow potential well of 0.1 eV. The results could be reproduced by classical trajectory calculations (Tully, 1981; Lucchese and Tully, 1984). Then, by invoking microscopic reversibility, the high-energy cut-off for the desorbing particles has been explained by the fact that the interaction potential was too weak to bind them.

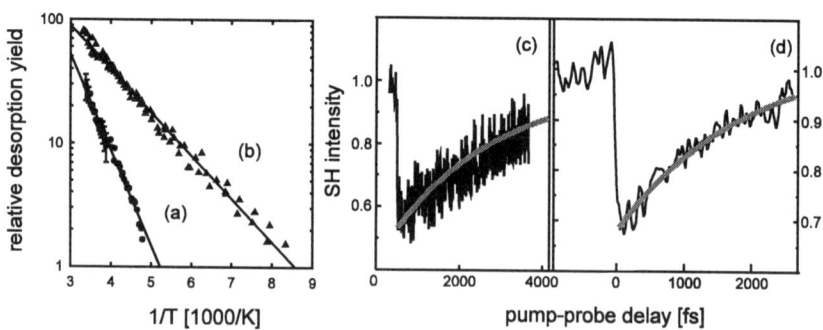

Figure 49 Substrate influence on Arrhenius-type desorption probability ((a) and (b)) and on picosecond temporal response ((c) and (d)) of rough Na cluster films adsorbed on insulators. The figures (a) and (c) correspond to mica as a substrate, (b) and (d) to lithium fluoride.

Obviously, the excited clusters are coupled to the supporting substrate. A significant influence of the surface structure (poly- vs single crystalline) on the desorption dynamics has been observed, for example, in ion-induced desorption of Na atoms from NaCl (Postawa et al., 1992) and in pulsed-laser sputtering from gold surfaces (Bennett et al., 1996) (see also Chapter 6, Section 6.2). In Fig. 49 the substrate influence is demonstrated for alkali clusters on insulators via changes in the picosecond temporal response and in the Arrhenius-type dependence of desorption yield on the substrate temperature T_s :

$$Y_{\text{tot}} \propto \exp\left(-\frac{E_A}{k_B T_s}\right), \qquad (2.17)$$

with E_A being different for desorption from clusters adsorbed on lithium fluoride as compared with mica surfaces.

Fig. 49a is the temperature dependence of the laser-induced yield (λ=500 nm, fluence 0.5 mJ/cm^2) of atoms desorbed from Na clusters with 120 nm radius, adsorbed on mica (Renger and Rubahn, 1993). The straight line corresponds to E_A=153 meV. Fig. 49b shows the same dependence for clusters adsorbed on lithium fluoride. Here, the straight line is for E_A=69 meV. Fig. 49c is a pump/probe measurement at Na clusters with mean radius 38 nm, adsorbed on mica (cf. Chapter 4, Section 4.1.3). The gray line corresponds to a decay time constant of 1.8(2) ps. Finally, Fig. 49d is for clusters adsorbed on lithium fluoride with the fit curve corresponding to 1.2(1) ps.

In all cases the coupling factor depends on the type of substrate. It has been found that electronic relaxation proceeds faster and desorption is more likely for clusters grown on lithium fluoride (LiF) as compared with clusters grown on mica. This could indicate an influence of the ionic nature of the LiF substrate on the relaxation processes or imply that the clusters grown on LiF are rougher (have more defects) than the cluster films grown on mica. The role of surface defects for such laser-induced desorption processes has been elucidated recently (Viereck et al., 1997a).

2.4 Diffusion

A laser technique for the study of diffusion processes on surfaces has been discussed already in Section 2.1 in the context of laser-induced thermal desorption processes. There, the desorption laser only served to generate a burn spot in the adsorbate, which was subsequently refilled by the diffusional motion of the adsorbates. The yield of particles that covered the initial burn spot was monitored by another laser-desorption pulse.

A more general method that also allows one to observe transient processes *on* the surface implies the use of laser-induced gratings. The coherent spatial overlap of two laser beams under a full angle Θ on a surface leads to interference phenomena, which result in a modulation of the light intensity $I(x) = A_1^2 + A_2^2 + 2A_1 A_2 \cos(2\pi x/\Lambda)$, the depth of which is given by the ratio of the amplitudes of the beams, A_1 and A_2, and the total laser power $(A_1^2 + A_2^2)$. The lattice constant of this laser-induced grating (Eichler et al., 1986) is

$$\Lambda = \frac{\lambda}{2\sin(\Theta/2)}, \qquad (2.18)$$

which amounts to 6.5 μm for a wavelength of $\lambda=1064$ nm (fundamental of a Nd:YAG laser, which is usually injection seeded to increase the coherence length) and $\Theta=10.4°$.

Materials diffusion

If the surface is covered by an adsorbate, then the laser-induced heating at positions of maximum field intensity might lead to desorption. The spatial intensity modulation then results in a two-dimensional grating[4] on the surface. An example is shown in (Fig. 50). On the left-hand side the initial grating (a) as well as the calculated lattice modulations for times $t+100$ s (b) and $t+500$ s (c) after laser irradiation are shown. Diffusional motion of the adsorbate molecules NH_3 at the surface Re(001) result in a smearing of the grating. The surface temperature for this calculation was 120 K, meaning a relatively slow diffusional motion. On the right-hand side scattering from this grating in the form of the measured second harmonic intensity is shown. An example for the temporal evolution of the signal is given in Fig. 52.

For read-out of the grating a second laser beam is diffracted linearly or non-linearly, i.e., via frequency doubling or second harmonic generation, SHG. The sharpness of the diffraction maxima and the observed intensity distribution allows one to draw conclusions about the state of the grating and thus about the diffusion processes at the surface. If a one-dimensional diffusional motion of the surrounding adsorbate atoms takes place into the voids that have been induced by the writing laser, then the diffraction intensity will decrease exponentially with time following Ficks second law since the grating will be washed out (Fig. 50) :

$$I(t) = I(t_0)\exp\left(\frac{-8\pi^2 Dt}{s}\right). \qquad (2.19)$$

Here, $I(t_0)$ denotes the initial intensity and D the macroscopic diffusion coefficient, which is assumed to be independent of coverage, but depends exponentially on the surface temperature T_s:

$$D(T_s) = D_0 \exp\left(\frac{-E_a}{k_B T_s}\right). \qquad (2.20)$$

E_a is an activation energy and D_0 the preexponential factor, which in a random walk model is directly related to the average hopping frequency

---

[4] In Chapter 6, Section 6.4 it is shown that even on the adsorbate-free surface a single impinging laser beam due to surface roughness might induce spatially periodic structures.

2.4. DIFFUSION

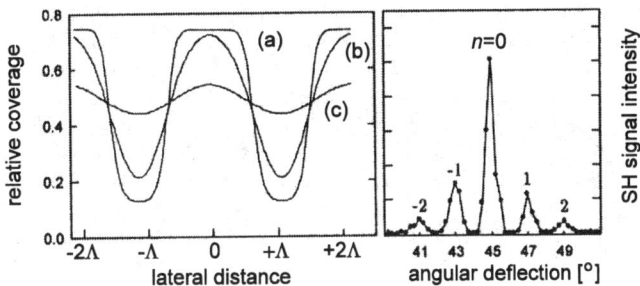

Figure 50 Left-hand side: Spatial variation of surface coverage Θ, normalized to the saturation coverage Θ_S as a result of laser-induced desorption. On the right-hand side the measured intensity of light is shown as a function of angle of incidence for the fully modulated grating. The light has been nonlinearly reflected (frequency doubled) at the ammonia molecules. Reprinted with permission from (Rosenzweig et al., 1993). Copyright 1993 American Institute of Physics.

ν_0 between neighboring surface sites (distance a_0).[5] For a surface with a quadratic unit cell one finds:

$$D_0 = \frac{a_0^2 \nu_0}{4}. \qquad (2.21)$$

The reader should note that the effective diffusion length is smaller than the grating constant, which is given by the intensity modulation. This is because the desorption rate, which leads to the grating formation, depends exponentially on the surface temperature. Hence the desorbed fractions of the adsorbate result from spatially narrower regions than the unaffected ones. In order to draw conclusions from the diffusion measurements a thorough knowledge of the grating profile that has been written into the adsorbate surface is thus absolutely necessary.

Since the surface potential is usually different along different surface directions (for example, parallel and perpendicular to atomic steps), one expects an anisotropy in the diffusional motion, i.e., different diffusion constants for different crystallographic directions. A huge advantage of the application of polarized laser light is the possibility to write a grating along a well-defined crystallographic orientation in the adsorbate. That way it is

---

[5] Here, we have assumed that adsorbate–adsorbate interactions can be neglected.

indeed possible to determine the activation energy for diffusion along or perpendicular to atomic rows on the surface.

A further advantage of the laser method is that laser light can be used to study surfaces *in situ* at points which are hardly accessible for other methods such as field electron microscopy, atom scattering or scanning tunneling microscopy. Thus it becomes possible to apply *simultaneously* to the laser method a second method in order to determine, for example, the change in crystallographic order during the diffusion measurement.

An example of this simultaneous measurement of several surface-specific quantities is the determination of average surface coverage and diffusional motion by the use of specularly reflected light and the light that has been diffracted into first order (Fig. 51). The set-up shown in Fig. 51 has been used to determine the diffusional motion of CO (coverage one monolayer) on Ni(111) at different surface temperatures via detection of the temporal variation in the frequency doubled light (SHG) (Zhu et al., 1988).

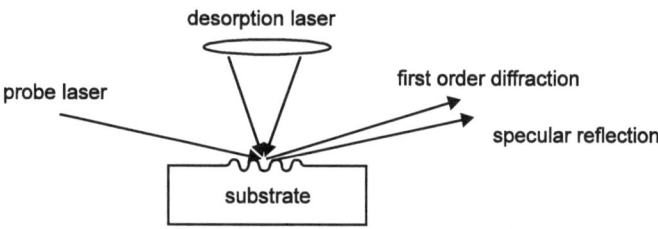

Figure 51 Principal set-up for the measurement of surface diffusion with lasers.

Fig. 52 shows that the SH intensity decreases exponentially as a function of time. If one assumes that the second-order nonlinear surface susceptibility increases linearly with adsorbate coverage and the diffusion constant depends exponentially on temperature (Eq. (2.20)), then the SH intensity in first-order diffraction should obey the exponential decrease of Eq. (2.19). From an exponential fit to the measured values in Fig. (52) and using the known grating constant $s=20$ μm one might deduce a value of D for each surface temperature. An Arrhenius plot then leads to values of the activation energy for diffusion (300 meV for CO/Ni(111)) as well as the pre-exponential factor ($D_0 \approx 10^{-5}$ cm^2 s^{-1}).

The SH method for the optical measurement of surface diffusion, although implemented for the study of rather fundamental systems such as the quantum diffusion of hydrogen on metals (Cao et al., 1997), relies on two important

2.4. DIFFUSION

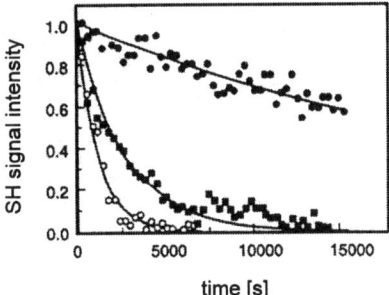

Figure 52 The frequency doubled light that has been diffracted into the first order of a grating that has been written into a CO monolayer on Ni(111) as a function of diffusion time. The symbols denote measurements at different surface temperatures: 219 K (•), 261 K (filled squares) and 273 K (o). The solid lines are exponential fit curves. Reprinted with permission from (Zhu et al., 1988). Copyright 1988 American Physical Society.

assumptions: (i) the linear dependence of the SH signal intensity on the adsorbate density; and (ii) a diffusion coefficient that is independent of surface coverage. If these conditions are not fulfilled, then the modulation depth of the grating within the adsorbate should be very small (less than 0.1 monolayer) in order to avoid falsification of the measurement via coverage dependencies, or interactions between the adsorbates such as lateral repulsion have to be taken carefully into account (Wei et al., 1996). Note that SHG is a nonlinear optical method which has an intrinsic high surface sensitivity but an overall sensitivity that is restricted to the observation of minimum coverages of about a tenth of a monolayer.

As a sensitive alternative one might use linear optical diffraction to read-out the grating, *e.g.*, via a HeNe laser (Xiao et al., 1992; Xiao et al., 1993). The main problem with this approach is given by the background signal intensity from light scattered off the bulk or off the surface roughnesses. If one modulates the impinging light of the HeNe laser between s- and p-polarization via a photoelastic modulator, then the diffracted signal will also be strongly modulated according to the angle between the laser polarization vector and the grating direction on the surface. On the other hand, the background signal will only be slightly modulated. Hence a phase sensitive detection technique ('lock-in' technique) allows one to attenuate the noise background down to 10^{-11} of the incoming light power (Xiao et al., 1992). Since the diffraction efficiency of the modulated grating of, for example, half a monolayer of CO on Ni(110) is about 5×10^{-7} of the incoming power (Xiao et al., 1992), variations in the grating modulation might be detected down to a small fraction of a monolayer. In that way an accurate determination of the coverage dependence of the anisotropic diffusion coefficient becomes possible (Xiao et al., 1993).

Limitations of all optical methods are given by the macroscopic dimensions of the generated gratings. In the case of linear gratings these dimensions are of the order of micrometers; in the case of nonlinear gratings, of fractions of the laser wavelength. The measured diffusion coefficient is thus always a *macroscopic* diffusion coefficient, and its determination makes it necessary to bring the system initially into a nonequilibrium-state, from which it relaxes into the equilibrium state that is given by the surface temperature. Hence one might expect that the *microscopic* tracer diffusion coefficient D^t, which includes the motion of individual particles on the surface, differs significantly from the macroscopic coefficient due to the defects and the microscopic surface morphology.

Thermal diffusion

Grating techniques are also used to investigate (anisotropic) lateral and vertical heat transport along the surface of bulk materials or through thin films (Käding et al., 1995). A transient thermal grating is induced in the material to be investigated by the absorption of two interfering pulses of a heating laser. The temporal evolution of the grating is monitored by reflection of a (cw, e.g., HeNe) probe laser. Spatial displacements of the surface (of the order of a tenth of a nanometer) result via the elastic properties of the material (i.e., the linear thermal expansion coefficient, Poisson ratio and thermal diffusivity κ)[6] in an angular deflection of the probe laser by a few μrad, which might be detected using a position-sensitive photodiode. The exact value of the change in displacement is governed by the temperature rise induced by the heating laser and might be described by the coupled differential equations for thermal diffusion (cf. Eq. (5.21)) and for thermoelasticity (Nowacki, 1986). Obviously, besides angular reflection of the probe laser beam, the spatially periodic heating also induces a change in signal amplitude via the change in surface reflectivity (tenths of a promille). Solving the coupled differential equations allows one to predict both intensity and deflection amplitude changes with a single set of thermal constants and thus provides a convenient cross-check of the results.

It has been shown (Käding et al., 1995) that the surface displacement decays at the surface of bulk materials with a characteristic time constant

---

[6] This approach neglects the possibility of exciting acoustic modes (Duggal et al., 1992), which oscillate on a much shorter time scale as compared with thermal diffusion phenomena; cf. Chapter 4, Section 4.2.2. It is noted that a combination of pulsed laser excited ultrasonic waves and cw laser detected waves has enabled noncontact surface-hardness measurements to be performed for industrial applications (Safaeinili et al., 1996).

2.4. DIFFUSION

$$\tau_{sd} = \frac{\Lambda^2}{4\pi^2 \kappa}, \qquad (2.22)$$

where Λ is the grating constant, given by Eq. (2.18). Within this time interval (the 'sensitivity time' of the method) the spatial extension of the material that is affected is given by the thermal diffusion length $L = 2\sqrt{\kappa \tau_{td}}$. Equating the time constants, one sees that

$$L = \frac{\lambda}{2\pi \sin(\Theta/2)}. \qquad (2.23)$$

Thus the thickness to which the grating method is sensitive to thermal properties of subsurface layers can be varied by varying the interaction angle Θ of the heating laser beams. This idea has been applied successfully to the measurement of effective diffusivities of a free-standing 76 μm thick diamond film as a function of depth inside the film (Käding et al., 1995), thus showing that the diffusivity of a thin film decreases due to an increase in growth defects.

3

Spectroscopy

In Chapter 1, Section 1.2.4 we discussed the spectroscopy of electronic and ro-vibrational states at surfaces using high energy photons (UV or X-ray photoemission), electrons (Auger, energy loss spectroscopy) or low-energy atomic beams (helium atom scattering). Those methods have been well-established during the last decades as versatile tools to characterize single-crystalline surfaces in ultrahigh vacuum environments (Duke, 1994). However, they lose significance if the solid under investigation is situated in ambient air, a liquid or a similar environment.

Here, optical methods gain importance, which include near-infrared, visible or ultraviolet photons. Using light, the surface under investigation can be investigated even through perturbing media, and the information of inerest can be transported by the reflected or transmitted light beam out of the medium. Widespread linear techniques are *absorption spectroscopy, laser-induced fluorescence, ellipsometry and reflection spectroscopy*. Nonlinear methods such as *optical second harmonic generation* or *surface enhanced stimulated Raman scattering* show, in contrast to the linear methods, in many cases high surface sensitivity.

3.1 Spectroscopic Methods

Laser light allows one to obtain a very high spectral resolution $\delta\nu/\nu \geq 10^{-15}$, i.e., to irradiate the system under investigation with photons of extremely small energy uncertainty. This good monochromasy of laser light thus enables one to determine the energies of the eigenstates of the system with high precision. In the case of a free molecule those are the quantized vibration of the molecular nuclei (Fig. 53), the eigenenergies of which might be expanded as a function of vibrational quantum number v:

$$E_v = \omega_e \left(v + \frac{1}{2}\right) - \omega_e x_e \left(v + \frac{1}{2}\right)^2 + ... \tag{3.1}$$

with the molecule-specific vibrational constants ω_e (harmonic) and $\omega_e x_e$ (anharmonic) as well as higher terms which take the anharmonicity of the molecular potential into account.

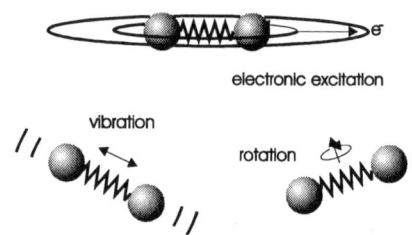

Figure 53 Elementary excitations of diatomic molecules.

An additional degree of freedom is the rotation of the molecule along the nuclear axis (Fig. 53), which results *within* a vibrational state in a splitting of the eigenenergies into quantized rotational states:

$$E_R = B \cdot J \cdot (J+1) + D \cdot (J \cdot (J+1))^2 + ..., \qquad (3.2)$$

where B and D are the harmonic and anharmonic rotational constants of the molecule and J is the rotatonal quantum number. The rotational eigenenergies are usually one to three orders of magnitude smaller compared with the vibrational energies (for the lithium dimer, *e.g.*, ω_e=351.43 cm^{-1} and B=0.673 cm^{-1} (Herzberg, 1950)).

In the case of direct laser excitation of molecular rotations or vibrations the photons couple with the molecule via the permanent dipole moment (infrared excitation) or via the nuclear motion induced, spatially variable polarizability (Raman excitation). However, the valence electron of the molecule might also be excited via a dipole transition (Fig. 53). Then the electronic term energy T_e is added to the rovibrational energies. This latter energy is usually an order of magnitude higher compared with the vibrational energy (for the lithium dimer the lowest electronically excited state $A^1\Pi$ has a term energy of $T_e = 14068.31$ cm^{-1} (Herzberg, 1950)).

All possible eigenenergies of a diatomic molecule might be plotted in the form of an adiabatic potential as a function of distance between the two nuclei (Fig. 54). The ground state (v=0, J=0) defines the vibration and rotationless molecule at an average equilibrium distance of the two nuclei. The potential energy of this state possesses, due to the zero-point energy, the finite value

3.1. SPECTROSCOPIC METHODS

$\frac{1}{2}\omega_e$.

Following excitation into higher excited rovibrational states within the electronic ground state (infrared transition) the equilibrium distance between the atoms of the molecule increases (Fig. 54). With increasing distance the anharmonicity is increasing, resulting in a decreasing distance between neighboring vibrational levels until the molecule dissociates. An excitation of the valence electron using a laser (visible or UV excitation) results in a new, energetically higher positioned electronic potential curve, the equilibrium distance of which usually is located at larger nuclear distances compared with the electronic ground state curve. This shift of the mean internuclear distances means that the relaxation of electronically excited molecules into the electronic ground state generates vibrationally highly excited molecules. Since the electronic relaxation time (nanoseconds) is short compared with the vibrational relaxation time in the electronic ground state (microseconds), the laser-excited molecules remain highly vibrationally excited. Hence one prepares a nonequilibrium distribution of energetically 'hot' molecules in the electronic ground state ('optical pumping' (Bergmann, 1988)).

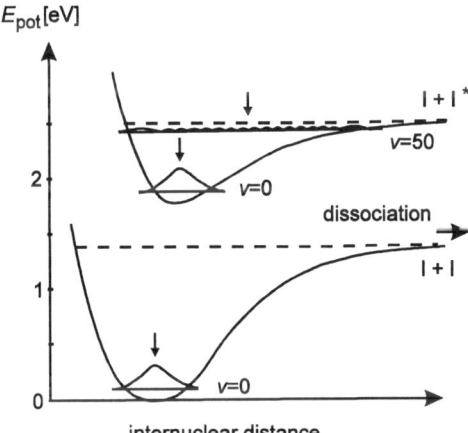

Figure 54 Electronic potentials of the iodine molecule. For the electronic ground state and $v=0$ the spatial probability wave function of both nuclei is plotted as well as that energy above which the molecule dissociates into its atomic fragments. The arrows denote the average nuclear distances. In the electronically excited state the molecule dissociates into a ground state (I) and into an electronically excited atom (I^*). Here, the spatial probability wave function for a highly vibrationally excited molecule ($v=50$) is shown.

If the molecule is adsorbed with its intramolecular axis oriented perpendicularly to the surface plane, then besides the perturbed internal vibration the excitation of other vibrational modes with respect to the surface plane becomes possible, namely the 'frustrated translation' (both atoms moving in the same direction) and the 'frustrated rotation' (the atoms are moving in opposite directions). This is demonstrated in Fig. 55, showing a CO molecule which is adsorbed on a metal surface in the 'bridge position' between two substrate atoms. The adsorption site determines the pointgroup symmetry of the adsorbed molecule (C_{2v} for the shown case). The low C_{2v} symmetry (Hamermesh, 1962) for the adsorption on a bridge site means that all N possible vibrational modes (N = number of atoms in the molecule) possess different frequencies. A measurement of the number of modes thus reveals information about the symmetry of the binding site.

The modes ν_1 (≈ 1930 cm^{-1} for CO/Ni(100) (Hähner et al., 1990)) and ν_2 (≈ 360 cm^{-1}) are the stretching modes of the CO molecule perpendicular to the surface. The mode ν_1 corresponds to the free stretching mode of the CO molecule and ν_2 to the metal–carbon stretching mode (frustrated translation in z-direction), which is on the low energetic side. The values of those two modes might be determined via infrared spectroscopy since they involve changes in the dipole moment perpendicular to the surface. The residual modes are bending vibrations, corresponding to dipole moment changes parallel to the surface. Thus they are not infared-active on metal surfaces. The modes ν_3 (≈ 223 cm^{-1}) and ν_6 (≈ 37 cm^{-1}) are frustrated translations along and perpendicular to the line connecting the metal atoms of the substrate, while ν_4 (≈ 660 cm^{-1}) and ν_5 (≈ 275 cm^{-1}) are frustrated rotations along those directions (Richardson and Bradshaw, 1979). A measurement of the energies of the frustrated rotations and translations (*e.g.*, via helium atom scattering for the modes with lowest energies, i.e., ν_6) provides detailed information on the molecule-surface potential.

The normal mode description of the adsorbate modes implies that the energy spectrum of the bound molecule is the spectrum of a harmonic oscillator: the adsorbate molecule is bound with elastic force constants to the substrate and to its neighboring molecules. Compared with the energy of a free molecule the eigenenergies of the adsorbate modes are usually smaller, since part of the electron density is transferred into the substrate, thus weakening the intramolecular bond. The energy of the ground state vibration of the free CO molecule is 2143 cm^{-1}, while the ν_1 mode in Fig. 55 has an energy that is about 10% smaller. Depending on the coverage of the substrate the frequency spectrum of the bound molecule is determined by the lateral interactions between the adsorbate molecules in addition to the coupling to the substrate phonons. This leads, in the case of a thick adsorbate layer, to the generation of phonon bands of the adsorbate, which replace the former discrete eigenenergy states.

3.1. SPECTROSCOPIC METHODS

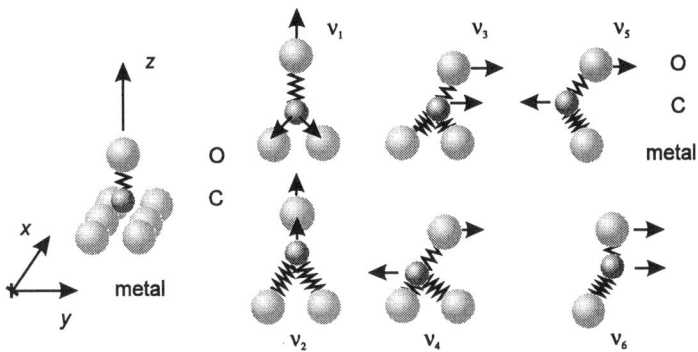

Figure 55 Normal mode description of the vibrations of CO molecules, which are adsorbed in bridge positions on a metal surface (C_{2v}-symmetry). The springs symbolize the elastic force constants between the atoms, the arrows denote the relative deflections of the atoms for the different normal modes. The symbols ν_1 and ν_2 denote the frequencies of the stretch modes perpendicular to the surface. The modes ν_3 and ν_6 are frustrated translations, parallel with respect to an atomic row, above which a bridge position has been occupied (ν_3), and perpendicular to it (ν_6). The modes ν_4 and ν_5 are the corresponding frustrated rotations.

3.1.1 Absorption spectroscopy

Fig. 56 shows the set-up for an experiment to determine the absorption spectrum of an adsorbate on a surface. The light intensity I_0 hits the surface perpendicularly and its transmission is measured behind the substrate using a photomultiplier. If one changes the frequency of the laser light and if the photon energy coincides with the energy of a transition between two eigenstates of the adsorbate, then the photons are adsorbed with a cross section σ and the photomultiplier detects a change $\Delta I = I_0 - I$ between incident (I_0) and detected (I) signal intensity:

$$\frac{\Delta I}{I_0} = \sigma \cdot m \cdot n = \sigma \cdot N. \tag{3.3}$$

Here, n is the surface density of the adsorbate and m the number of adsorbate layers. With increasing total number of molecules adsorbed per area, N, the change in signal intensity is also increasing. If far more than one adsorbate layer is deposited on the surface ($m \gg 1$), then the absorption is described by Beer's law:

$$\frac{\Delta I}{I_0} = 1 - \exp(-\sigma \cdot N). \tag{3.4}$$

Thus a systematic change of the laser frequency results in a *spectrum* of possible energetic differences between the eigenstates of the adsorbate. The absorption lines reflect the distribution of the internal degrees of freedom of the molecule in the ground state as well as the selection rules for the excitation with light. The *depth* of the absorption lines is given by the number density of the molecule in the selected state and the excitation cross section σ, which depends on the optical transition. The *width* of the lines is given by a convolution of the laser linewidth $\Delta\nu_L$ [1] and the intrinsic linewidth $\Delta\nu_i$ of the molecule. The lower limit of this linewidth $\Delta\nu_i$ is given by the lifetime $\tau_\infty = 1/(2\pi\Delta\nu_\infty)$ of the excited state, which is the Fourier transform of the natural linewidth $\Delta\nu_\infty$. The natural linewidth is inversely proportional to the transition frequency, $\tau_\infty \propto 1/\nu^3$ (Condon and Shortley, 1957). Hence, infrared-excited states have long (millisecond) and UV-excited states short (nanoseconds) lifetimes.

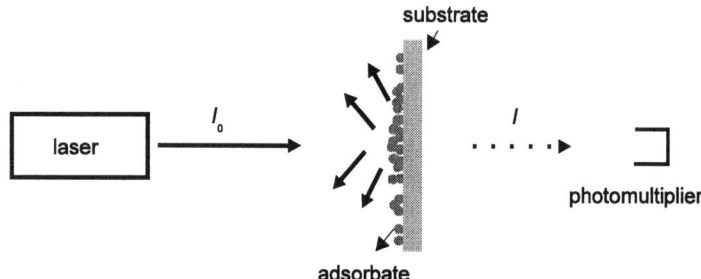

Figure 56 Absorption spectroscopy of adsorbates on surfaces.

The long lifetimes of IR-excited states correspond to narrow linewidths (in the Hz to kHz range). Thus the linewidths that are detected in IR spectroscopy of molecules, which are adsorbed on surfaces, are dominated by the damping of the vibration by coupling to substrate or adsorbate phonons (and/or electron–hole pair excitations in the case of metallic substrates). These processes as well as perturbations of the phase of the internal molecular vibration

---

[1] If one uses a lamp for excitation and a spectrometer for the spectral analysis (*e.g.*, for FTIR), then the spectral width of the spectrometer has to be deconvoluted from the measured linewidth.

3.1. SPECTROSCOPIC METHODS

via thermal excitation of the molecule–substrate vibration (dephasing or T_2 processes) result in a homogeneous line broadening (Lorentz profile).[2] An inhomogeneous broadening (Gauß or Voigt profile (Demtröder, 1996)) is expected if the excitation and emission probabilities of the photons are different for different molecules, i.e., if some of the molecules are adsorbed at defects or other energetically preferred places or if different adsorbate domains have been created.

Besides the linewidth the *position* of the transitions might differ from the spectral position of lines in the free molecule due to interactions with the substrate or among the adsorbed molecules. A precise measurement of the changes using infrared absorption spectroscopy (IRAS) (Hoffmann, 1983) or Fourier transform infrared spectroscopy (FTIR) (Griffiths and Haseth, 1986; Chabal, 1988) results in detailed information on adsorbate densities, adsorbate structure and orientations.

The measurement of the frequencies of the modes of the adsorbed molecules allows one to chemically identify the adsorbed species. The sensitivity of the method is about 5×10^{-3} monolayers with a spectral resolution of better than 0.4 cm^{-1}. This resolution is similar to the value of molecular rotational constants and allows one to resolve even low-energy adsorbate modes.[3] If the molecules are only weakly bound to the surface (physisorbed), then the resonance frequency will be changed with respect to the unperturbed molecule by only a few percent due to electrostatic interactions. Chemisorbed adsorbates show changes of more than 10% due to covalent interactions. Usually the intramolecular binding is weakened by the substrate binding, resulting in a red shift of the transition frequency, the amount of which providing information about the binding site.

On metals due to the generation of image dipoles IRAS or FTIR allows one to excite solely modes with dipole moments, which are oriented perpendicular to the surface (cf. Chapter 3, Section 3.2). This limitation is no longer valid for semiconductors or insulators. Since the absorbed light intensity is proportional to the square modulus of the transition dipole moment and the field strength vector of the incoming lightwave, i.e., proportional to the angle between the dipole moment and the field vector, measurements using s- or p-polarized light[4] can be used to determine the angle of the molecular transition dipole moment with respect to the surface normal. Since the transition dipole

---

[2] For this dephasing to occur a thermal excitation is necessary. Hence it is possible to separate the dephasing part of the total linewidth from other damping processes by performing temperature-dependent measurements.
[3] If one is interested in scanning the excitation energy over several eV instead of several 1000 cm^{-1}, then electron-energy loss spectroscopy (EELS) is the method of choice. Via short-range, 'impulsive' electron–adsorbate collisions, modes can be excited which are not IR active. The disadvantage of the method is its much smaller spectral resolution of about 40 cm^{-1}.
[4] Polarized perpendicular or parallel to the plane of incidence.

moment (at least for diatomic molecules such as CO) has a well-defined angle with respect to the nuclear axis, one is able to determine even the tilt angle of the molecule with respect to the surface. This is an important quantity for a structural analysis of adsorbates.

The evaluation of the spectra has to take into account that part of the incoming light is absorbed in the substrate and part gets lost via scattering at the adsorbate or the substrate. Therefore the spectra have to be corrected by reference measurements, which define the so-called 'apparatus function'. In the case of clusters adsorbed on surfaces (c.f., Chapter 4, Section 4.1.3) the value of the scattering cross section at the adsorbate can be similar to the value of the absorption cross section. Also, the cross sections show a pronounced wavelength dependence. In that case the measured quantity is the *extinction*, i.e., the sum of absorption and scattering. Fig. 57a shows a typical measurement of the extinction of light by the distribution of sodium clusters, adsorbed on a dielectric substrate. The comparison with calculated spectra reveals (Figs. 57b and 57c) that absorption and extinction show very different behavior. In the presented case the scattering cross section is about a factor of three larger than the absorption cross section.

3.1.2 Laser-induced fluorescence

The light intensity that is emitted by a laser usually shows, even in the case of a stationary lasing mode, fluctuations in the percent range due to amplitude fluctuations of the pump source. This sets limits for the absorption spectroscopy of thin adsorbate layers or of adsorbates with small cross sections. An additional problem for spectroscopy on surfaces is reflected laser light, which is overlayed over the signal of interest. An appealing way to solve this problem is to observe spectrally shifted light emitted by the molecules of interest. For example, if one optically excites a diatomic molecule from a rovibrational state of the electronic ground state into a rovibrational state of an electronically excited state, then the excited electron will return to the electronic ground state via emission of a photon. If the average nuclear distances of the electronic states are different (which is usually the case), then the excited molecule returns to energetically higher or lower lying states in addition to the initial rovibrational state. The emitted photons possess another energy (color) compared with the exciting photon and can be easily separated from the initial photons by use of a narrow-band spectral filter. The energetic difference between excitation and fluorescence energy is transferred into internal excitation or de-excitation of the molecule.

Figure 58 exemplifies the principle of this 'laser-induced fluorescence' (LIF) for lithium molecules. A typical fluorescence transition is indicated by arrows in the potentials. The observed fluorescence intensities are given by

3.1. SPECTROSCOPIC METHODS

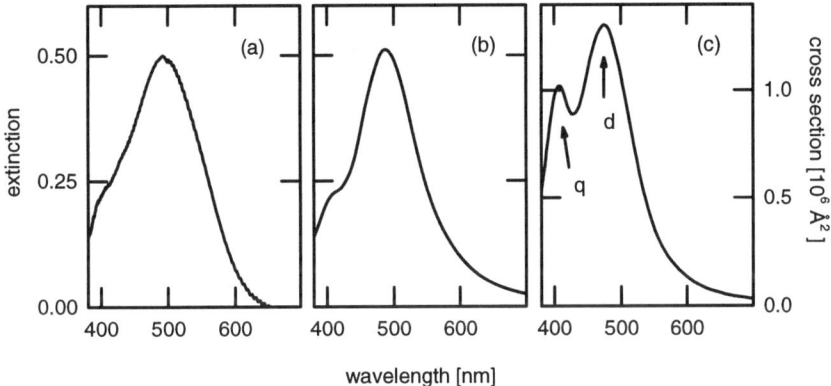

Figure 57 Measured extinction (a), calculated extinction (b) and absorption spectra (c) for a size distribution of sodium clusters, adsorbed on a dielectric substrate (mica). The average radius of the clusters is 40 nm and the half width of the distribution is 50% of the mean cluster radius. The calculation is on the basis of Mie theory (Bohren and Huffman, 1983). The maxima, which can be seen quite well in the absorption spectrum, result from the dipole ('d') and the quadrupole resonance ('q') of the collective motion of the electrons with respect to the lattice.

$$I_{\text{LIF}} = A \cdot N' \cdot \nu^4 \cdot \left| \int \psi'^*(v) \cdot \psi''(v) \mathrm{d}r \right|^2 \cdot \frac{S_{\text{rot}}}{2J'+1}, \qquad (3.5)$$

where the proportionality constant A includes the detector sensitivity, the observed solid angle and the electronic transition dipole moment, N' is the density of excited particles and $\psi(v)$ are the vibrational wavefunctions for ground (") and excited states ('). The integral (the transition moment between vibrational states) is called the 'Franck–Condon factor', S_{rot} (the transition moment between rotational states) is called the 'Hönl–London factor', and $2J'+1$ is the degeneracy of the excited state. The total transition moment is thus factorized into a part that depends on the electronic state of the molecule and a part that is determined by the nuclear motion (rotation and vibration). This implies that the nuclei do not move during the electronic excitation process: the optical transition occurs 'perpendicular' in the Born–

Oppenheimer picture of adiabatic potentials ('Franck–Condon principle'); see Figs. 54 and 58.

Due to the efficient elimination of background light by a color filter, LIF is a sensitive detection method even close to highly reflecting surfaces. If one wants to apply the same method to atoms in the electronic ground state the problem is that exciting and emitted radiation have the same wavelength. Here, excitation via a two-photon process (Göppert-Mayer, 1931) and observation of blue-shifted fluorescence (Fig. 58, right-hand side) is a possible way out. The resulting fluorescence spectrum is shown in Fig. 59b for the example of laser-excited sodium atoms.

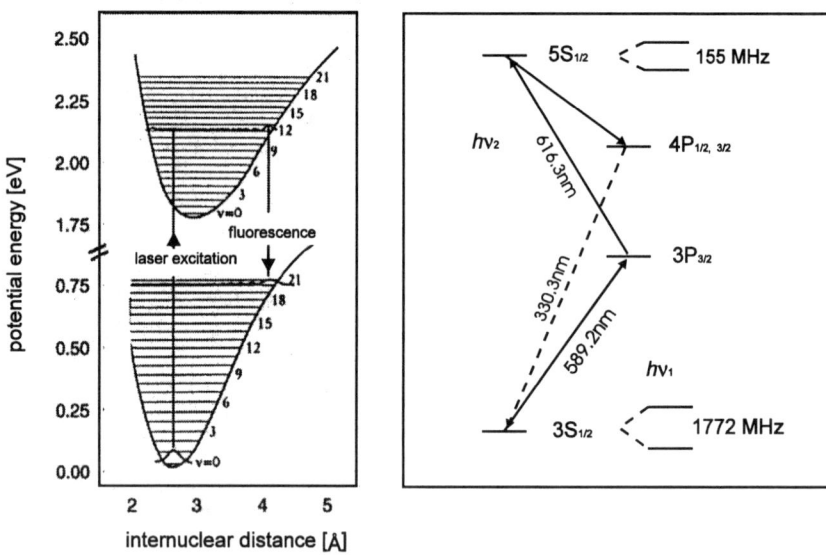

Figure 58 Two methods for the nearly background-free fluorescence detection of particles. Left-hand side: Laser-induced fluorescence demonstrated for electronic excitation ($X^1\Sigma \to A^1\Pi$) of Li_2. As indicated in the figure, the electronic transitions between the potential curves are 'perpendicular': the nuclear motion is much slower as compared with the electronic motion. The fluorescence is red-shifted compared with the excitation wavelength.
Right-hand side: term scheme for two-photon excitation between electronic states in sodium atoms. Here, a possible fluorescence wavelength (4P → 3S) is blue-shifted to 330 nm.

Two-photon laser-induced fluorescence (TPLIF) has the additional

3.1. SPECTROSCOPIC METHODS

advantage that Doppler broadening of the spectral lines can be avoided. The Doppler broadening is sometimes even larger than the lifetime broadening, which characterizes, besides dephasing broadening, the interaction with the surface. It is an inhomogeneous broadening process since every single molecule has its own excitation probability: depending on the velocity components v_i with respect to the wavevector $k_i = 2\pi/\lambda_i$ only blue- or red-shifted photons $(\omega_0 \pm k_i \cdot v_i)$ compared with stationary molecules (ω_0) can be absorbed. The resonance condition for the two-photon transition is thus

$$\Delta E = \hbar(\omega_1 + \omega_2) - \hbar v \cdot (k_1 + k_2). \tag{3.6}$$

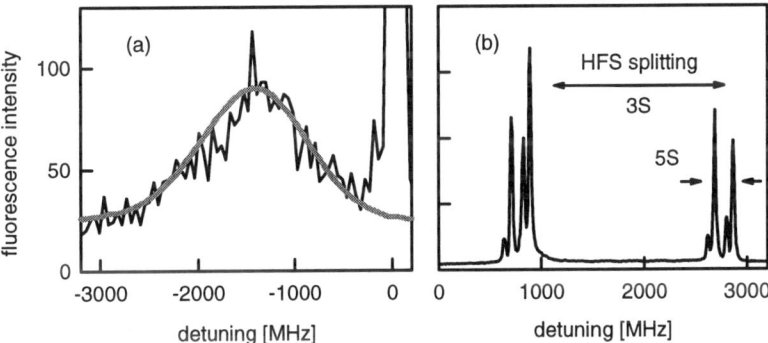

Figure 59 Fluorescence spectra of sodium atoms near dielectric interfaces and excited within the evanescent wave of a prism. (a) Spectrally resolved one-photon fluorescence. The broad line results from light that is emitted by desorbing atoms. Reprinted from *Optics Communications*, (Bordo et al., 1997), Copyright 1997, with permission from Elsevier Science. (b) Two-photon fluorescence following anticollinear irradiation by a second laser. The lines are due to the hyperfine structure splitting of ground (3S) and excited (5S) states. The dynamic Stark effect results in an additional splitting of all the lines.

With increasing mobility of the molecules (i.e., increasing temperature of the gas) the value of the Doppler broadening is increasing. However, if one uses for the two-photon excitation process photons with wavevectors of opposite sign, $k_1 = -k_2$, then the velocity-dependent term in Eq. (3.6) cancels and all molecules or atoms contribute at the same frequency to the transition. Hence, a significantly higher spectral resolution follows.

In Fig. 59a one-photon fluorescence is shown, spectrally resolved by an etalon while the laser $h\nu_1$ (Fig. 58, right-hand side) was fixed at the $3S_{1/2}(F=2) \rightarrow 3P_{3/2}(F)$ transition. The strong, sharp line on the right-hand side is elastically scattered light, whereas the broad line is due to the $F=2$ hyperfine component of desorbing atoms. The grey line is the result of a rigorous theoretical approach. In Fig. 59b two-photon fluorescence 4P→3S is plotted, following irradiation by a second laser $h\nu_2$, which has been directed anticollinearly to the first laser. The laser $h\nu_1$ has been detuned over the 3S→3P transition, while $h\nu_2$ was fixed at 16227.17 cm^{-1}. The maxima correspond to the hyperfine structure splitting of the 3S state (distance between line centers 1772 MHz) as well as that of the 5S state (distance between line centers 155 MHz). The additional splitting of all the lines is due to the high laser intensity, giving rise to the optical Autler–Townes effect (Delsart and Keller, 1978).

The interpretation of fluorescence spectra of adsorbates becomes difficult in the neighborhood of an interface (the substrate surface), since the spontaneous emission, which follows the excitation, does not occur in free but in confined space, i.e., in space with boundary conditions. The changes in optical properties, however, can be predicted rather accurately by the use of electrodynamic calculations (see Section 3.2).

One-photon LIF is mainly used for the state-selective determination of product distributions of desorbing molecules. In that case the particles are a few millimeters apart from the surface following the desorption process. Therefore the surface influence on the spectra can be safely neglected.

3.1.3 Multiphoton ionization

An alternative to the detection of desorbing particles via fluorescence spectroscopy is ionization via a resonant intermediate state ('resonance enhanced multiphoton ionization', REMPI). The first photon excites the molecules into an electronically excited state, and the subsequent photons from the same or a different laser ionize it. In Fig. 60 results of an experiment are plotted, where aniline molecules close to a gold surface have been ionized by 260 nm pulses with 1 μJ energy. This experiment was aimed at obtaining information about the surface binding of volatile molecules. For this purpose one might determine the time-of-flight distribution (and thus the distribution of kinetic energies) of the ions with a time-of-flight mass spectrometer (TOF-MS).

Since the laser light is reflected from the surface and since the molecules are part of a gas *above* the surface and only partially adsorbed on the surface, the measured signal is generated both by gas- and surface-bound molecules. In order to distinguish between these two possibilities the hypotenuse of a quartz prism, which has been coated with an approximately 5 nm thick gold film, is used as the surface. Irradiation with the desorption-ionization laser from the vacuum side results in an interference pattern between incoming and reflected light, which has a node (i.e., vanishing field strength) directly on the surface. Consequently, the aniline gas is ionized with the same spatial distribution of

3.1. SPECTROSCOPIC METHODS

a standing wave. The first maximum (the minimum distance to the surface at which ionization takes place), occurs at $\lambda/2\Theta$, with Θ the angle of the incoming beam with respect to the surface plane (Chai and Reilly, 1984). For grazing incidence ($\theta < 1°$) and UV light (λ=260 nm) this distance amounts to more than 10 μm.

The resulting time-of-flight distribution of the ions is plotted in Fig. 60a. The ions, which are generated by the laser via REMPI, are accelerated by an electrically charged aperture and are detected by an open electron multiplier. The spatial density modulation of the desorbing particles is thus transformed into a temporal modulation via velocity and flight path.

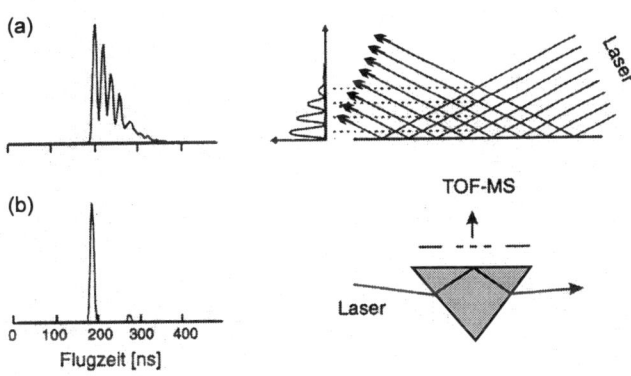

Figure 60 Multiphoton ionization of desorbing molecules in a light field that is reflected from a surface (a) and in an evanescent wave field (b). On the right-hand side the excitation geometry is sketched. On the left-hand side the observed intensity distributions are plotted. (a) Reprinted with permission from (Yang and Reilly, 1990). Copyright 1990 American Chemical Society. (b) Reprinted from *Optics Communications*, (Yang and Reilly, 1989), Copyright 1989, with permission from Elsevier Science.

Since a field strength node occurs on the surface, one will detect mainly aniline molecules from the gas phase. If, however, the laser irradiates in the geometry for total internal reflection (Fig. 60b), then the molecules will be ionized in the evanescent wave close to the surface (for the experiment of Fig. 60 within the initial 50 nm above the surface, compare the following chapter). As a consequence the TOF maximum appears at a shorter flight time and no further maxima are observed (Fig. 60b). The tiny second maximum in Fig. 60b results from the carbon isotope ^{13}C, which due to its higher mass has a longer flight time for the same velocity. By the use of the evanescent wave it thus becomes possible to monitor the desorption dynamics of molecules, which are in an adsorption/desorption equilibrium close to a surface.

The detection of molecules via REMPI has two major advantages compared with LIF detection: firstly, a severe limitation is no longer valid, namely that the excited molecules have to fluoresce. Secondly, ions can be detected with much higher efficiency as compared with photons: a strong electric field extracts nearly all generated ions and focuses them onto an open particle multiplier, which creates an electron avalanche from each impinging ion. In contrast, the quantum efficiencies of photomultipliers are of the order of 20%, and additional losses occur due to the imaging optics and due to the small solid angle (about 10% of the total solid angle (Hefter and Bergmann, 1988) can be covered).

A severe disadvantage of REMPI is the fact that it is a multiphoton process. Even for a two-photon process the signal intensity depends in the saturation-free case quadratically on the irradiating laser intensity. Since ionization with the second photon usually excites the electron into a continuum, the excitation probability is even smaller compared with a resonant transition. An additional problem close to the surface are spurious ions, which are generated by the laser at the surface itself and can form a fairly large background signal.

3.1.4 Spectroscopy with evanescent waves

If one irradiates a prism in vacuum with laser light of wavelength λ at an angle Θ_i with respect to the surface normal, which is larger than the critical angle $\Theta_c = \sin^{-1}(1/n_1)$ (with n_1 the refractive index of the prism), then the incoming light will be totally reflected at the interface between the prism and vacuum.

The electromagnetic boundary conditions make it necessary that an exponentially decaying electromagnetic field exists perpendicular to the surface with a decay length β of the order of a fraction of the wavelength (Eq. (2.12)). The corresponding evanescent wave traverses along the surface without taking energy out of the prism. However, if the prism is situated inside an absorbing medium or if an absorbing film has been evaporated onto the prism hypotenuse, then energy can be transferred into the medium (Fig. 61). In that way it becomes possible that surface plasmons are excited in the adsorbed metal film, which is not possible under the usual conditions since the transverse electromagnetic light wave cannot couple to the longitudinal plasmon oscillation (see Chapter 1, Section 1.2.2). A phase matching between the incoming photon and excited plasmon oscillation is achieved if the angle of incidence of the light, Θ_i, fulfills the condition:

$$\sin\Theta_i = \sqrt{\frac{\epsilon_1 \epsilon_2}{\epsilon_p(\epsilon_1 + \epsilon_2)}}, \qquad (3.7)$$

3.1. SPECTROSCOPIC METHODS

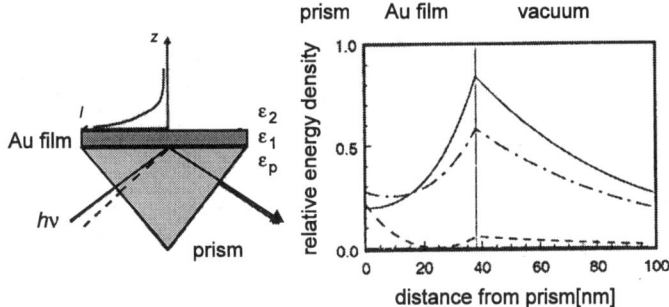

Figure 61 Left-hand side: Total internal reflection of a light beam at a prism surface and generation of an evanescent wave. The dielectric functions of a vacuum, gold film and prism are denoted by ϵ_2, ϵ_1 and ϵ_p. Right-hand side: Energy densities of the electromagnetic field of light that hits a glass prism, coated with a 38 nm thick gold film, for three different angles of incidence (42°(- - -), (46°(—) and (50°(- · -)). The plasmon resonance angle is 46°. Reprinted with permission from (Lee et al., 1993). Copyright 1993 American Physical Society.

where $\epsilon_p, \epsilon_1, and \epsilon_2$ are the complex dielectric functions of the prism, metal film and medium on top of the metal film, which of course depend on the wavelength of the exciting light. In the case of resonance, i.e., for the correct angle of incidence at a given wavelength, the photon energy is absorbed in a plasmon excitation, resulting in a sharp minimum in the reflected light as a function of angle of incidence (Fig. 19). If the prism/metal combination is not changed, then the resonance angle of incidence is a sensitive function of the dielectric function of an attached medium. In that way sensors for biochemically relevant molecules with sub-nanomolar sensitivity have been constructed (Harris and Wilkinson, 1995). It is also an elegant method to determine the thickness of the evaporated film (Knoll, 1998). For example, the angle of total reflection for light from a HeNe laser, which hits a glass prism that is coated by a 50 nm thick silver film and Langmuir–Blodgett (LB) multilayers of Cd-arachidic acid (see Fig. 23 for the LB method) changes per 5 nm organic film thickness by 0.5 to 1.1° in the thickness range 5 – 60 nm (Swalen et al., 1980). Expanding the laser beam by a cylindrical lens and adsorbing an increasing number of LB layers in the form of a wedge (Fig. 62, right-hand side) allows one to detect the change in angle as a function of film thickness directly by a photographic plate (Fig. 62, left-hand side).

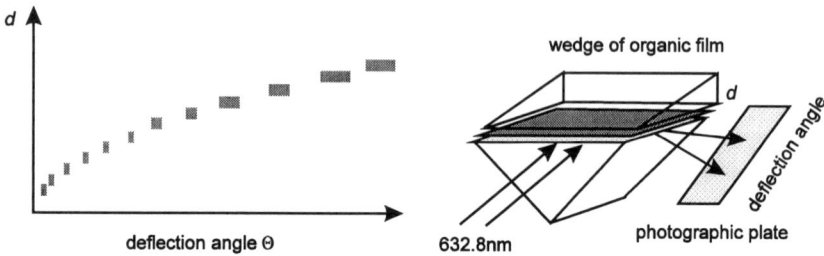

Figure 62 Change of the angle of total internal reflection for a HeNe laser beam (632.8 nm), which hits a prism that is coated by a thin metal film, as a function of thickness d of adsorbed Cd-arachdic acid films.

Another widespread optical method for the determination of film thicknesses is ellipsometry (Azzam and Bashara, 1992; Tompkin, 1993) (Fig. 63). Depending on the thickness and the optical constants of the reflecting layer phase and amplitude of the reflected (r) wave as compared to the incident (i) wave change while traversing the layer. Let $\delta\phi$ be the phase difference between the parallel (p) and perpendicular (s) components of the electromagnetic waves and δE the ratio of reflected to incident field amplitudes. Then one measures the phase change $\Delta = \delta\phi_i - \delta\phi_r$ between incident and reflected waves as well as an angle Ψ, related to the change in amplitudes via Fresnels formulas: $tan\Psi = |\delta E_p|/|\delta E_s|$. If the optical constants of the film are well known, then application of Fresnels formulas (Eq. (5.1)) provides one an accurate measure of film thickness from the measured changes in Δ and Ψ.

Let us finally remark that the 'selective reflection spectroscopy' was founded already at the beginning of this century on the basis of evanescent waves. It has been used to deduce atomic dispersion spectra via detection of the frequency dependence of the light that has been reflected from the dielectric interface between a window and an atomic gas (Wood, 1909). Recently, besides the 'attenuated total reflection spectroscopy' (ATR, Chapter 1, Section 1.2.2)) the 'Doppler-free evanescent wave spectroscopy' (Simoneau et al., 1986) has been developed, implementing two lasers. The advantage of evanescent wave spectroscopy is its sensitivity to atoms which travel along the surface (Ducloy, 1993). In that way the van der Waals interaction between the atom and dielectric surface becomes measurable (Section 3.2). If one excites the atoms with a circularly polarized laser in the presence of a strong magnetic field and probes their angular momentum substates with a second probe laser, then the interaction of spin-polarized atoms with dielectrics can be investigated,

3.1. SPECTROSCOPIC METHODS

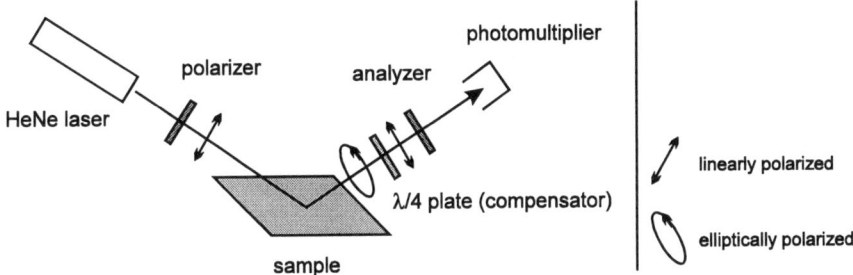

Figure 63 Set-up for the ellipsometric determination of optical constants of thin films. If the optical constants are known, then the rotation of the polarization angle by the film allows one to determine the thickness of the film with high precision.

thus providing information about wall relaxation processes (Grafström et al., 1996).

Fig. 61 shows the electromagnetic energy density inside a 38 nm thick gold film, which has been evaporated onto a prism, as a function of distance from the prism surface (Lee et al., 1993). The curves have been calculated for different angles of incidence with respect to the surface normal. At the plasmon resonance angle (46°) a strong enhancement of the energy density at the gold-film vacuum boundary layer is observed as compared with the gold-prism interface. If one attaches the molecules that are to be spectroscopically investigated onto such a gold layer, then a strong enhancement of the electromagnetic energy density can be obtained, resulting in a more sensitive detection via, for example, fluorescence methods.[5] A combination of this approach with intracavity total internal reflection via evanescent wave cavity ring-down[6] spectroscopy (Pipino et al., 1997) results in a sensitivity of 0.04 monolayers for a strong absorber (iodine molecules).

---

[5] A combination of fiber-shaped evanescent wave field sensors with bioaffinity assays, using fluorescent tracer probes, obviously results in even higher sensitivity down into the sub-picomolar range (Duveneck et al., 1996).

[6] Cavity ring-down spectroscopy relies on injecting a laser pulse into a cavity, monitoring the leaking of the intra-cavity intensity out of the cavity and sensing the change of the decay curve due to the existence of an absorbing species inside the cavity. From that the absorption spectrum of the species can be determined with high accuracy since the effective absorption length is largely increased and the amplitude fluctuations of the light source have only minor influence on the signal intensity.

3.1.5 Two-photon photoemission

As discussed in Chapter 1, Section 1.2.2, surface states are generated on metals if an excited electron induces a polarization charge in the metal and thus a Coulombic potential well ($V_C = -e^2/4z$). The energy of those image potential states is fixed with respect to the vacuum level, and is given to first order by the binding energy of the electron to the hydrogen atom with an effective nuclear charge of $Z_{eff}=1/4$, i.e., $E_b \approx 0.9$ eV. The states are situated energetically between the Fermi level and the vacuum level, i.e., they are unoccupied states (Figs. 16 and 64).

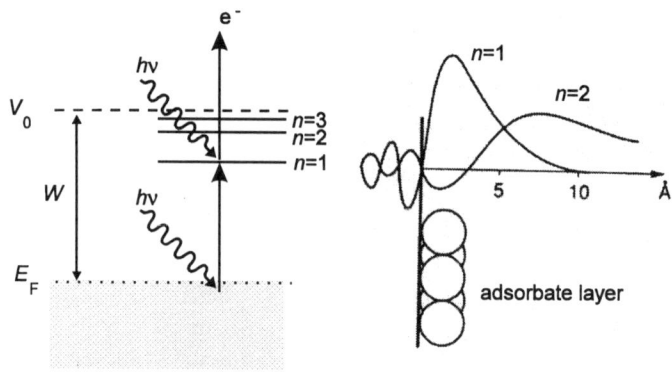

Figure 64 Left-hand side: Term scheme for the detection of image potential states via two-photon photoemission. The first laser generates the image state (e.g., $n=1$), and the second one ionizes it. The Rydberg series of image states converges against the vacuum-energy V_0. The difference between V_0 and the Fermi energy is the work function W. Right-hand side: Wavefunctions of the first two image potential states on a metal surface, compared with the thickness of a xenon adsorbate layer. Reprinted with permission from (Padowitz et al., 1992). Copyright 1992 American Physical Society.

If one solves the Schrödinger equation for an electron in the Coulombic potential perpendicular to the surface (i.e., $k_\parallel = 0$), then one obtains a Rydberg series of states with quantum numbers n and binding energies $E_b = -0.85/n^2$ eV relative to the vacuum level V_0. The binding energies are about 0.1 to 0.3 eV smaller due to the modification of the ideal Coulombic potential by the real crystal potential. For the first two image potential states the corresponding wavefunctions are shown on the right-hand side of Fig. 64 in comparison with the thickness of a typical adsorbate layer (xenon for this

3.1. SPECTROSCOPIC METHODS

example). The maximum of the electron density distribution is found a few Ångstrom outside the surface. The $n=2$ image potential state has its maximum probability already outside the adsorbate layer. Taking this large distance to the substrate surface into account it follows that image potential states exist even if there are highly polarizable adsorbates such as Xe atoms on the surface. Changes in image potential states compared with those of an ideal surface reflect the properties of the adsorbate–substrate and the adsorbate–vacuum interfaces (Padowitz et al., 1994; Hotzel et al., 1998).

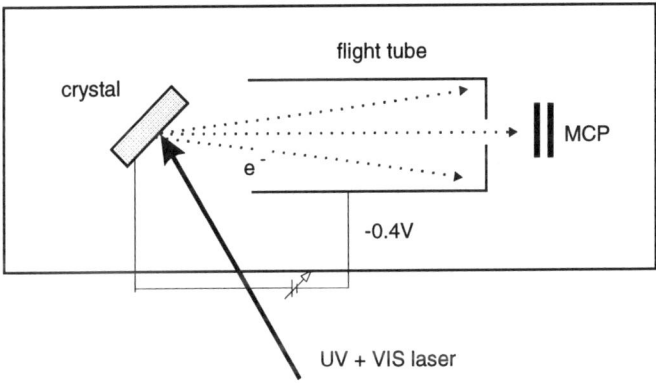

Figure 65 Laser TOF set-up for the determination of image potential states via two-photon photoemission spectroscopy. The photoelectrons, which are generated by coaxially irradiating UV (323 nm) and visible (VIS, 646 nm) photons, are detected after a flight path of 13.5 cm by a multi-channelplate amplifier (MCP). Reprinted with permission from (Padowitz et al., 1994). Copyright 1994 Society of Photo-Optical Instrumentation Engineers (SPIE).

Another description of image potential states results from the observation that the long-range interaction of the solid with an external charge occurs mainly via excitation of the collective electronic states (surface plasmons in metals if excitable). It has been shown that the image charge that is generated by the excited electron is identical to the distribution of surface plasmon frequencies ω_s (Zangwill, 1988). The binding energy of the image potential states is then directly related to the plasmon frequency via (Quiniou et al., 1994)

$$E_b \propto -1/(\omega_s^4 n^2). \tag{3.8}$$

This means that the properties of the surface plasmon frequency of the investigated metal (for example the temperature dependence) are reflected in the energetic position of the image states. Since the position of the image states is fixed with respect to the vacuum energy V_0, shifts of V_0 e.g., via adsorbate-induced changes in the work function, should be reflected directly in shifts of the image state maxima. This has indeed been observed for oxygen adsorption on copper surfaces (Quiniou et al., 1994).

Due to their Rydberg-like character the image states have a long lifetime and the corresponding linewidth is only of the order of a few tens of meV. Therefore spectroscopic methods with high resolution such as two-photon photoemission (Rudolf and Steinmann, 1977; Steinmann, 1989) are needed to observe spectral shifts. Fig. 65 shows a typical set-up for the measurement of image states (Padowitz et al., 1994). The liquid-helium-cooled crystal is irradiated simultaneously with visible light pulses from a mode-locked Nd:YLF synchronously pumped dye laser (λ=646 nm) and with frequency doubled light, which has been generated in a nonlinear optical crystal. The repetition rate of the laser is 2 MHz and the pulse energy 50 nJ with 6 ps pulse duration. These short light pulses generate only one photoelectron per 1000 laser pulses, which, however, generate a satisfactorily strong signal on a multichannel-plate amplifier following temporal averaging. The field-free flight path of about 10 cm is passed by these electrons within a few hundred nanoseconds. This is a factor of 10^5 longer as compared with the initial pulse length, and thus the theoretical energy resolution is about 0.01 meV. Delays, induced by the TOF electronics, result in a practical energy resolution of 5 meV. As an alternative to this TOF method hemispherical energy analyzers might be used to energetically resolve the photoelectrons, which, however, usually have an even worse energy resolution.

In Fig. 66 as an example we show the spectra of the $n=1$ image state of a silver (111) surface, which has been covered by different amounts of cyclohexane. With increasing cyclohexane coverage a decrease in binding energy of the image state by up to 0.2 eV and simultaneously a splitting into five maxima is observed. Apparently islands are formed which have different cyclohexane coverages. Thus the electrostatic potential varies between different positions on the surface, leading to peaks at several spectral positions. The spectra of the higher image states $n=2$ to $n=4$ are shifted only marginally with increasing cyclohexane coverage since the maximum of the electron wavefunction is far outside the surface (Fig. 64). By the use of this method changes in the interaction of the electron with the adsorbate can be detected on a subnanometer scale. It is also obvious that the method can be used in a more general sense to study surface morphology with high spatial resolution (Fischer et al., 1993).

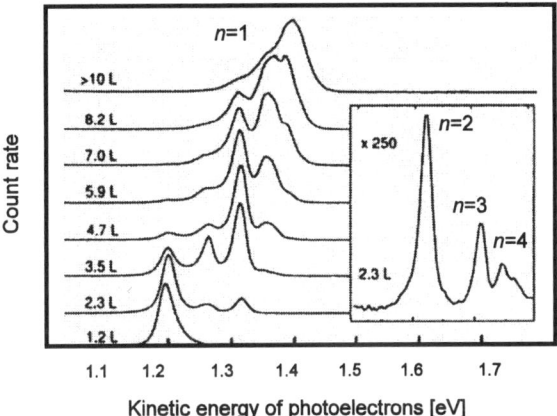

Figure 66 Photoelectron spectra of the image potential states of Ag(111) as a function of coverage with cyclohexane (in Langmuir). The large figure shows the $n=1$ state, the small one the $n=2$, 3 and 4 states. Only the maxima in the $n=1$ state shift with increasing cyclohexane coverage. Reprinted with permission from (Padowitz et al., 1992). Copyright 1992 American Physical Society.

For a more detailed discussion of the combination of ultrashort laser pulses with this method see Chapter 4, Section 4.1.1 .

3.2 Fluorescence Spectroscopy Near Interfaces

An interface close to a laser-excited atom or molecule significantly alters its response to an electromagnetic wave (Ford and Weber, 1984; Metiu, 1984; George and Arnoldus, 1990). Comprehensive reviews are given in (Chance et al., 1978; Waldeck et al., 1985; Cavanagh et al., 1994; Barnes, 1998). This matter has found renewed interest in recent years since it influences directly the development of, for example, biosensors (Leung et al., 1994) or cell–silicon junctions (Weis et al., 1996), where the distance of the living cell from the silicon surface could be accurately determined by the modified fluorescence rate (Lambacher and Fromherz, 1996). A thorough knowledge of surface-induced changes in the fluorescence rate is also important for imaging and spectroscopy of isolated molecules in scanning near field microscopy (SNOM) (Betzig and Chichester, 1993; Ambrose et al., 1994b; Bian et al., 1995; Pagnot et al., 1997). Note that recent possibilities to tailor and characterize the surfaces on nanoscale dimensions (Drexler, 1992; Jersch et al., 1998) have made the investigation of the influence of the *microscopic roughness* (Leung and George, 1989) of the surface an interesting subject too.

The lifetime of the excited state, τ_∞, is inversely proportional to the homogeneous linewidth $\Delta\nu_\infty$. Thus a measurement of the linewidth as a function of distance to the interface results directly in values of the changed lifetime. The forces, which act via the metal surface on the ground and excited states of an atom, result in energetic shifts of the influenced energy levels. Due to the usually different polarizabilities of excited and ground states, the shifts have different magnitudes and thus the optical transition frequency between the energy levels is changed compared with the case of the free particle. This transition frequency shift can also be determined. An accurate calculation of this shift is possible via quantum electrodynamics, e.g., (Hinds, 1991; Hinds, 1994). Approximate, but for most purposes sufficiently accurate values of changed lifetimes, can be calculated also in a more simple, classical picture, which describes the radiation field of a dipole antenna above a surface (Sommerfeld's radiation theory (Sommerfeld, 1909)).

Level shifts
An atom in the *ground state* experiences, at small distances $z \leq 50$ nm from the surface, the van der Waals potential (Lennard-Jones, 1932)

$$V_{\text{vdW}} \propto \frac{\mu^2}{z^3}. \tag{3.9}$$

The fluctuating electric dipole moment μ of the ground state atoms is *instantaneously* coupled to the mirror image in the metal. Hence the mean value of the interaction energy does not vanish as a function of averaging over all spatial orientations of the dipole. A calculation on the basis of perturbation theory results in the z^{-3} dependence of a dipole–dipole interaction potential (Israelachvili, 1992).

In order to calculate the frequency shift we assume the atom to be a point-dipole (or 'antenna'), embedded in a spacer layer with dielectric constant ϵ_1 and driven by its own reflected radiation field $E_R(\omega)$ (Fig. 67). This corresponds to configuration (a) in Fig. 70. The classic shift of the optical dipole transition frequency (real part of the reflected field) for an atom in front of a flat surface is given by (Chance et al., 1975)

$$\Delta\omega_a^{\text{cl}} \propto -\frac{3}{4} q \frac{1}{\tau_\infty} \frac{\epsilon_1}{k_1^3} \cdot \text{Re} G^R(\omega). \tag{3.10}$$

Here, $G^R(\omega) = E_R(\omega)/\mu_0$ is a Green's function, k_1 is the emission wave vector within the dielectric spacer layer, q the quantum efficiency for radiation and μ_0 the amplitude of the dipole moment. The challenge is to find the reflected field E_R at the dipole position, which can be done via an energy

3.2. FLUORESCENCE SPECTROSCOPY NEAR INTERFACES

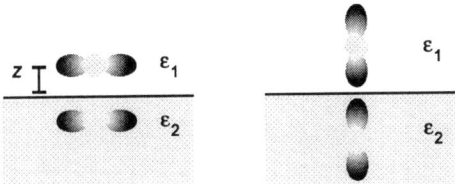

Figure 67 Point-dipole above a metal surface. Two possible geometries and the radiation of the dipole parallel and perpendicular to the surface are indicated.

flux method (Chance et al., 1978) or, as suggested by Eq. (3.10), by applying a dyadic Green's function method. The topography of the interface and its optical properties might be introduced via Fresnel's reflection coefficients into the explicit form of the Green's function. From Eq. (3.10) the shift is expected to become large if the lifetime τ_∞ of the unperturbed, excited atom inside the spacer layer is small. Hence atoms with nanosecond lifetimes of the excited states and well-defined locations of the transition dipole moments are much more favorable as probes of this shift compared with, for example, large dye molecules.

If the distance between atom and surface is so large that the time which the electric field of the atom takes to be reflected from the surface is comparable eith the dipole period itself, then the orientation of the initial dipole and its image in the metal are no longer instantaneously coupled. The orbital frequency of an electron is of order of 3×10^{15} s^{-1}. The critical distance is given by the distance the light travels during the orbit time (about 100 nm in vacuum). Since in that case no optimum coupling between the atom and its mirror image occurs, the attractive interaction is weakened. As a consequence, the retarded van der Waals or 'Casimir–Polder' potential depends with higher order on the distance to the surface (Casimir and Polder, 1948):

$$V_{\mathrm{CP}} \propto \frac{\alpha}{z^4}. \qquad (3.11)$$

Here, α denotes the static dipole polarizability of the atom. The attractive long-range force of the Casimir-Polder potential results mainly in a change in the gradient of the vacuum field, which is caused by the conductive surface due to the spatially varying Lamb shift.[7] This effect is of a quantum-electrodynamical nature. Experiments that measure directly its consequences are the topic of cavity quantum electrodynamics (CQED) (Berman, 1994). In a recent experiment sodium atoms have been expanded through a micron-sized parallel plate cavity (Sukenik et al., 1993) and the deflection of the atoms has been measured for cavity widths of the order of a micrometer. This experiment verified the d^{-4} dependence of the Casimir–Polder potential, but the corresponding level shift was only of the order of 100 kHz and thus too small to be measured.

If one employs sodium atoms which have been excited into Rydberg states and expands them between two parallel metal plates of microscopic (but not atomic) dimensions (μm) (Sandoghdar et al., 1992; Sandoghdar et al., 1996) then the van der Waals shift can be determined, which in that case is of the order of a few hundred MHz. Alternatively the van der Waals induced frequency shifts between excited atoms and dielectrics can be deduced from selective reflection spectroscopy by the use of evanescent waves, e.g., (Chevrollier et al., 1991; Ducloy and Fichet, 1991; Chevrollier et al., 1992; Ducloy, 1993; Fichet et al., 1995; Gorris-Neveux et al., 1997); see also Section 3.1.4. If one is interested in the Casimir–Polder shift, however, and applies large distances between atom and surface, the d^{-1} dependence of the charge–conductor interaction dominates and retardation effects cannot be observed directly. Also, the microscopic surface structure (viz., roughness) does not influence the optical spectra.

Lifetime damping

The same classic model as used above for calculation of the frequency shift allows one also to calculate the lifetime of an excited atom close to a metal surface (Kuhn, 1970; Chance et al., 1978). The classic lifetime τ of the dipole is changed compared with the lifetime τ_∞ of the unperturbed, excited atom by the amplitude $E_R(\omega)$ of the field at the position of the dipole, which is reflected from the metal (*c.f.*, Eq.3.10):

$$\frac{\tau_\infty}{\tau} = 1 + \frac{3}{2} q \frac{\epsilon_1}{k_1^3} \mathrm{Im}(G^R(\omega)). \tag{3.12}$$

Thus compared with the free dipole the lifetime of the dipole in front of the lossless mirror is altered by the absorbance or enhancement of the field amplitude $E_R(\omega)$ due to its mirror image in the metal. Depending on whether the dipole *oscillates* perpendicular or parallel to the surface plane, its decay

---

[7] The Lamb shift describes originally the lifting of the degeneracy between the $2S_{1/2}$ (shifted by μeV upwards) and the $2P_{1/2}$ level of hydrogen due to the quantization of the electromagnetic field (Lamb and Retherford, 1947). Level shifts of this kind, which result from the exchange interaction with the vacuum field, are generally called 'Lamb shifts'.

3.2. FLUORESCENCE SPECTROSCOPY NEAR INTERFACES

rate will increase with decreasing distance to the metal by a factor of two compared with the value of the free dipole or vanish. In the former case the dipole and mirror–dipole *radiate* in parallel to each other with respect to the surface plane and enhance each other, in the latter case they radiate in opposite directions and the spontaneous emission is suppressed via destructive interference.[8]

Since the lifetime is reciprocal proportional to the decay rate, the lifetime of the parallel dipole decreases close to the surface and that of the perpendicular dipole goes to infinity. Of course, this is only the case for a perfect mirror having the only property to reflect the radiation at the surface with a phase shift and not to absorb any radiation. For a real mirror with complex dielectric function ϵ_2 both lifetimes of the parallel and perpendicular dipoles vanish with decreasing distance to the metal surface, since the dipole dissipates energy into the metal and the electromagnetic field strength is damped (Fig. 68). For example, ϵ_2 enters in the reflected field in the case of a dipole which is embedded in a vacuum ($\epsilon_1=1$) via $(\epsilon_2-1)/(\epsilon_2+1)$ (Balzer et al., 1997a), and for $\epsilon_2=-1$ there is a resonant energy transfer to surface plasmon excitations in the metal. A recent experimental study of the lifetime of europium ions close to a thin (less than 100 nm thick) silver film nicely demonstrates the importance of surface plasmon polariton excitations as a decay channel (Amos and Barnes, 1997).

Using Sommerfeld's radiation theory one finds for the change of lifetime with distance the behavior that is shown in Fig. 68 (Chance et al., 1978). Here, the ratio τ/τ_∞ is plotted for unpolarized light of wavelength λ=616 nm. As seen, the lifetime of the dipole oscillates if one approaches the surface due to interferences between the initial and the phase-shifted reflected dipole field. In this distance regime far away from the surface ($z > \frac{c}{\omega} = \frac{\lambda}{2\pi} \approx 100$ nm, i.e., within an order of magnitude of the wavelength of the emitted light) the radiative interaction of the dipole with its own image in the metal results in an acceleration or deceleration of the original dipole due to the finite value of the speed of light. Hence the lifetime is either decreasing or increasing as a function of distance. Since the reflected field is weak far away from the surface, the emitted *frequency* is close to its value without surface. Also, in this regime the microscopic surface structure does not influence the optical properties.

For small distances to the surface (but still far enough away that the probability for electron transfer and chemical bondings is small) the excitation spectrum of the surface itself becomes important and the lifetime is quickly decreasing, $\tau \propto z^3$, i.e., the damping increases. The corresponding energy loss results from excitation of surface plasmons (Morawitz, 1978; Weber and Eagen, 1979; Pockrand et al., 1980; Gruhlke et al., 1986)(if the excitation frequency is high enough) and from the excitation of electron–hole pairs

---

[8] This explains the strict selection rules for IRAS from adsorbates on metal surfaces, namely that the dipole moment has to be oriented parallel to the surface normal.

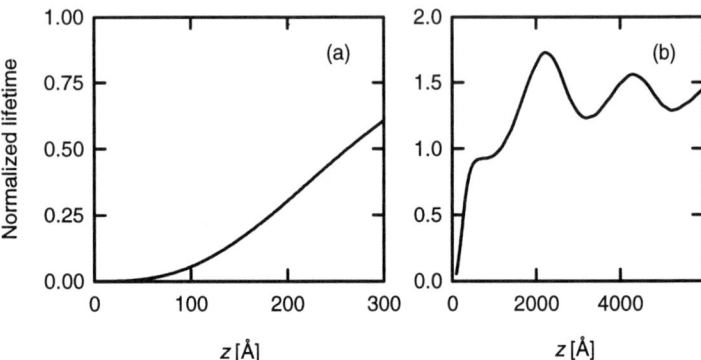

Figure 68 Lifetime of a point-dipole, embedded in a dielectric and driven by unpolarized light, as a function of distance to a platinum surface. The lifetime for given distance z to the surface is plotted with respect to the lifetime τ_∞ of the dipole, which is located infinitely far away from the metal surface, but still embedded in a dielectric. The plot on the left-hand side is an enlargement of the plot on the right-hand side.

(Whitmore et al., 1982a). Fig. 69 shows the importance of the different mechanisms both on the lifetime and on the transition frequency as a function of z for a model system (sodium atoms near a flat gold surface).

For very small distances ($z \leq 1$ nm) the classic electrodynamic description assuming a local dielectric function breaks down. Both the non-locality of the interaction between the electromagnetic field and the surface and the surface roughness have to be taken into account. In a microscopic picture, the lifetime of the excited dipole depends explicitly on the form of the dipole–surface potential (Head-Gordon and Tully, 1992; Morin et al., 1992b).

Experimental investigations on the radiation damping of optically excited dye molecules as a result of interaction with a metal surface were performed more than 30 years ago (Drexhage et al., 1968; Kuhn et al., 1972; Drexhage, 1974), albeit for microscopically not very accurately defined systems. The dye molecules were bound chemically to long-chain fatty acid molecules, which were adsorbed on metallic surfaces in the form of monomolecular multilayers using the Langmuir–Blodgett technique (Blodgett and Langmuir, 1937) (Chapter 1, Section 1.2.3). In that way the distance between the dye molecules and the surface could be varied between ten and a few hundred nanometers. The fluorescence lifetimes of the molecules, which were excited

3.2. FLUORESCENCE SPECTROSCOPY NEAR INTERFACES 107

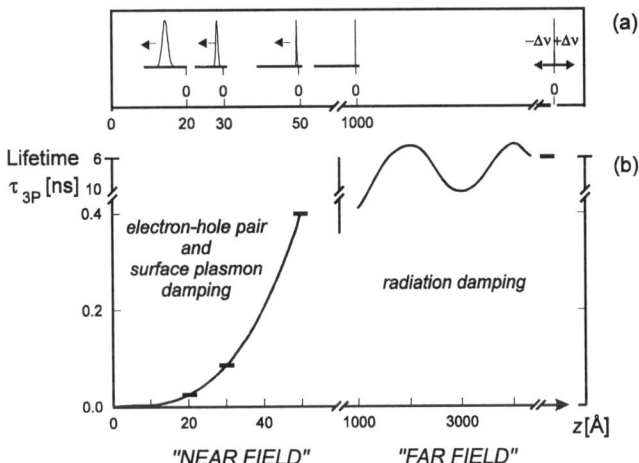

Figure 69 Distance dependence of the 3S–3P transition frequency (a) and lifetime of the 3P state (b) for laser-excited sodium atoms as a result of their interaction with a flat gold surface. The physical effects that are mainly responsible for the corresponding changes are indicated in the figure.

by a pulsed light source, showed for large distances the expected oscillations (Fig. 69) and thus agreed well with the predictions of the classic dipole theory (Harris et al., 1985).

Another way to realize a given distance z between dipole and metal is to use as the dielectric spacer layer a self-assembled ultrathin organic film (Bammel et al., 1993; Balzer and Rubahn, 1995a; Kuhnke et al., 1997) (see Chapter 1, Section 1.2) or a rare gas matrix at low temperatures of the order of a few degrees Kelvin (Rossetti and Brus, 1980).

The energy transfer between excited molecules and the surface *very close* to a silver surface (down to a few nanometers) was observed by Harris and coworkers (Whitmore et al., 1982b). These results too could be understood by applying classic electrodynamics without the need to take the microscopic surface structure into account (Waldeck et al., 1985). However, near the surface (closer than 10 nm) quantitative deviations between experiment and theory have been observed (Alivisatos et al., 1985), which addressed the *roughness* of the surface and the assumption of a local dielectric function. The local approximation of classic electrodynamics means that the induced polarization at a given point does not depend on the field and the polarization of the surroundings of this point.

Calculations of the lifetime assuming a small, Gaussian distributed roughness (Arias et al., 1982) showed a stronger dependence of lifetime on distance compared with that expected from the approximation for a smooth surface, $\tau \propto z^n, n \geq 3$. The exact value of the coefficient n depends on the degree of roughness. The lifetime of the atom close to the rough surface, τ_{rough}, as compared with the lifetime near the flat surface, τ_{flat}, can be estimated for oscillation wavelengths λ far away from the surface plasmon resonance and for $\delta \ll \lambda$ as (Arias et al., 1982)

$$\tau_{\text{rough}}/\tau_{\text{flat}} \approx 1/(1 + 6(\delta/\lambda)^2). \tag{3.13}$$

Hence significant effects are expected for roughnesses with average heights δ of larger than 2 nm (see Section 3.3) and for distances to the surface smaller than 5 nm.[9] In this distance regime the potential between atom and surface has significant influence on the damping of the excited dipole. Electron–hole pairs are excited not only in the bulk, but also at the surface. From momentum conservation one finds that the excited electrons in the substrate, which are responsible for the energy loss, have to experience collisions within the solid.

The probability of undergoing such energy-consuming collisions increases with the decreasing mean free path λ of the electrons. For transition metals such as nickel with a high density of states and thus a small mean free path close to the Fermi level and with excitations of a few eV ($k_F \lambda \approx 1$) bulk losses play an important role even very close to the surface. Here, the z^3 distance dependence is expected to be valid. For free-electron metals such as aluminum or sodium and for noble metals with excitation energies below interband transitions one has $k_F \lambda \approx 100$. In that case the electrons have to interact with the surface potential in order to undergo energy losses. The de-excitation occurs via surface electron–hole pair excitation and the lifetime changes with a z^4 distance dependence (Persson and Lang, 1982). This effect has been invoked to explain the significant deviations between the electronic lifetimes of N_2^* on Al(111), measured via EELS, and predictions of classic electrodynamics (Avouris et al., 1983). Lifetime measurements of biacetyl molecules, separated by ammonia spacer layers from silver surfaces (Alivisatos et al., 1985), as well as measurements of the C_{60} exciton lifetime in the close neighborhood (less than 3 nm) of silver surfaces (Kuhnke et al., 1997) have successfully demonstrated this d^4 distance dependence.

Calculations, taking explicitly into account the surface roughness, suggest that both the lifetimes (Xiong et al., 1995; Palasantzas, 1997) and the *frequency shifts* (Leung et al., 1988; Leung and Hider, 1993) of a dipole in front of a rough surface differ significantly from those of a dipole in front of

---

[9] As noted in (Leung et al., 1987; Leung, 1997), the electrostatic image potential approach (i.e., the long-wavelength limit, $z < \lambda$, meaning instantaneous interactions of dipole and reflected field) used in (Arias et al., 1982) indeed implies $z \ll \delta_{\text{Skin}}$, where δ_{Skin} is the skin depth of the metal ($\delta_{\text{Skin}} \approx 3$ nm for gold and $\lambda = 600$ nm).

3.2. FLUORESCENCE SPECTROSCOPY NEAR INTERFACES

a smooth surface. This theoretical finding agrees with recent measurements of lifetimes and transition frequencies of sodium atoms close to rough metal surfaces (Bammel et al., 1993; Balzer et al., 1993a). As demonstrated in Fig. 70 for the case of lifetime changes, the measured values are significantly smaller close to an experimentally rough surface as compared with the suggestions of classic image dipole theory (dashed line in Fig. 70).

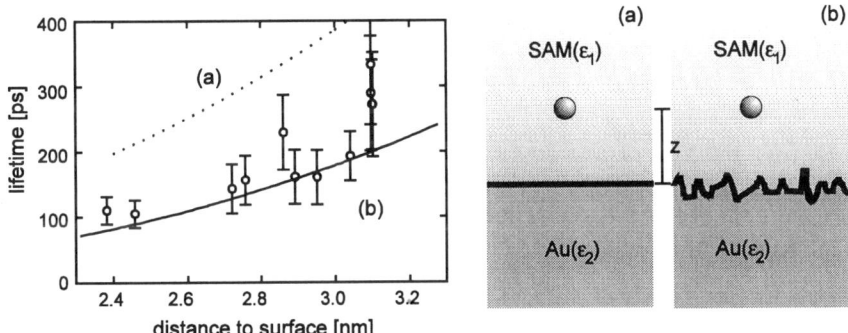

Figure 70 Dependence of the lifetime of sodium (Na) atoms on the distance to a rough gold (Au) surface. The atoms are embedded in a self-assembled organic film (SAM). The dashed line is the result of classic dipole image dipole theory (Chance et al., 1978), corresponding to configuration (a) on the right-hand side. The solid line (configuration (b)) is from a quantum-mechanical linear optical response formalism, taking into account measured surface roughness and nonlocality of the light–matter interaction. Reprinted with permission from (Balzer et al., 1998a). Copyright 1998 Society of Photo-Optical Instrumentation Engineers (SPIE).

This effect becomes even more pronounced if one measures *shifts* of the energies of the eigenstates of the atom as it approaches the rough surface (Balzer et al., 1997a). Frequency shifts of laser-excited dipoles close to (but not *on*) a conducting surface have previously been determined for a thin silver island film, separated via an insulator of known thickness d from another silver film (Holland and Hall, 1984). The surface plasmon resonance frequency of the silver island film (cf. Chapter 4, Section 4.1.3) was measured to be shifted as a function of thickness of the insulator to longer wavelengths for 20 nm$\leq d \leq$30 nm and to shorter wavelengths for 30 nm$\leq d \leq$75 nm as compared with the resonance frequency of the free film.[10] While the frequency shift could be reproduced for large distances by calculations on the basis of the classic dipole theory (Eq. (3.10)) deviations between experiment and theory were detected for smaller distances ($d \leq 25$ nm). This disagreement is not that surprising since the surface plasmon frequency is influenced by several effects which are not included in the theory, for example the collective excitations of higher order surface plasmons.

In order to elucidate the possibility of resonant excitations in the metal surface a jellium model for the electrons is assumed in a first-order approximation and the dielectric response is represented by the Drude equation

$$\epsilon_2(\omega) = 1 - \frac{\omega_p^2}{\omega(\omega + i\Gamma)} \qquad (3.14)$$

with ω_p the bulk plasmon frequency, Eq. (5.8), and Γ the bulk damping constant.

The external field $E(\omega)$ induces a surface charge

$$\sigma(\omega) = \left(\frac{\epsilon(\omega) - 1}{\epsilon(\omega) + 1}\right) \frac{E(\omega)}{2\pi} \qquad (3.15)$$

in the metal (in vacuum). Obviously, $\sigma(\omega)$ diverges for $\epsilon(\omega) = -1$, which, together with Eq. (3.14), means that $\omega_{sp} = \omega_p/\sqrt{2}$ represents a resonance in the surface charge ('surface plasmon frequency', cf. Chapter 4, Section 4.1.3). In the corresponding spectral range the frequency shift shows a dispersion-like behavior, which could give rise to significantly smaller values of the frequency shift compared with that expected without elementary excitations in the metal.

The solid line in Fig. 70 results from a model that includes roughness and nonlocality of the optical response (Balzer et al., 1997a). This model is based on a quantum mechanical linear response formalism, which was developed for excited atoms close to flat surfaces (Wylie and Sipe, 1984; Wylie and Sipe, 1985). This approach takes into account the multilevel nature of the atoms by use of all relevant real and virtual transitions and the microscopic surface structure (i.e., the statistical roughness; cf. Section 3.3, Eq. (3.24)) via modified Fresnel coefficients. The nonlocality of the optical response is described in the framework of bulk–selvedge coupling theory (Sipe, 1979; Sipe, 1980).

Note that one has to be aware of resonance effects since the effective values of ω_{sp} can significantly deviate from the values of the plain metal surface

---

[10] A similar red shift for small distances has been observed recently for large sodium clusters, separated from a gold surface by self-assembled organic films of varying thickness (Balzer et al., 1998a).

depending on the metal film thickness and structure (Abe et al., 1982; Fischer and Pohl, 1989) or on an adsorbed overlayer on the metal surface. Additional resonances might result from excitations in the selvedge region of the metal. We denote by 'selvedge' that region parallel to the surface plane where electronic polarization and external electromagnetic field are no longer locally connected and the conduction electrons produce nonlocal optical responses (Sipe, 1979). The corresponding optical response function has resonances (called 'multipole surface plasmons' (Liebsch, 1997) due to their dipolar character, which contrasts the usual surface plasmon with its monopolar sheet of charge), which can be calculated from a hydrodynamic dielectric function. Resonances of this kind were proposed a long time ago (Bennett, 1970) and have been observed previously in photoemission (Levinson et al., 1979; Barman et al., 1998) and electron energy loss experiments (Tsuei et al., 1991; Moresco et al., 1996; Chiarello et al., 1997). They can be observed even for perfectly flat surfaces because their dispersion curve extends to values of the parallel wavevector $q_\|$ smaller than the light line $\omega = q_\| c$ (Fig. 18). The MSP modes are responsible for significant modifications of ATR spectra (Sipe, 1979; Lang and Fukui, 1994). They are also expected from time-dependent density-functional calculations of the nonlinear response of alkali overlayers on aluminum (Liebsch, 1989; Ishida and Liebsch, 1992) to be responsible for the local-field-enhanced second harmonic generation of alkali–metal adsorbates (Tom et al., 1986). Usually the ratio between MSP and bulk plasmon frequency is $\omega_{\mathrm{MSP}}/\omega_\mathrm{p} \approx 0.8$ at $q_\|=0$ (Liebsch, 1991; Tsuei et al., 1991).

In the case of a statistical rough surface the thickness of the selvedge region (i.e., the region of strongly varying electromagnetic field strength) is approximated by $l=2\,\delta$ with δ the statistical roughness parameter[11] (Eq. (3.24)). Typically, δ is of the order of a nanometer (cf. Fig. 73)

The introduction of the selvedge region allows one to treat frequency shifts and relaxation rates nearly independent of each other. This is because the relaxation of the electronic optical response arises mainly from mutual electron collisions. Due to the extremely small thickness of the selvedge region (a few Fermi wavelengths) one can neglect the electron collisions inside it. Thus the *bulk* region gives the main contribution to the lifetime damping rate, while the *selvedge* region gives the main contribution to the frequency shift.

The hydrodynamic selvedge dielectric function using only a single multipole surface plasmon (MSP) excitation is written as (Balzer et al., 1997a)

$$\epsilon_s(k_1,\omega) = 1 + \frac{\omega_p^2 l_1}{l(\omega_1^2 - \omega^2 - i\omega\Gamma_1)} \qquad (3.16)$$

with the bulk plasmon frequency ω_p, the MSP frequency ω_1, its damping

---

[11] In the case of a smooth, flat surface the thickness would be of the order of a few Fermi wavelengths.

constant Γ_1, wavevector k_1 and $l_1 = f_1/n_0$ a length that characterizes the effective oscillator strength per unit area f_1 (n_0 is the bulk electron density). This selvedge dielectric function implies neglecting the spatial dependence of the external field within the selvedge and averaging the selvedge polarization over the z-coordinate, which is equivalent to setting $k=0$ (long-wavelength limit). Hence the track of nonlocality are the MSP frequencies that have been obtained in the framework of the nonlocal hydrodynamic model.

Within this quantum mechanical approach the metal-induced energy shift of atomic state $|a\rangle$ is induced by the polarization energy of the atom in a fluctuating electromagnetic field and is described as the sum of two terms (Wylie and Sipe, 1984; Wylie and Sipe, 1985)

$$\Delta\omega_a = \Delta\omega_a^{vdW} + \Delta\omega_a^{cl} , \qquad (3.17)$$

with a van der Waals and a classic term. The van der Waals term is given by

$$\Delta\omega_a^{vdW} = -\frac{1}{2\pi} \int_0^\infty d\omega G^R(\omega)\alpha_a(\omega), \qquad (3.18)$$

with the atomic polarizability

$$\alpha_a(\omega) = \frac{2}{\hbar} \sum_n \frac{\omega_{na} D_{an}}{\omega_{na}^2 - \omega^2} \qquad (3.19)$$

and D_{an} the squares of the dipole moments for the relevant transitions ω_{an}, calculated from the oscillator strengths. The classic term is given by Eq. (3.10).

The selvedge contributions to the classic level shift and relaxation rate are determined by the function G^R at the frequencies of the relevant atomic transitions, which is the expectation value of the field generated by the oscillating dipole (in the nonretarded limit, $(c \to \infty)$):

$$G^R(\omega) \propto \frac{1}{\epsilon_1} \int_0^\infty dk k^2 R(\omega,k)\exp(-2kd). \qquad (3.20)$$

The thickness of the selvedge region, l, might be introduced into the theoretical approach by a waveguide-type Fresnel coefficient for the interface between the metal and the dielectric spacer layer

$$R(\omega,k) = \frac{r_{1s} + r_{s2}\exp(-2kl)}{1 + r_{1s}r_{s2}\exp(-2kl)} , \qquad (3.21)$$

which includes the reflection Fresnel coefficients for p-polarized light between spacer layer ('1'), metal bulk ('2') and selvedge region ('s') (the coefficients for s-polarized light are zero in the nonretarded limit):

3.2. FLUORESCENCE SPECTROSCOPY NEAR INTERFACES

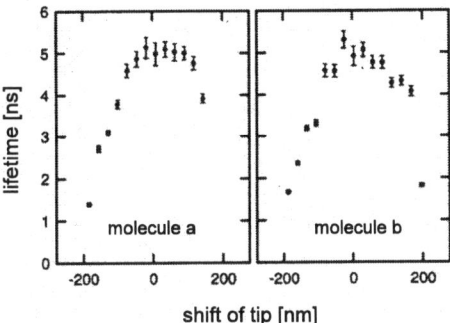

Figure 71 Dependence of the lifetime of *individual* Rh6G molecules ('a' und 'b') on a quartz substrate on the lateral distance to the tip of a SNOM. The measurement has been performed with near-field microscopy and using a mode-locked Ar$^+$ laser (514.5 nm). The far-field lifetime of the molecules is about 3.7 ns. Reprinted with permission from (Ambrose et al., 1994b). Copyright 1994 Society of Photo-Optical Instrumentation Engineers (SPIE).

$$r_{ij} = \frac{\epsilon_j - \epsilon_i}{\epsilon_j + \epsilon_i}. \tag{3.22}$$

As seen in Fig. 70, taking into account the measured surface roughness and choosing an appropriate value for the MSP frequency results in satisfactory agreement between the measured and calculated values of lifetime damping (and also frequency shift; see (Balzer et al., 1997a)). A more elaborate inspection of the theoretical approach reveals that especially the frequency shift changes dramatically with increasing roughness of the surface, whereas the lifetimes vary only of the order of a few tens of percent.

Experiments with surface probe microscopes or microminiaturized optical sensors show that changed fluorescence lifetimes close to a surface are not just of academic interest. After demonstrating the possible nanometric resolution of fluorescence measurements, the fluorescence lifetime of *single* dye molecules (Xie and Trautman, 1998) has been observed by a near-field microscope (SNOM, Chapter 1, Section 1.2.4) both in the spectral domain (Trautman et al., 1994) and in the time domain (Xie and Dunn, 1994; Ambrose et al., 1994a). The influence of the aluminum tip on the lifetime has been studied too (Ambrose et al., 1994b) (cf. Fig. 71).

Since the effect might result in an increase, and also in a decrease, in the lifetime (depending on the position of the tip (Pagnot et al., 1997)), a quantitative understanding (Bian et al., 1995) is important if one is interested in measuring the fluorescence lifetimes of individual molecules independent of the detection instrument. Whether the observed signal intensities are due to the optical response of a single, isolated molecule, or whether there are alterations in the optical response due to heterogeneous broadening (i.e., different environments for different molecules) might be decided from time-correlated, single-photon counting measurements of the fluorescence decay times (Enderlein et al., 1997). By the use of confocal laser scanning optical microscopy (CLSM, laser focus area 0.36×0.6 µm), the expected antibunching behavior of an individual quantum system on a surface (i.e., a rhodamine 6G molecule on a borosilicate sample) has been demonstrated recently (Ambrose et al., 1997).

3.3 Surface-Enhanced Raman Scattering

In the course of absorption or fluorescence spectroscopy electromagnetic dipole radiation is employed to excite polarizable molecules from a ground state into an excited state. Photons are emitted subsequently between two real states with the same or different energy. However, they also can be elastically or inelastically *scattered*. The elastic 'Rayleigh' scattering (i.e., scattering without energy transfer from or to the molecule) provides information about the scattering medium since the scattering cross section depends strongly on the wavelength ($\propto \nu^4$).[12] More information can be gained from inelastic scattering, i.e., from the existence of Raman lines, which are shifted with respect to the elastic line by the difference between the energies of two real states (Fig. 72).

The cross section for inelastic scattering is, just as in the case of particle scattering, much smaller compared with that for elastic scattering. It is determined by the Raman tensor $d\alpha/dQ \neq 0$. In contrast to infrared absorption, which occurs only if the molecule possesses a permanent dipole moment, here only the polarizability α of the molecules has to change in the course of a variation in the nuclear coordinates Q, *e.g.*, during vibration. The intensity of the Raman lines is, for the excitation of an ensemble of particles of density n with frequency ν and laser intensity I_L, given by (Long, 1977)

$$I_{\text{Raman}} \propto \nu^4 I_L \cdot n \cdot \left| \sum_z \frac{\mu_{fz}\mu_{zi}}{\Delta \nu - i\Gamma} \right|^2. \qquad (3.23)$$

Here, μ_{fz} and μ_{zi} are the Raman transition probabilities between final state $|f\rangle$ and intermediate state $|z\rangle$ and between intermediate state and initial state $|i\rangle$. The letter Γ summarizes possible relaxation processes, *e.g.*, via

---

[12] This enhanced scattering of high-frequency radiation, which reflects the radiation behavior of a Hertzian dipole, leads to the blue daylight firmament.

3.3. SURFACE-ENHANCED RAMAN SCATTERING

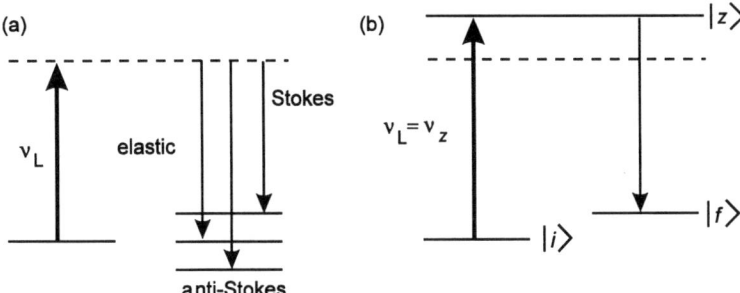

Figure 72 Elastic, Stokes and ant-Stokes Raman lines due to nonresonant (a) and resonant (b) light scattering at a molecule. The symbol ν_L denotes the laser frequency, while ν_z means a transition frequency between two real states, i and z, within the molecule. Real states are characterized by solid lines, virtual states by dashed lines.

electromagnetic radiation or via collisions, while $\Delta\nu = \nu_L - \nu_z$ describes the deviation of laser frequency from the transition frequency into the real state $|z\rangle$. If the laser hits a real state ($\Delta\nu=0$) and the damping Γ is small, then a high Raman intensity is expected. The same is the case for large densities n of the scattering medium. Molecules, which are adsorbed on surfaces, are expected to provide only weak or no Raman signals at all due to their low density, which is about a factor of 10^5 smaller compared with that of molecules in the gas phase. However, it has been found that under certain circumstances in the case of molecules that are adsorbed at rough metal surfaces, very high Raman intensities can be found, even in the case of a spontaneous Raman process ('SERS', surface enhanced Raman scattering) (Fleischmann et al., 1974). The enhancement compared with a smooth surface is between 10^4 and 10^6. As an explanation of this effect one might invoke the changed binding sites (as compared with a smooth surface), which might change the Raman polarizability ('chemical effect'). But the roughness of the surface itself might also lead to resonance enhancement of the electromagnetic field at the surface since surface plasmons might be excited in the rough metal surface ('electromagnetic effect'). Consequently, a change in surface roughness might be used for optimization of the SERS effect (Nitzan and Brus, 1981b).

In order to change the surface roughness at will, it is necessary first to find a quantitative description. A first-order approach is to use a Gaussian function

$$G(\vec{r},\vec{r}') = \delta^2 \exp\left(-\frac{|\vec{r}-\vec{r}'|^2}{\lambda_0^2}\right), \qquad (3.24)$$

which describes the height–height correlation between different points of the surface profile $S(\vec{r})$. The function thus describes in which way the heights at different points of the surface are correlated as a function of their mutual distance $|\vec{r}-\vec{r}'|$. In Eq. (3.24) $\vec{r}=(x,y)$ means the position at the surface, λ_0 the transverse correlation length ($\lambda_0 \to 0$ means the statistical distributed roughness) and δ the depth of corrugation of the surface or 'mean height', $\delta = \sqrt{\overline{S^2}}$ (Fig. 73). Values for δ and λ_0 for small $\delta \leq 2$ nm might be deduced optically via the efficiency of the surface–plasmon coupling (Raether, 1988; Tajima et al., 1995) (Section 3.1.4) or directly by the evaluation of scanning probe microscopy pictures (Raether, 1984).[13] Fig. 73 shows the topography of a rough gold surface as measured using a force microscope, which can be described by $\delta = 1.67$ nm and $\lambda_0 = 23.5$ nm over a wide range of correlations (up to $|\vec{r}-\vec{r}'| = 40$ nm).[14]

A macroscopic method for the determination of surface roughnesses is the measurement of the reflected light-intensity I as a function of the reflection angle Θ_o for different angles of incidence Θ_i of an laser beam of wavelength λ (Fig. 74). For smooth surfaces with $\delta \ll \lambda$ the laser light is reflected specularly ($\Theta_o = \Theta_i$), and the angular dependence of the reflected light intensity is described as

$$I_{\text{refl.}} = I_0 \exp\left(-\left(\frac{4\pi\delta}{\lambda}\cos\Theta_i\right)^2\right). \qquad (3.25)$$

The value of I_0 for a perfectly smooth surface can be calculated from the optical constants of the reflecting material via Fresnel formulas (Born and Wolf, 1975). Measurement of the intensity distribution then results in a value of δ.

In the case of a moderately rough surface ($\delta \approx \lambda$) diffuse scattering also becomes important as well as elastic scattering. With increasing Θ_i and increasing ratio δ/λ the fraction of diffusely scattered light also increases (Ogilvy, 1991). Since diffuse scattering depends on the correlation length λ_0, several measurements using different angles of incidence and different polarizations of the light allow one to obtain values for λ_0 and δ. For very rough surfaces ($\delta \geq \lambda$) the specularly reflected light intensity is of minor importance,

---

[13] Ellipsometry (Azzam and Bashara, 1992) also provides information about the roughness of the reflecting layer.

[14] For larger values of $|\vec{r}-\vec{r}'|$ the Gaussian distribution, Eq. (3.24), has to be shifted by a finite value in reciprocal space. This means for the height-height correlation function that an additional term $cos(|\vec{r}-\vec{r}'|b)$ has to be introduced with fitting parameter b (Tong and Williams, 1994).

3.3. SURFACE-ENHANCED RAMAN SCATTERING

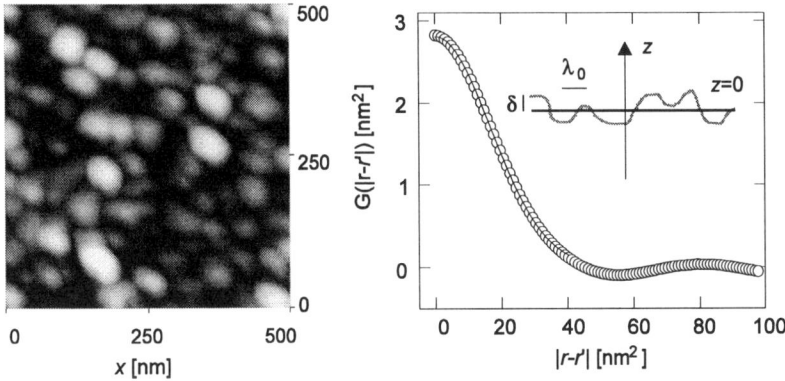

Figure 73 Scanning force microscopy picture of a rough gold film, evaporated on polycrystalline silicon, and corresponding correlation function. The points have been taken from an evaluation of the picture, the curve is a Gaussian fit and the inset defines the parameters in Eq. 3.24.

and the emission characteristics is a cosine-distribution with respect to the angle of reflection, nearly independent of Θ_i (Lambert's diffuse scatterer).

In Fig. 75 the intensity of the 1597 cm^{-1} Raman line of PNBA (p-nitro-benzol-acid) is shown as a function of mass thickness of the silver film on which it has been adsorbed. The dielectric substrate has been shaped via plasma-etching into the form of tall needles[15] (100 nm long, with 60 nm diameter) before evaporating the silver on it (Fig. 75). With increasing mass thickness the average diameter of the silver clusters, which are evaporated onto the needles, is increasing, resulting in an increase in surface roughness and (for given correct wavelength) of the field enhancement.[16]. As can be seen from the figure, the Raman signal intensity can be enhanced by a well-defined change in the surface roughness.

The different symbols in Fig. 75 represent measurements taken under

---

[15] An effect similar to the generation of cones by laser ablation, cf. Fig. 143.
[16] Recent measurements seem to reveal that not only the average cluster size, but also the distance between the clusters and their aggregation into larger units (Chen et al., 1995) as well as their fractal dimension (Douketis et al., 1995b) influence the SERS activity. However, it has also been noted that the fractality alone is insufficient to fully account for the SERS activity (Douketis et al., 1995a). Near-field optical microscopy (SNOM) studies suggest that SERS is not uniquely coupled to specific surface sites (Zhang et al., 1998).

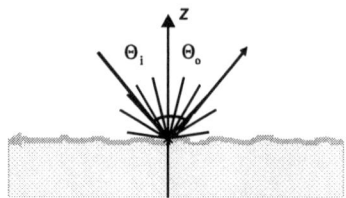

Figure 74 A light beam that hits a rough surface at an angle of incidence Θ_i is specularly reflected at Θ_o, but also has components from diffuse scattering, depending on the surface roughness.

various angles of incidence, which reproduce approximately the trend given by the dashed curve. Following a maximum at a thickness of about 100 nm, the intensity decreases for larger thicknesses since for larger clusters radiation damping reduces the enhancement effect.

Note that coupling to the surface plasmon field also changes the fluorescence properties of the adsorbates (cf. Section 3.2). In contrast to the Raman process for fluorescence intensities, radiationless energy losses into the metal give important contributions besides enhanced absorption and emission, since they influence the *spontaneous* emission, which is independent of the field and does not exist for Raman processes. The fluorescence yield of a molecule close to a rough surface, which might be described approximately via an array of spheres with radius r, is given by the molecular lifetime multiplied by the absorption and emission rates. The lifetime of the molecule decreases with decreasing distance z to the surface as z^3, while the enhanced electromagnetic field, which results in a changed absorption and emission rate, decreases as $(r+z)^{-3}$ with *increasing* distance. These two effects result in a maximum of the fluorescence yield, the spatial position of which depends on the average roughness of the surface (Fig. 76) (Aussenegg et al., 1987; Kümmerlein et al., 1993). The existence of strong damping mechanisms for molecules that are adsorbed directly onto metallic surfaces, which compete with the field enhancement effects, means that, for example, for surface enhanced photochemistry a dielectric spacer layer (or a functionalized self-assembled monolayer film (Ye et al., 1997)) has to be brought between adsorbate and substrate in order to obtain optimum enhancement conditions.

It is noted that the 'optimum' enhancement could be even further increased due to positive electromagnetic interference effects by placing a spacer layer in between the rough, adsorbate-covered metal surface and an additional metal surface. The latter one then serves as a mirror which significantly enhances the local fields. As a spacer layer one might use quartz films of variable thickness

3.3. SURFACE-ENHANCED RAMAN SCATTERING

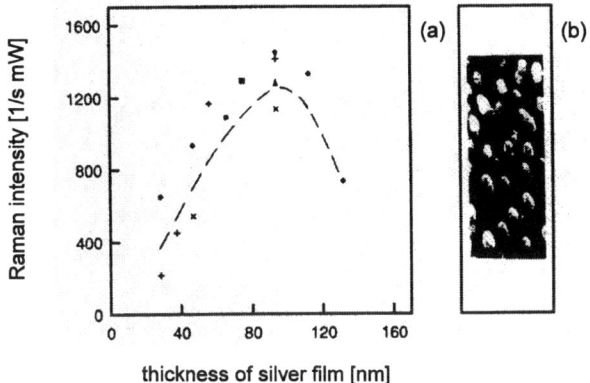

Figure 75 Raman intensity of a PNBA film, adsorbed onto silver and as a function of silver layer thickness. For 100 nm thickness a picture of the distribution of silver-covered needles using electron microscopy is shown on the right-hand side. Reprinted with permission from (Wokaun, 1985). Copyright 1985 Taylor & Francis.

(Bingler et al., 1995).

The importance of SERS is that it enables one to perform Raman spectroscopy with extremely small particle densities in chemical and biological important environments. The investigation of surface-specific spectral features allows one to determine the physical and chemical properties of the adsorbates such as geometry and orientation, the kind of chemisorption binding and the binding sites or the existence of catalytic chemical reactions, for example at the surface of electrodes (Moskovits, 1985). SERS has been used as a versatile method for chemo-optical sensing of organic trace compounds at interfaces in air or water (Hill et al., 1995), but also to study single molecules in colloidal solutions (Kneipp et al., 1997) or to two-dimensionally map rough surfaces (Zhu et al., 1997). Drawbacks include problems of adsorbing the species to be investigated at the rough surface (a major problem for nonpolar solutes) and the impossibility of analyzing quantitatively signal strengths. The former problem might be overcome by the use of organic film stabilized colloids (Chumanov et al., 1995) or organic film-coated metallic nanospheres (Freunscht et al., 1997). The latter problem is more difficult to handle since it involves unclear conditions in the generation of the rough, SERS-active samples as well as a lack in understanding of the SERS mechanism itself.

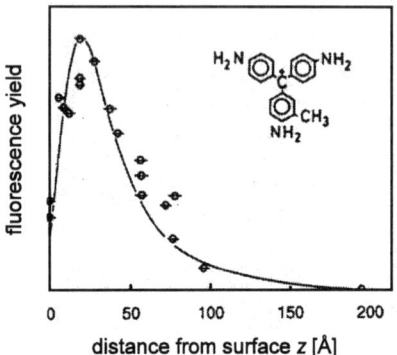

Figure 76 Fluorescence yield of the chromophore Basic Fuchsin (chemical structure shown in the upper right corner), adsorbed close to a 4 nm thick, rough silver film, as a function of distance to the film, which has been adjusted in increments of 0.5 nm by the use of silicon oxide spacer layers. The solid line is the theoretical expectation for silver ellipsoids with large semiaxes of 30 nm and 14 nm. Reprinted with permission from (Wokaun et al., 1983). Copyright 1983 American Institute of Physics.

Given the present possibilites of real-space investigations and manipulations via surface scanning techniques, one might hope to see significant progress in this area within the next few years.

3.4 Second Harmonic Generation

A light wave that hits a polarizable medium deflects the electrons from their equilibrium positions. The macroscopic polarization $\vec{P}$, which is induced by these oscillating dipoles, and which leads eventually to the refracted or reflected light wave, is proportional to the incoming electromagnetic field strength $\vec{E}$ as long as the deflection of the electrons can be described within the framework of an harmonic oscillator approximation. The proportionality constant χ depends on the irradiated material and is called the 'susceptibility'. The harmonic approximation is equivalent to an independence of susceptibility and, for example, absorption coefficient on the strength of the exciting electromagnetic field.

In a more general description, which is not restricted to the linear range of

3.4. SECOND HARMONIC GENERATION

small[17] field strengths, the polarization which is induced by the light wave, might be written as a Taylor expansion[18] in the electric field

$$\vec{E}(\vec{r},t) = \mathrm{Re}\vec{E}(\vec{r})\exp(-i\omega t) = \frac{1}{2}(\vec{E}(\vec{r})\exp(-i\omega t) + (\vec{E}^*(\vec{r})\exp(i\omega t)), \quad (3.26)$$

namely (Jackson, 1975):

$$\vec{P} = \epsilon_0[\chi^{(1)}\vec{E} + \chi^{(2)}\vec{E}\vec{E} + \chi^{(3)}\vec{E}\vec{E}\vec{E} + ...] \quad (3.27)$$

with the dielectric constant of the vacuum, $\epsilon_0 = 8.86 \times 10^{-14}$ As/V·cm. Here, the (anisotropic) coupling between light and matter is described by the susceptibility tensors $\chi^{(i)}$. While the tensor components of $\chi^{(1)}$ are of order 1 in solids, $\chi^{(2)}$ is about ten orders of magnitude smaller, i.e., $\approx 10^{-10}$ cm/V and $\chi^{(3)} \approx 10^{-17}$ cm^2/V^2. Thus the exciting fields $\vec{E}$ have to be strong to observe nonlinear optical effects. Laser excitation of a strongly nonlinear crystal such as beta bariumborate (BBO, β-BaB$_2$O$_4$, with $\chi^{(2)} \approx 2.2 \times 10^{-10}$ cm/V) results in a spectrum of light that contains, besides the fundamental frequency, sidebands with light of multiple frequency or — in the case of excitation with two or more different frequencies — of light with mixed frequencies. These nonlinear optical phenomena (Shen, 1984; Butcher and Cotter, 1990) of frequency multiplication or frequency mixing have extended the available spectral range of coherent radiation far into the infrared and ultraviolet ranges. Back in 1961, one year after the description of the first optical laser, optical second harmonic generation was reported the first time (Franken et al., 1961).

The optical nonlinearities of the investigated media can also be used for spectroscopic analysis, resulting in numerous new spectroscopic methods such as saturation spectroscopy, stimulated Raman spectroscopy, multiphoton absorption, frequency doubling, sum frequency generation, four-wave mixing, etc. The potential of those methods, which may be applied in most cases also to surfaces, becomes nowadays increasingly explored due to the availability of ultrashort pulse lasers (viz., femtosecond pulses) and lasers with wavelengths from the infrared (between 1 and 10 μm) to the VUV ($\lambda \leq$ 197 nm) spectral range.

---

[17] The external field strengths are called 'small' if they can be neglected compared with the intra-atomic field strengths. The electrostatic field that acts on a ground state electron in the hydrogen atom, for example, has a strength of 5.14×10^9 V/cm. This field strength can be achieved in the 10 μm focus of a laser with 10 ns pulse length and a pulse energy of 0.2 J. For more complex media the intra-atomic field strengths are smaller (typically 10^8 V/cm). Since the ratio between nonlinear polarization of $(n+1)$th order to polarization of nth order is about equal to the ratio between external to intra-atomic field strength (Shen, 1984), a field strength of 10^3 V/cm is usually sufficient to obtain a significant nonlinear signal intensity.

[18] Accurately spoken, this power expansion is valid only for $\chi^{(2)}\vec{E}\vec{E} \leq \chi^{(1)}\vec{E}$. This condition might be not fulfilled, for example, for resonance enhanced frequency-doubling or very strong laser fields.

Figure 77 shows the set-up for an experiment to observe optical second harmonic generation, SHG, at surfaces.[19] A pulsed laser beam of wavelength 1064 nm (fundamental of a Nd:YAG laser) irradiates the substrate under an angle of 45° with respect to the surface normal and is reflected. The generated SH signal at 532 nm is separated from the fundamental light by use of a monochromator and a filter and is detected by a photomultiplier. The polarization direction of the incoming laser beam is either parallel (p-polarized) or perpendicular (s-polarized) with respect to the plane of incidence. The SH signal too might be observed p- or s-polarized using a polarizer. The restriction to defined polarization combinations means that only selected components of the $\chi^{(2)}$ tensor contribute to the signal intensity and allows one to deduce information about symmetries at the surface (see below).

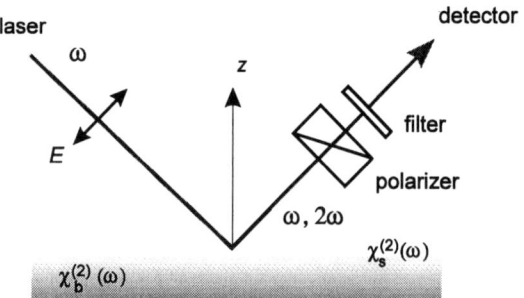

Figure 77 Set-up for the observation of optical frequency doubling at surfaces with surface susceptibility $\chi_s^{(2)}(\omega)$. The solid has the bulk susceptibility $\chi_b^{(2)}(\omega)$. The filter transmits solely the frequency-doubled light. The arrow drawn perpendicular to the line of the incoming laser beam characterizes the direction of polarization (p-polarized, i.e., parallel to the plane of incidence).

A necessary condition for the generation of nonlinear polarization of even order is the lack of inversion symmetry in the nonlinear medium. In media with inversion symmetry the even electric multipole terms (dipole, quadrupole and so on) in the multipole expansion of the electromagnetic fields are forbidden due to conservation of parity: under a parity operation the fields $\vec{E}$ change sign, $\vec{P}(-\vec{E}) = -\vec{P}(\vec{E})$. This condition is to be fulfilled for Eq. (3.27) only if the even terms vanish. In media with inversion symmetry thus only nonlinear

---

[19] A detailed account of the principles of surface SHG is given in (Brevet, 1997).

3.4. SECOND HARMONIC GENERATION

terms of odd order exist (beginning with $\chi^{(3)}$) as well as higher order bulk contributions (magnetic dipole, electric quadrupole, etc.). These terms are usually small compared with the dipole term.

At the surface, however, the inversion symmetry of the bulk is broken. Electric dipole contributions to the nonlinear polarization become possible due to the spatial structure of the surface and due to the discontinuity of the normal component of the electric field at the surface. In the case of a *nonlocal* interaction the polarization at a given position $\vec{r}$ depends on the external field of the surroundings (i.e, 'spatial dispersion'). In most cases, however, one assumes for the sake of simplicity a *local* interaction, in which case the susceptibility is independent of the polarization in the surroundings. If one assumes that the main reason for this nonlinear polarization is the generation of a strong (static) dipole field at the surface, then it becomes clear that it should be localized in the uppermost atomic layers down to a depth of 0.5 – 1.5 nm (Sipe et al., 1987). Hence in contrast to SHG in the bulk at surfaces, due to the thin nonlinear optically active layer phase matching is automatically fulfilled under all angles of incidence.

In the case of metals, the damping of the incoming electromagnetic light wave by Friedel oscillations provides the most important length scale (Song et al., 1988). The wavelength of the oscillations is π/k_F with the Fermi wavevector (the largest possible wavenumber of the free electron gas) $k_F = 2\pi/\lambda_F = (3\pi^2 \bar{n})^{1/3}$ and $\bar{n}$ the average density of the positive charge of the ion cores.[20] If one uses for $\bar{n}$ the value for gold (5.9×10^{22} cm^{-3}), then one obtains $k_F = 1.2$ Å^{-1} and $\lambda_F = 5.24$ Å . The damping thus mainly occurs over a depth of a few nanometers. At the interface with the vacuum the electronic charge density is smeared out. This so-called 'spill out' of the electrons leads to an electrostatic dipole layer (Fig. 78).

The additional electric quadrupole and magnetic dipole contributions from the bulk can be separated from the real surface contributions only under special conditions: for example, if one modifies the surface layer in the form of an evaporated thin film and if one extracts the bulk contributions by calculating the interface contributions (Koopmans et al., 1993). The bulk contributions are proportional to the field gradient, meaning that a zone of about $\lambda/2\pi \approx$ 5...10 nm contributes to the total signal intensity.

Determination of coverages

If one takes these restrictions into account, SHG is, for media with inversion symmetry, an extremely surface-sensitive method. This class of media covers fcc (gold, etc.) and bcc (sodium, etc.) metals, diamond, silicon, germanium, all gases and liquids or glasses. Depending on the absolute value of the

---

[20] Often one characterizes the average density by the 'Wigner–Seitz radius', which is given by $r_S = (3/(4\pi\bar{n}))^{1/3}$. The Wigner–Seitz cell is the unit cell of the reciprocal lattice.

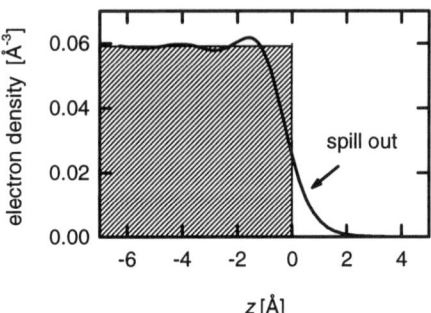

Figure 78 Electron density distribution as a function of distance to a simple metal surface ($z=0$), calculated using the jellium model (Lang and Kohn, 1970) with an average charge density $\bar{n} \approx 0.06$ Å^{-3}. The shaded box in the figure denotes the positive charge background of the ion cores.

components of $\chi_s^{(2)}$, adsorption and desorption processes might be observed in the submonolayer regime. This is demonstrated in Fig. 79 for CO molecules, adsorbed on Ni(111) (Zhu et al., 1985) and alkali films on Si(111) (Boneß et al., 1998). The signal intensity is given by (Shen, 1986)

$$I_{2\omega} = \frac{32\pi^3 \omega \sec^2 \Theta_i}{c^3 \hbar \epsilon(\omega) \sqrt{\epsilon(2\omega)}} \cdot |\chi_s^{(2)}|^2 I_\omega^2 \cdot F \cdot \tau. \qquad (3.28)$$

Here, Θ_i is the angle of the incoming laser beam with respect to the surface normal. For a Nd:YAG laser pulse (1064 nm) of duration $\tau = 10$ ps, intensity $I_\omega = 10^9$ W/cm^2 and area $F=1$ mm^2 (corresponding to a pulse energy of 100 µJ), which irradiates a monolayer of molecules with a surface susceptibility of $\chi_s^{(2)} = 10^{-15}$ esu, one calculates from Eq. (3.28) about 10^5 photons per laser pulse.[21] This signal intensity can be measured without problems with a photomultiplier. The total susceptibility of the monolayer is given by the nonlinear polarizability of each single molecule, $\alpha^{(2)}$, and the number density of the surface molecules, N_s,

---

[21] In the CGS system the susceptibility is defined via $P^{(n)} = \chi^{(n)} E^{(n)}$. The conversion factor between the two systems is $\chi^{(n)}[(\text{cm/V})^{n-1}] = \chi^{(n)}[\text{esu}] \cdot 4\pi/(3 \times 10^2)^{(n-1)}$ (Butcher and Cotter, 1990). For the presented example this means $\chi^{(2)} \approx 4 \times 10^{-17}$ cm/V. This value is six orders of magnitude smaller compared with that for a BBO crystal!

3.4. SECOND HARMONIC GENERATION

$$\chi_s^{(2)} = N_s \cdot \alpha^{(2)}, \qquad (3.29)$$

if one averages over the orientations of the molecules and neglects their mutual interactions.[22] For the surface coverage of a monolayer ($\approx 5 \times 10^{14}$ cm^{-2}) the value $\chi_S^{(2)} = 10^{-15}$ esu corresponds to a molecular polarizability of 2×10^{-30} esu. The nonlinear polarizabilities of metals are often more than an order of magnitude larger; those of dye-doped monofilms can be even higher. In Table 1 exemplary measured values of $\chi_s^{(2)}$ are summarized for different excitation wavelengths and the coverage of a monolayer.[23]

Table 1 Selected nonlinear surface susceptibilities. The values are from: a: (Wang and Duminski, 1968); b: (Heinz et al., 1982); c: (Kelly et al., 1991); d: (Chen et al., 1981); e: (Bloembergen et al., 1968); f: (Chen et al., 1973); g: (Marowsky et al., 1988).

Material	$\chi_s^{(2)}$ [10^{-17} esu]	λ [nm]	Reference	Remark
Lithiumfluoride	5	694	a	insulator
Rhodamine 110	150	670	b	dye
Silicon	160	1064	c	semiconductor
Pyridine	800	1064	d	dye
Silver	900	1064	e	metal
Sodium	2700	694	f	metal
Hemicyanine film	50000	1064	g	fatty acid film

Of course, upon reflection from the surface, the local field strength tensor **L** has to be taken into account and the above $\chi_s^{(2)}$ accordingly modified:

$$\chi_s^{(2)} \to \mathbf{L}(2\omega)\chi_s^{(2)}\mathbf{L}(\omega)\mathbf{L}(\omega). \qquad (3.30)$$

In the case of an ideal smooth surface **L** is mainly represented by the Fresnel factors $f_i(\omega, 2\omega)$ (explicit values for ω and 2ω are given, for example, in (Mizrahi and Sipe, 1988)). These factors depend strongly on the angle of incidence of the light and average over the optical properties via a general dielectric function of the medium down to the penetration depth of the light. Hence they are not surface sensitive in the same way as $\chi^{(2)}$ is. For a rough surface the electromagnetic field enhancement might be much stronger due to resonance effects (see below, Fig. 80 and also Section 3.3). For example,

---

[22] Mutual interactions of the nonlinear optically active particles modify the local fields (Ye and Shen, 1983).
[23] Recently, by generating Langmuir-Blodgett films (see Fig. 23) of supramolecular-nested chiral molecules second order nonlinear susceptibilites of 10^{-11} esu have been achieved (Verbiest et al., 1998).

for pyridin molecules, adsorbed onto a rough silver electrode, the effective surface susceptibility is about a factor of 50 higher compared with that for a smooth surface, namely $\chi_S^{(2)} \approx 4 \times 10^{-13}$ esu (Boyd et al., 1986). This strong nonlinearity makes the measurement of SH signals, even with continuous lasers and a power of a few tens of milliwatts, possible; the good spatial resolution of the laser beam (focal diameter a few μm) should allow one to perform nonlinear surface microscopy via SH generation (Boyd et al., 1986; Smilowitz et al., 1997). In recent research by exploiting magnetic field influences on SHG (see later in this chapter) magnetic domains could be imaged via SHG microscopy (Kirilyuk et al., 1997b; Kirilyuk et al., 1997c).

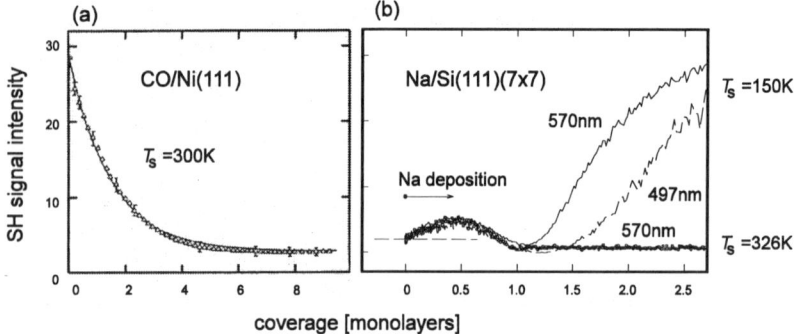

Figure 79 (a) Coverage dependence of the intensity of frequency doubled light, induced by irradiating a Ni(111) surface with 1064 nm laser light. The surface has been covered with CO molecules. Reprinted from *Surf. Sci.*, (Zhu et al., 1985), Copyright 1985, with permission from Elsevier Science. (b) Coverage dependence for an alkali-film-covered Si(111)(7×7) surface. Solid and dashed lines are obtained for $T_S = 150$ K, open circles for $T_S = 326$ K. The intensity from the plain Si surface is denoted by a dashed horizontal line. Reprinted from *Surf. Sci.*, (Boneß et al., 1998), Copyright 1998, with permission from Elsevier Science.

Figure 79 shows changes in the SH signal intensity from metal (left-hand side) and semiconductor surfaces (right-hand side) as the surfaces are covered by CO molecules and sodium atoms, respectively. In the latter case above $\Theta = 1$ ML the abscissa has to be compressed by a factor of ten for T_S=326 K due to the strong decrease in sticking coefficient. The SH signal intensity was induced by 60 fs pulses at 570 nm (solid line, circles) and at 497 nm (dashed line) in 60° reflection geometry and in pp-orientation. The experimental set-up is shown in Fig. 105.

3.4. SECOND HARMONIC GENERATION

Obviously the SH intensity of the plain Ni(111) surface is decreasing with increasing coverage by CO molecules since CO is an electron acceptor, which binds the initially nearly free electrons of the nickel surface. In contrast, alkali atoms are electron donors and thus increase the magnitude of the nonlinear surface signal intensity (Tom et al., 1986). Hence the initial increase in SH intensity with coverage for $\Theta < 0.5$ ML is dictated by the generation of Na/Si induced surface dipoles and thus is independent of the surface temperature and also of the wavelength of the fundamental beam. A further increase in coverage leads to a depolarization of the dipoles due to mutual interactions and an accompanying decrease in the binding energy between Na and Si. This is accompanied by a structural transformation of the Na overlayers since above $\Theta = 0.5$ ML the lattice sites for covalent bindings are occupied.

At room temperature the strong decrease in the sticking probability for Na on Si(111) (Papageorgopoulos and Kamaratos, 1992) prevents a subsequent increase in signal intensity above 1 ML. At significantly lower surface temperatures the following increase in SH signal intensity is due to collective electronic excitation since the adsorbate layer becomes metallic at a coverage close to 1 ML (Soukissian et al., 1989), corresponding to the formation of a conduction band within the Na overlayer (Jeon et al., 1992). With increasing alkali coverage the surface electron density increases strongly, shifting the surface plasmon frequency of the adsorbate from the value of the 3S–3P transition for free Na atoms ($\hbar\omega_{3S-3P} = 2.1$ eV) towards the value of the bulk-terminated sodium surface ($\hbar\omega_{sp} \approx 4.1$ eV) at coverages of the order of 2 ML (Jostell, 1979). Hence a near-resonance enhancement of the nonlinear signal intensity is expected since $\hbar\omega_{sp}$ approaches the value of the second-harmonic frequency, $2\hbar\omega = 4.35$ eV. If one increases the fundamental beam frequency ω and consequently 2ω, then the slope of the signal increase is reduced due to the enlarged difference between 2ω and ω_{sp} (dashed line in Fig. 79).

Figure 79 also demonstrates the possibility of SHG to discriminate between different growth modes of ultrathin films on surfaces, namely layer growth (at room temperature) and 'Stranski-Krastanoff' growth (layer growth, followed by three-dimensional island growth) (cf. Chapter 1, Section 1.2.3, Fig. 20 (Reichelt, 1988)). If one exchanges the substrate and uses, for example, a dielectric instead of the semiconductor surface, then the alkali films will show a three-dimensional island growth mode ('Volmer–Weber' growth mode) from the beginning. The measured SH intensity from an array of spherical islands is given by (Chen et al., 1983)

$$I(2\omega) \propto I^2(\omega)|\chi^{(2)}|^2 Q^2(\omega)Q(2\omega)r_0^4 \qquad (3.31)$$

and thus depends on the cluster radius r_0. The local field tensors **L** in Eq. (3.30) are represented by the factors $Q(\omega)$, which describe the mean local electric field at the cluster surface in units of the geometrical cross section:

$$Q(\omega) = \frac{R^2}{\pi r_0^2} \int_0^{2\pi} \int_0^{\pi} E_{\text{int}} E_{\text{int}}^* \sin\Theta \mathrm{d}\Theta \mathrm{d}\Phi |_{R=r_0}. \quad (3.32)$$

The validity of this approach can be verified by growing alkali islands on dielectrics such as mica or lithium fluoride (see also Chapter 4, Section 4.1.3). As demonstrated in Fig. 80 by comparison of calculated[24] and measured (Müller et al., 1997) relative SH intensities as a function of cluster radius, the coverage-dependent SH signal intensity shows pronounced resonances at well-defined values of mean cluster radius that depend on the excitation wavelength. These resonances are the nonlinear optical analog to the resonances observed in linear extinction spectra of cluster films (Fig. 100) and can be predicted via classic Mie theory by resonances in the local field factors. They in turn are due to a collective electronic or 'surface plasmon' excitation (cf. Sections 1.2 and 4.1.3). Thus SHG in this case provides both a unique fingerprint of the Volmer–Weber growth mode and a means of determining the mean cluster radius.

In the past, strongly enhanced second-harmonic signals due to local field enhancement have been observed for thin alkali films in the ATR (attenuated total reflection) geometry (Simon et al., 1975) and later also in reflection geometry on rough silver and gold films (Wokaun et al., 1981; Chen et al., 1981). Finding a strong relationship between enhanced SHG and enhanced Raman scattering (SERS) (Boyd et al., 1984), a phenomenological treatment of the problem for small spherical particles (Chen et al., 1983) was provided. More accurate treatments using Green's function formalisms (Hua and Gersten, 1986) and nonlinear Mie scattering (Östling et al., 1993; Dewitz et al., 1996) were applied subsequently, permitting a qualitative prediction of the SH enhancement even for larger spheres. So far, in all theoretical investigations of resonance enhancement an accurate calculation of $\chi^{(2)}$ from an electronic theory has not been provided. As a consequence, $\chi^{(2)}$ is always assumed to be independent of cluster size[25] and the clusters are assumed to be noninteracting spheres (with the exception of calculations in (Garcia-Vidal and Pendry, 1996)). These assumptions have also been made for the calculations that led to Fig. 80.

---

[24] Here, $\chi^{(2)}$ is assumed to be independent of cluster size in analogy to density matrix calculations for the size-dependence of $\chi^{(3)}_{\text{Au}}$ (Hache et al., 1986). This assumption is reasonable for clusters with radii larger than about 5 nm (Östling et al., 1993) since the surface plasmon frequency of the clusters reaches the value for the infinitely extended bulk for clusters already containing not more than 100 atoms (Parks and McDonald, 1989).

[25] For flat metallic surfaces nonlocal response theory has recently shed some light on the resonance mechanism (Liebsch, 1989; Jensen et al., 1997). However, it has been shown that due to a large contribution of bulk SHG to the total SH signal from adsorbed clusters, the SHG from metal films and cluster films cannot be easily compared (Aussenegg et al., 1995).

3.4. SECOND HARMONIC GENERATION

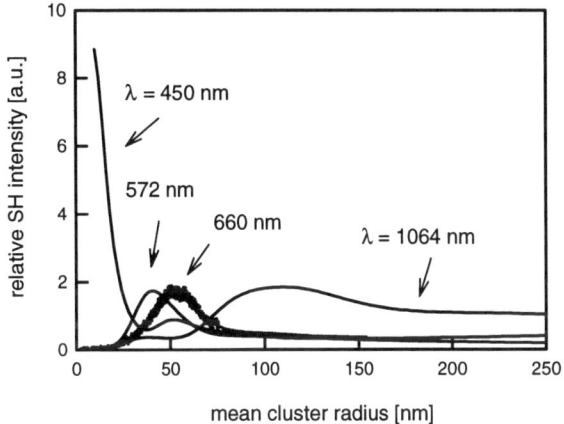

Figure 80 Calculated relative SH intensities as a function of fundamental wavelength λ and mean cluster radius for sodium clusters adsorbed on dielectrics. Typical experimental values are indicated by dots.

SH signals following excitation perpendicular to the surface normal can only be observed if the solid has order over a range which is comparable with the wavelength of the generated second harmonic light. This means that in the case of long-range order (*e.g.*, via the correlation of steps on a surface) a SH signal can be observed, while microscopic methods with smaller coherence length (*e.g.*, LEED) no longer detect order.

The sensitivity of the method increases for given laser energy if one uses higher irradiances or shorter pulses since it follows from Eq. (3.28) that

$$I^{2\omega} \propto \frac{1}{\tau}, \qquad (3.33)$$

with τ being the pulse duration. Limits are given by melting of the surface, which usually occurs for pulsed excitation with fluences of the order of J/cm^2 (cf. Chapter 5, Section 5.1). For excitation with continuous lasers the thermal diffusion coefficient of the surface (the time constant at which heat is transported out of the radiation zone) determines the maximum power that can be applied (about 1 W for a strongly focused laser beam on silicon).

Linear-optic methods for the determination of adsorbate coverages are, for example, ellipsometry (Azzam and Bashara, 1992) (Fig. 63) or the measurement of changes $\Delta R/R$ in the surface reflectivity (Dvorak et al., 1997).

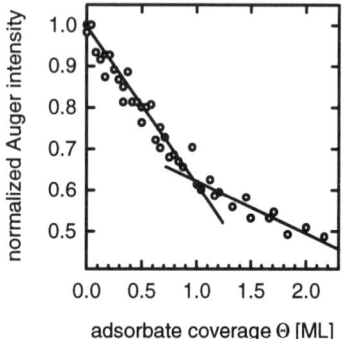

Figure 81 Normalized Auger intensity of the substrate as a function of coverage with an adsorbate. Reprinted from *Surf. Sci.*, (Boneß et al., 1998), Copyright 1998, with permission from Elsevier Science.

In the case of chemisorption (for example, oxygen on Mo(100) (Anderson et al., 1973)) one measures reflectivity changes of up to 1% for an adsorbate-saturated surface. The changes in reflectivity result probably from the fact that the chemisorbed adsorbates create new surface states which contribute to the light absorption. Using this method, one can determine, under suitable circumstances, changes of coverage of less than 0.04 monolayer of oxygen.

The SH method has the same sensitivity as methods using linear reflectivity changes and it has no major restrictions concerning the observable adsorbates. However, while coverage changes can be deduced straightforwardly from the measured SH values, the evaluation of *absolute* coverages is only possible by the use of reference measurements (*e.g.*, thermal desorption measurements (Chapter 2, Section 2.1) (Zhu et al., 1985) or Auger measurements (Chapter 1, Section 1.2.4) (Jordan et al., 1995)). For example, in Fig. 81 a measurement of the intensity of a substrate Auger line (here Si(LVV)) is plotted as a function of increasing adsorbate coverage. The adsorbate attenuates the Auger intensity linearly until the surface is fully covered (one monolayer). For higher coverages the relative attenuation is weaker since now the Auger signal stems mainly from the bulk of the substrate, which obviously is less influenced by an adsorbate coverage. As a consequence, a pronounced kink appears which, in combination with SHG measurements, can be used for a coverage calibration.

3.4. SECOND HARMONIC GENERATION

Symmetries

The $\chi^{(2)}$ tensor reflects the symmetries of the surface via the symmetry properties of the electronic surface states (Chapter 1, Section 1.2.1). Since it is a tensor of rank three, only three- or lower-fold surface symmetries can be resolved.[26]

The second order nonlinear component of Eq.3.27 is given by:

$$P_l(2\omega) = \sum_{m,n} \chi^{(2)}_{lmn}(-2\omega;\omega,\omega)E_m(\omega)E_n(\omega), \qquad (3.34)$$

where $\chi^{(2)}_{lmn} = |\chi^{(2)}_{lmn}|e^{i\phi_{lmn}}$ couples the mth and nth components of the fundamental with the lth components of the generated nonlinear polarization (ϕ is the phase). The $\chi^{(2)}$ tensor, due to the condition $\omega_1=\omega_2=\omega$, might be contracted to 18 components ('piezoelectric contraction'), which — depending on the symmetry of the surface — are not independent of each other (Yariv, 1989). If the exciting laser irradiates along the surface normal (z-direction), then the emitted SH intensity is the following function of the angle Θ between the electric field vector of the linearly polarized laser and preferred directions on the crystal (labelled x and y) (McGilp, 1987),

$$I_x(2\omega) \propto |(\chi^{(2)}_S)_{xxx}\cos^2\Theta + (\chi^{(2)}_S)_{xyy}\sin^2\Theta + (\chi^{(2)}_S)_{xyx}\sin2\Theta|^2 \qquad (3.35)$$

and

$$I_y(2\omega) \propto |(\chi^{(2)}_S)_{yxx}\cos^2\Theta + (\chi^{(2)}_S)_{yyy}\sin^2\Theta + (\chi^{(2)}_S)_{yxy}\sin2\Theta|^2. \qquad (3.36)$$

For laser irradiation along the surface normal one expects SH signals by the symmetry classes 1, 1m and 3, 3m.[27] For 1m-symmetry the nonlinear susceptibilities vanish in the directions xyx, yyy and yxx, and thus in the y-direction the observed intensity I_y is proportional to $\sin^2 2\Theta$. In Fig. 82 the measured SH intensities along the y-direction (upper part) and x-direction (bottom part) are shown for two differently reconstructed silicon crystals: the (2×1) reconstruction, which results from cleaving of the single crystal in vacuum, and the (7×7) equilibrium reconstruction, which can be obtained by annealing to surface temperatures of usually above 1000°C. The solid lines in the y-direction are fits to the measured points assuming a 1m-symmetry.

---

[26] At least in dipole approximation ($L=1$). If it is possible to measure higher order multipole-contributions, then the highest resolvable rotational symmetry is for an Nth order nonlinear technique given by $(N + L)$ (Koopmans et al., 1992).

[27] This corresponds in the 'Schönflies-notation' (Schönflies, 1891) to the crystal classes C_1 (single rotation axis, triclinic), C_s (single mirror plane, monocline), C_3 (triple rotation axis, rhombohedric) and C_{3v} (triple rotation axis and three vertical mirror planes, rhombohedric) (Hamermesh, 1962).

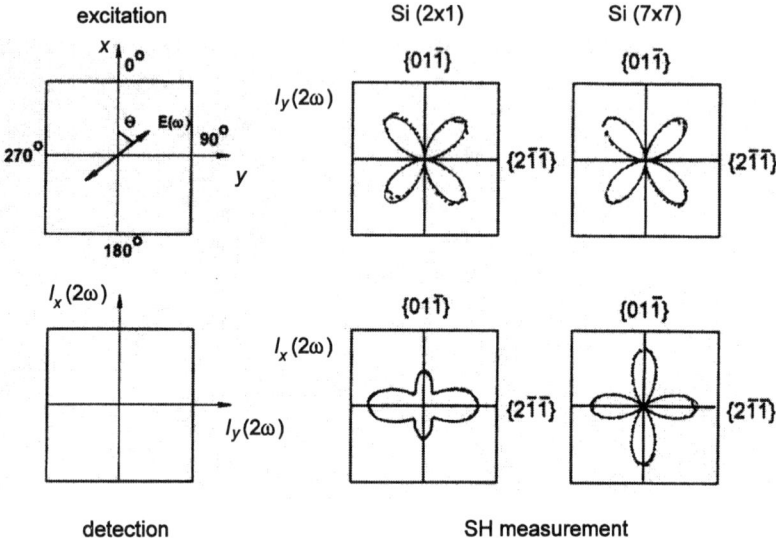

Figure 82 Symmetries of the reconstructed silicon (2×1) and (7×7) surfaces, observed via rotation of the polarization vector of the exciting laser (field strength $E(\omega)$) within the crystal plane and measurement of the second harmonic intensity $I(2\omega)$ along the $(2\bar{1}\bar{1})$ and $(01\bar{1})$ crystallographic directions, respectively. Reprinted with permission from (Heinz et al., 1985). Copyright 1985 American Physical Society.

In the x-direction one expects to observe in the case of the higher 3m-symmetry ($-xxx = xyy = yxy; xyx = 0$) an intensity dependence proportional to $\cos^2 2\Theta$, since $\cos^2\Theta - \sin^2\Theta = \cos 2\Theta$. Experimentally, this intensity dependence is observed only for the (7×7) reconstructed surface, while the (2×1) surface shows a dependence in the x-direction, which can be reproduced only via three independent tensor elements $-xxx$, xyy and yxy. Hence the SH measurements directly prove that the (2×1) surface shows the lower 1m-symmetry, while the (7×7) surface has three mirror planes.

The deduced symmetries are consistent with LEED results. In contrast to LEED studies, however, which do not allow one to exclude the existence of a higher order (*e.g.*, sixfold) symmetry, this is possible via SH measurements since at normal incidence a surface with such high symmetry would show no SH signal intensity at all.

Via SHG it is possible to watch the heat-induced increase in symmetry of

3.4. SECOND HARMONIC GENERATION

the silicon surface if one observes the vanishing of the signal in a direction in which only the (2×1) surface induces SH signal intensity (Fig. 82). It is seen that the phase transition from the lower to the higher symmetry starts already around 550 K surface temperature (Heinz et al., 1985).

Measurement of the rotationally anisotropic second harmonic yield has been shown to be a versatile tool for a nonintrusive, relatively simple determination of interface symmetry. Besides plain surfaces it of course has also been applied to the determination of adsorbate symmetries, which turns out to be a difficult task in the case of adsorption on insulating surfaces. For example, the adsorption symmetry of molecular water on alkaline-earth halides such as CaF_2 has been deduced as a function of coverage (Zink et al., 1992), demonstrating oriented initial adsorption. This orientation is subsequently lost for higher coverages.

In the case of an isotropic distribution in the (x, y)-plane no rotational anisotropies are observed in the SH plots at normal incidence. However, if one irradiates with the laser under a fixed angle of incidence Θ_{in} with respect to the surface normal for a given polarization of the electric field vector with respect to the plane of incidence and varies the angle α of a given crystallographic direction on the surface (i.e., by azimuthal rotation of the sample with respect to the surface normal) (Jordan et al., 1995), then one is able to deduce different components of the effective $\chi^{(2)}$ tensor. The information content increases further if one also varies the angle of incidence (Bratz and Marowsky, 1990; Ying et al., 1993).

Orientations and chirality

The orientation of molecular adsorbates on surfaces might also be deduced from the polarization dependence of the SH signal (Andrews and Hands, 1996). Let us assume that a monomolecular film has been adsorbed on the surface. The transition dipole moments of the molecules, which are responsible for the SH generation, define a preferred axis in the molecules, which is tilted with respect to the surface normal by the angle Θ. Let the molecules be isotropically distributed in the (x, y)-plane. Their microscopic polarizability in the z-direction, $\alpha^{(2)}_{zzz}$, determines the nonlinear optical response. Then the orientation angle Θ can be determined from the measurement of the two macroscopic tensor components (zxx) and (zzz) (Heinz, 1991):

$$\sec^2\Theta \approx 2\frac{\chi^{(2)}_{zxx}}{\chi^{(2)}_{zzz}} + 1. \tag{3.37}$$

For this to hold, a knowledge of the absolute values of $\chi^{(2)}$ or of the surface density of the molecules is not necessary, assuming that the influence of local fields on the surface (especially the interaction between the molecules) can be neglected.

Especially for the application to biological relevant surfaces (*e.g.*, adsorbed proteins) the sensitivity of polarized SHG to the chirality of the molecules (the 'handedness' of their structures) is important (Verbiest et al., 1998). Recently, the second-order nonlinear optical analog to circular dichroism and optical rotatory dispersion spectroscopy has been successfully developed (Yee et al., 1994; Byers et al., 1994).

Surface magnetization

The investigation of the magnetic properties of surfaces, thin films and layered structures is of practical importance for the optimization of magnetic data storage techniques. With respect to fundamental research, the investigation of the exchange interactions between thin films, of quantum well states, the spin flip dynamics or the growth of magnetic domain structures are especially interesting.

The magnetic properties of surfaces can be investigated by spin-polarized electrons (Feder, 1985) or via the magnetooptic Kerr effect (SMOKE, surface magnetooptic Kerr effect) (Qian and Wang, 1990). These linear techniques suffer from their large penetration depth (about 50 nm in the case of SMOKE), meaning that their surface sensitivity is provided solely by the properties of the investigated material, *e.g.*, the thickness of the evaporated film. The magnetism of interfaces between layers made of different materials ('buried interfaces') cannot be investigated independently of the magnetic properties of the bulk of the layers.

Since magnetization does *not* break the inversion symmetry of the investigated material ($\vec{M}$ is an axial vector with even parity with respect to the inversion operation), one does expect to preserve the large surface sensitivity of SHG even in the case of magnetic interfaces. On the other hand, the existence of the magnetization changes the symmetry of the bulk and surface. Thus additional elements of the electric dipole tensor $\chi_s^{(2)}$ become non zero compared with the nonmagnetized surface (Pan et al., 1989). Of special interest are the odd elements that change sign as a result of changing the direction of $\vec{M}$. This results in a change in the total SH-intensity

$$I_{+,-} \propto |\chi_{\text{nonmagnetic}}^{(2)} \pm \chi_{\text{magnetic}}^{(2)}|^2. \tag{3.38}$$

Depending on the direction of $\vec{M}$, the intensity should increase or decrease. Hence magnetic-field-induced SHG (MSHG) exists only if the surface is SH active even without a magnetic field, and $\chi_{\text{magnetic}}^{(2)}$ depends linearly on the magnetization. However, the nonlinear Kerr rotation might be much stronger compared with the linear one, as demonstrated recently by the use of a multilayer of 2 nm Fe on 2 nm Cr on a quartz substrate, which possesses only a very small linear Kerr rotation (Koopmans et al., 1995).

In a classic macroscopic picture the nonlinear Kerr effect can be explained as follows. The reflection of an incoming light wave at the interface is related to the nonlinear generation of an electron flux which is affected by the surface

3.4. SECOND HARMONIC GENERATION

magnetization via spin–orbit coupling and exchange interaction. The resulting Lorentz force $\vec{F}=---e\vec{v}\times\vec{B}$ acts perpendicularly to the magnetic induction and the velocity vector of the electrons and thus rotates the direction of the emitted electromagnetic field vector with respect to the incoming light vector. This is shown in Fig. 83 for p-polarized light, which hits a Fe(110) crystal that possesses a magnetization in the {001}-direction.

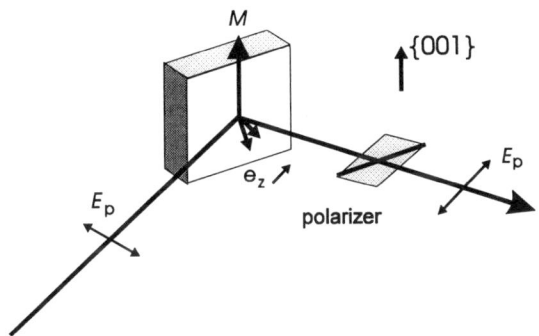

Figure 83 Set-up for the investigation of the influence of surface magnetization on the generation of frequency-doubled light from a Fe(110) crystal (light in pp-orientation). The symbol e_z denotes the Lorentz-force-induced rotation of the electron oscillation perpendicular to the surface. Reprinted with permission from (Reif et al., 1991). Copyright 1991 American Physical Society.

The projection of the incoming electric field vector induces an electron oscillation perpendicular to the surface, which is rotated counterclockwise by the magnetic field. If the magnetization is directed in the negative {001}-direction, then the generated SH light will be rotated clockwise. If one now observes p-polarized frequency-doubled light along a fixed direction, then this magnetic-field-induced rotation results in a decrease or an increase in the signal intensity, depending on the direction of the magnetic field. This behavior has indeed been observed in the course of pp-measurements under 45o angle of incidence (λ=532 nm, τ=6 ns) from a clean iron crystal, which has been mounted in an ultrahigh vacuum between the pole faces of an electromagnet (Reif et al., 1991). The observed ratio between magnetic-field-induced SHG and nonmagnetic SHG was about 0.25. This large effect makes the method a promising tool for other systems too. The surface sensitivity is demonstrated by observing that the signal intensity decreases exponentially with increasing coverage with background gas (mainly dissociative adsorption of CO) (Fig. 84a).

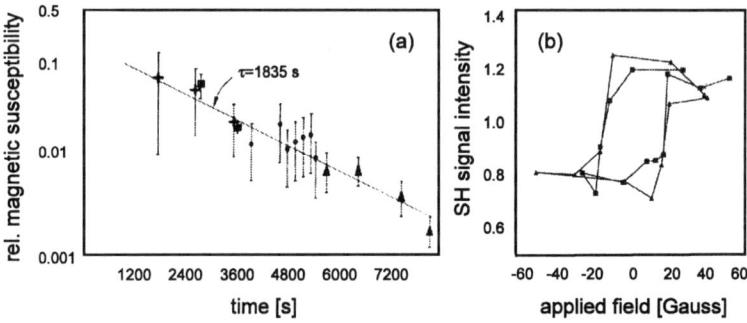

Figure 84 (a) Relative magnetic surface susceptibility $\chi^{(2)}_{mag.}/\chi^{(2)}_{nonmag.}$ as a function of CO coverage. Reprinted with permission from (Reif et al., 1991). Copyright 1991 American Physical Society. (b) Magnetic-field-induced SHG from a 50 nm cobalt/5 nm gold on a quartz substrate. Reprinted from *Surf. Sci.*, (Spierings et al., 1993), Copyright 1993, with permission from Elsevier Science.

In Fig. 84b the sensitivity of the method to buried interfaces is further demonstrated by a measurement on a cobalt/gold/quartz layered system (Spierings et al., 1993). The magnetic-field-induced effect is observed *only* if a single (or an odd number of) cobalt/gold interfaces does exist and in that case shows a significant hysteresis. For a longitudinally magnetized PtMnSb(111) surface, circular dichroism[28] has been observed via SHG, which provides information about the population of spin-up and spin-down states in the conduction band (Reif et al., 1993).

The versatility of the MSHG method is mainly due to its high sensitivity, which results partially from local-field enhancement effects.[29] Consequently,

---

[28] In order to perform this experiment one irradiates the surface with right- or left-hand circularly polarized light, corresponding to photon spins oriented parallel or antiparallel with respect to the magnetization direction. The different SH yields for the different directions of rotation result in a change in ellipticity of the nonlinearly reflected light as compared with the incoming light.

[29] Obviously, the sensitivity can be further enhanced by inducing the second-harmonic signal in the spectral neighborhood of a surface plasmon resonance of the thin magnetic film.

3.4. SECOND HARMONIC GENERATION

the method has found a large number of applications, including the investigation of quantum well oscillations (Wierenga et al., 1995; Kirilyuk et al., 1997a) or femtosecond time-resolved spin dynamics (Scholl et al., 1997; Hohlfeld et al., 1997a).

Spectroscopy

The spectroscopic possibilities of SHG lie in the resonance enhancement of the signal, namely if the second harmonic frequency, 2ω, coincides with the frequency of a dipole-allowed transition ω_{31} in the irradiated material. Here, '1' means the ground state and '3' an excited state. Possible resonance enhancement can be directly deduced from the microscopic expression[30] for $\alpha^{(2)}$:

$$\alpha^{(2)}(2\omega) \propto \sum_{1,2,3} \frac{M_{13}M_{32}M_{21}}{(2\omega - \omega_{31} - i\gamma_{31})(\omega - \omega_{21} - i\gamma_{21})}, \quad (3.39)$$

where M denotes the matrix element for dipole transitions between two states ('2' means a real intermediate state) and γ denotes the half width of the transition, corresponding to the characteristic relaxation time (Shen, 1984). Obviously the hyperpolarizability is increasing strongly if the difference between 2ω and ω_{31} or ω_{21} becomes small ('resonance').[31]

Figure 85 shows the measured SH signal from a silicon (100) surface, which has been covered by a 700 nm thick SiO_2 layer, as a function of laser energy. A prominent resonance is seen at 3.3 eV. Adsorption of oxygen on the surface reduces the height of the resonance not significantly. This means that the signal is not generated by the surface states of silicon. However, comparison of the energy dependence of the linear susceptibility $\chi(2\omega)=\epsilon(2\omega)-1$ of bulk silicon (solid line) with the measured data shows good agreement concerning the positions of the resonance. Hence the resonance might be induced by the direct valence–conduction band transition in silicon. Since the signal can be observed even from below a 700 nm thick silicon oxide layer, it is probably generated at the interface between the silicon and SiO_2, which does not possess inversion symmetry. The slight red shift of the measured as compared to the calculated resonance provides evidence that the Si–Si bindings are slightly elongated at the surface as compared with the bulk.

In contrast to the 3.3 eV resonance at the silicon surface the 4.1 eV resonance on a Cu(111) surface (Fig. 85b) reacts sensitively on the adsorption

---

[30] This equation describes the molecular hyperpolarizability via second-order perturbation theory assuming a single particle excitation. Collective phenomena are not taken into account.
[31] At the spectral position of resonance, however, the expansion equation (Eq. (3.27)) might no longer be valid and the transition probability is dominated by additional dynamic effects such as photon echoes, free induction decay, self-induced transparency, etc. (Mandel and Wolf, 1995). It is thus difficult to model quantitatively the absolute signal intensity at resonance.

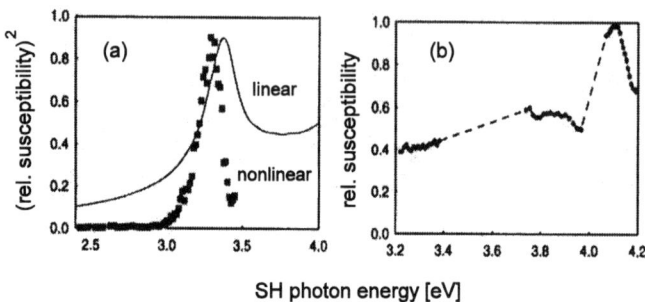

Figure 85 (a) Measured nonlinear susceptibility of second order of Si(100)/SiO$_2$. The solid line is the calculated linear susceptibility. Reprinted with permission from (Daum et al., 1993). Copyright 1993 American Physical Society. (b) Measured nonlinear susceptibility for Cu(111) using angle of incidence $\Theta=67°$ and azimuthal angle $\psi = 30°$. Reprinted with permission from (Lüpke et al., 1994). Copyright 1994 American Physical Society.

of oxygen. This is a hint that in that case transitions into *surface states* are responsible for the enhancement of the nonlinear signal intensity. The most probable candidate is a transition between a surface state close to the Fermi energy and the ($n=1$) image potential state (Chapter 1, Section 1.2.1). The energetic position of the resonance is $E = E_a - E_b = 4.92 - 0.82$ eV $= 4.1$ eV. Here, E_a is the work function of Cu(111) and E_b is the binding energy of the image potential state. The long lifetime of the Rydberg-like image potential states increases the possible population density and thus the probability for the existence of nonlinear processes between them and the surface states.

Frequency-dependent SH measurements at Ag(110) have detailed different resonances in the energetic range between 1.6 eV and 2.1 eV (fundamental of the laser), which can be traced back partially to interband transitions and partially to transitions between occupied and nonoccupied surface states (Urbach et al., 1992). In the regime of interband transitions (for silver $2\omega=3.8$ eV) the corresponding field enhancement results in an enhancement of the SH signal intensity. A similar effect occurs if 2ω corresponds to the transition between an occupied surface state below the Fermi level and an unoccupied state in the gap. Since transitions from bulk states of the same symmetry as the occupied surface state (but deeper below the Fermi level situated) might also occur, the resonance maximum usually has a broad tail to higher energies.

Up to now the experimentally observed resonances in the SH signals, which provide characteristic information about the electronic band structures and sometimes are truly surface-specific (i.e., have no analog in the bulk, e.g. (Erley and Daum, 1998)), could not be reproduced fully by theoretical approaches. Since the second-order nonlinear response of the surface to an incoming lightwave reacts sensitively to the charge density profile of the ground state (Murphy et al., 1989), a detailed understanding of the electronic surface structure is important. SH theory, on the other hand, mostly uses a jellium model for the surface structure. In the framework of the jellium model the ionic cores are represented by a homogeneous, spatially averaged positive background charge (Fig.78). It is not surprising that totally neglecting the lattice potential only in the case of surfaces with low electron densities and in the absence of surface defects (i.e., in the case of the homogeneous distribution of ions) provides qualitatively correct predictions of the nonlinear optical response function of the surface.

3.5 Sum Frequency Generation

An obvious disadvantage of using second harmonic generation for the spectroscopy at adsorbate-covered surfaces is the missing molecule specificity. This problem can be overcome if one mixes a fixed frequency laser pulse with a pulse of variable frequency ('sum frequency generation', SFG). Since $\chi^{(2)}_{SFG}$ also vanishes in dipole approximation in media with inversion symmetry, this method too has high surface sensitivity and shows resonance enhancement.

In most cases the second harmonic of a Nd:YAG laser (λ_{vis}=532 nm) irradiates the surface at a given angle Θ_{vis} with respect to the surface normal (Fig. 86). This light is mixed with infrared light (ω_{IR}), which might have been generated in an optical parametric oscillator with variable wavelength. The sum frequency $\omega_{SF} = \omega_{vis} + \omega_{IR}$, which is generated in the substrate, is reflected under an angle Θ_{SFG} with respect to the surface normal, which is given by (Hunt et al., 1987)

$$\omega_{SFG}\sin\Theta_{SFG} = \omega_{vis}\sin\Theta_{vis} + \omega_{IR}\sin\Theta_{IR} \qquad (3.40)$$

The sum-frequency light might be separated from the irradiating light either by virtue of its different frequency or spatially. If one tunes the infrared light, then in the case of a resonance (e.g., by excitation of an adsorbate vibration at the surface) the nonlinear signal intensity is strongly increasing, just as in case of SHG (Eq. (3.39)). The enhanced molecule specificity of the method results from the use of IR light, which allows one to directly excite intramolecular motions. As an example, in Fig. 86 SFG spectra for methanol (CH_3OH) and glycol ($C_2H_4(OH)_2$) are shown, which have been adsorbed on a quartz substrate (Hunt et al., 1987). The symmetric (s) and antisymmetric (a) CH stretch vibrations are clearly identified. For comparison, Raman spectra from the solution are shown by dashed lines. Their maxima are strongly shifted due

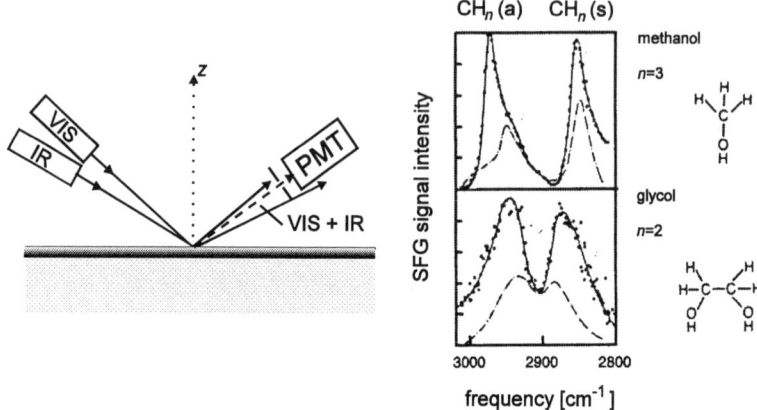

Figure 86 Sum frequency generation at surfaces. On the left-hand side the experimental set-up is shown schematically. An infrared (IR) and a visible (VIS) laser beam are mixed in a thin methanol adsorbate film and the resulting sum frequency signal is spatially separated and detected by a photomultiplier (PMT). If one tunes the wavelength of the IR laser, then adsorbate-spectroscopy can be performed (right-hand side). Reprinted from *Chem.Phys.Lett.*, (Hunt et al., 1987), Copyright 1987, with permission from Elsevier Science.

to the mutual interactions of the molecules.

An experimental set-up for the spectroscopy of silicon–hydrogen stretch vibrations via SFG is shown in Fig. 87. A mode-coupled Nd:YAG laser pumps a dye laser, which generates tunable 70 ps light pulses. Part of the YAG fundamental light (1064 nm) is amplified by a regenerative amplifier and is frequency doubled. This light then serves as the visible mixing pulse for the SFG, having 100 ps duration and 50 mJ energy. In addition it is used also to amplify the tunable dye laser pulse to 6 mJ. The dye laser pulse is Raman-shifted to the infrared spectral regime (4.4–5.6 μm) inside a cell which is filled with cesium vapor. This procedure results in about 40 μJ tunable IR light. Both pulses together allow one to determine the energy of the Si-H stretch vibration at a smooth surface (2084 cm^{-1}). The set-up shown in Fig. 87 can be used also to generate an IR pump pulse. In that way an adsorbate mode can be excited and the dynamics of the following relaxation, induced by the interaction with the substrate or with the surrounding adsorbed atoms, can be investigated by applying a temporal delay between pump and probe pulses. Recently, the identification of short-lived chemical species produced

3.6. HIGHER ORDER WAVE MIXING

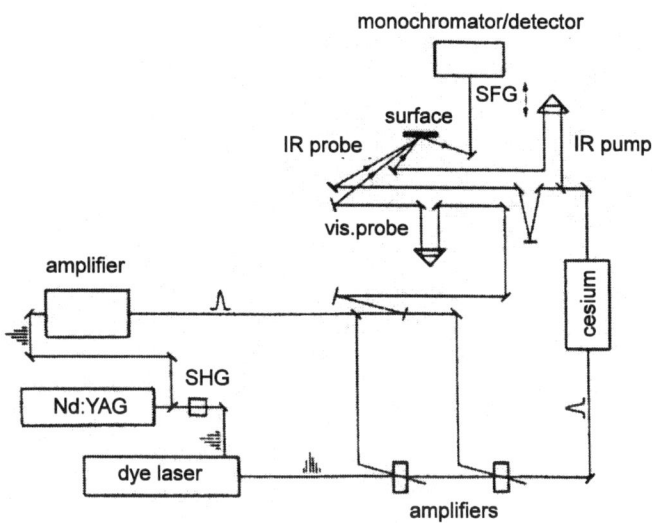

Figure 87 Set-up for IR pump/SFG probe, time-resolved nonlinear vibrational spectroscopy. Reprinted with permission from (Morin et al., 1992a). Copyright 1992 American Institute of Physics.

on a surface (Domen et al., 1998) but also at buried interfaces (Cremer et al., 1995) became possible via SFG vibrational spectroscopy.

While the spectral position of the resonance can be determined straightforwardly using SFG, evaluation of the line profile with respect to the origin of possible broadenings or the absolute densities of the adsorbates becomes difficult. The basic problems with all nonlinear spectroscopic techniques (SHG, SFG, CARS, etc.) are (i) a wavelength dependence of the local field strength tensors $\mathbf{L}(\omega, 2\omega)$ (Eq. (3.30)) and (ii) the non-resonant background, which is given by the additive nonresonant nonlinear susceptibility. Both factors have to be taken into account for an accurate profile analysis via extended fit procedures.

3.6 Higher Order Wave Mixing

Optical second harmonic and sum frequency generation are three-wave mixing processes. The nonlinear polarization, which is necessary for the generation of the third photon, can be described by the second term in Eq. (3.27). If one takes additional photons into account, multi-wave mixing processes and especially four-wave mixing takes place. The coupling constant between irradiating fields and induced polarization then is the nonlinear susceptibility of third order, $\chi^{(3)}$. Fig. 88 shows term schemes for some important four-wave mixing processes, which have been deduced by invoking energy conservation: coherent anti-Stokes Raman scattering (CARS), frequency tripling (THG) and degenerate four-wave mixing (DFWM).

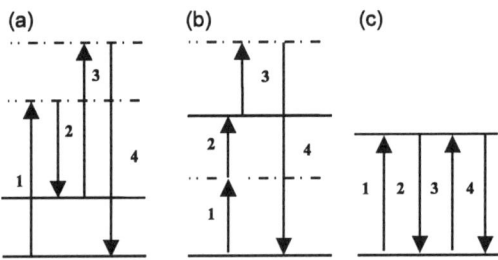

Figure 88 Possible four-wave mixing processes. Real states are plotted by solid lines, virtual states by dash-dotted lines. 1,2,3 and 4 denote the frequencies ω_1 to ω_4. (a) Coherent anti-Stokes Raman scattering (CARS), $\omega_1 = \omega_3$. (b) Third harmonic generation (THG), if $\omega_1 = \omega_2 = \omega_3$, otherwise 'up-conversion'. (c) Degenerate four-wave mixing (DFWM), $\omega_1 = \omega_2 = \omega_3 = \omega_4$.

As seen, the term scheme for DFWM is especially simple: all four photons induce transitions between real states. This 'resonance enhancement' results in a higher signal strength as compared with, for example, CARS. Momentum conservation determines the direction of the resulting signal wave. Since the phase-matching condition is (for the notation see Fig. 90):

$$\vec{k}_{\text{fp}} + \vec{k}_{\text{bp}} - \vec{k}_{\text{p}} - \vec{k}_{\text{s}} = 0, \qquad (3.41)$$

$\vec{k}_{\text{fp}} + \vec{k}_{\text{bp}} = 0$ and the energies of all involved photons are equal, the phase conjugate signal wave will counterpropagate to the probe wave (phase-conjugate (PC) geometry; Fig. 89).

DFWM (Fisher, 1983) is a real-time variant of optical holography, which has been known since the 1960s: two laser beams (here, forward pump and object beam) are overlapped coherently under a small angle Θ. They induce in a nonlinear medium with susceptibility $\chi^{(3)}$ an interference pattern (grating), which contains information about the amplitude and phase relations between the contributing waves. This information is recovered via Bragg scattering

3.6. HIGHER ORDER WAVE MIXING

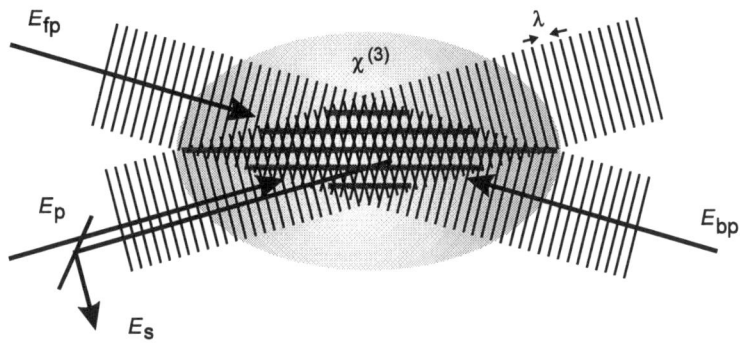

Figure 89 Generation of a holographic grating in a medium with nonlinear susceptibility $\chi^{(3)}$ by interference of a forward propagating incoming light wave (field strength E_{fp}) of wavelength λ, a backward propagating wave (E_{bp}) and an object or probe wave (E_{p}). The generated signal wave (E_s) is directed into the detector using a beam splitter.

using a third beam (backward pump) and generates a phase conjugate signal wave. In contrast to conventional holography, in the case of DFWM generation and recovery processes occur simultaneously.

'Phase conjugation' in this context means that the signal wave has the same wavefronts and phase relations as the object wave. Only the sign of the wavevector $\vec{k}$ has changed. For example, the original light wave might have passed through a phase-disturbing medium (e.g., an adsorbate) onto the nonlinear medium (e.g., a surface film). As long as the irradiated signal wave travels through the same phase-disturbing medium, the information content is not diminished.

In the case of a resonant transition involving population transfer the generated grating might be a density grating. But even without strong absorption the coherent superposition of the laser beams results in a modulation of the complex index of refraction of the medium (amplitude- or polarization-grating), which leads to the generation of a phase-conjugate signal wave. The signal intensity is a measure of the depth of modulation of the grating just as in the case of diffraction of an external laser beam from the transient grating structure. One might name the whole process also 'transient grating scattering' (cf. Chapter 2, Section 2.4). The lattice constant Λ (and thus the number of lattice rods within the overlap volume of the laser beams, i.e, the sensitivity of the method) depends on the crossing angle Θ via

$\Lambda = \frac{\lambda}{2\sin(\Theta/2)}$ (Eq. (2.18)). Thus a smaller crossing angle results in a more sensitive optical detection.

While the break of inversion symmetry at the surface makes nonlinear optical processes of second order intrinsically surface-sensitive in the case of centro-symmetrical solids, this is no longer the case for a nonlinear process of third order. Hence one has to induce surface-sensitivity externally. This might be done, for example, by performing nonlinear optics at adsorbates on the surface (Balzer and Rubahn, 1995c; Balzer and Rubahn, 1998), such as rough alkali cluster films (cf. Chapter 4, Section 4.1.3). A possible set-up for an experiment that exploits simultaneously second harmonic generation and four-wave mixing is shown in Fig. 90.

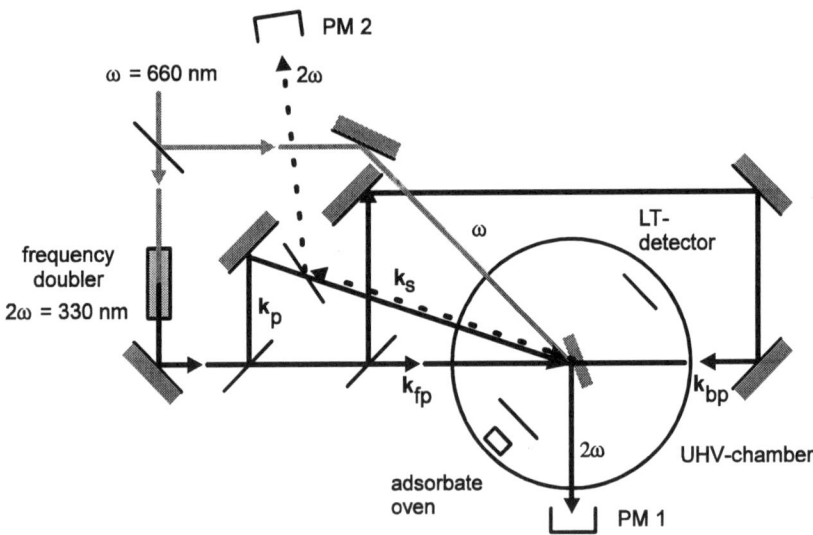

Figure 90 Set-up for the simultaneous observation of SHG (using ω-light) and DFWM (using 2ω-light from the same laser source) in the PC geometry from thin cluster films on transparent substrates. The sample is situated in an ultrahigh vacuum chamber and alkali films are deposited from an alkali oven while monitoring the frequency-doubling and the four-wave mixing signals induced by the growing adsorbate film. The deposition rate of adsorbates is calibrated using a Langmuir–Taylor hot-wire detector. PM1 and PM2 are two photomultipliers, which monitor solely 330 nm light. Mirrors are denoted by thick lines and beam splitters by thin lines.

3.6. HIGHER ORDER WAVE MIXING

The nonlinearly optically active medium is an array of spheroidal alkali clusters with index of refraction n, which gives rise to a DFWM signal intensity (Fisher, 1983)

$$I_{\text{DFWM}} = \frac{\mu_0 \omega^2}{\epsilon_0^3 n^4 c^2} l^2 |\chi_{\text{eff}}^{(3)}(-\omega_s; \omega_{\text{fp}}, -\omega_p, \omega_{\text{bp}})|^2 \times I_{\text{fp}} I_p I_{\text{bp}}. \tag{3.42}$$

To predict the signal intensity one needs a value of the macroscopic susceptibility $\chi^{(3)}$, which is given by the microscopic hyperpolarizability $\alpha^{(3)}$ and the volume of the metallic sphere V:

$$\chi^{(3)} = \alpha^{(3)}/V. \tag{3.43}$$

Again, as in the case of $\chi^{(2)}$ (Eq. (3.39)), the hyperpolarizability will show large values in the case of resonance between the laser frequency ω_L and the interband transition frequencies. Numerical values of the electric dipole $\chi^{(3)}$ might be calculated via density matrix theory as exemplified for gold spheres in (Hache et al., 1986). For larger particles the observed nonlinear optical signal is dominated close to electronic resonances by local field enhancement effects. The enhancement can be so strong that one is able to obtain DFWM signal intensities even for irradiation with cw lasers (Balzer and Rubahn, 1995c). A local field enhancement of a factor of 900 compared with a flat silver surface was found via FWM for an ordered array of equally shaped silver ellipsoids with semi-axes $a=50$ nm and $b=150$ nm (Chemla et al., 1983).

In Fig. 91 the measured DFWM signal intensity is presented from a cluster film with a mean radius of 50 nm, adsorbed on mica and irradiated by a pulsed laser at 330 nm. The dependence of the signal intensity on the forward laser power is shown with backward and probe power increasing simultaneously (the total value of the backward power was twice that of the forward power). The fit curve reveals a cubic dependence on laser power, as expected from Eq. (3.42).

Another way to make a four-wave mixing process surface-sensitive is to use the evanescent part of the electromagnetic field at interfaces for spectroscopic purposes. Here one takes advantage of the field enhancement that takes place at the surface of a prism, which is coated by a thin metallic film ('Kretschmann configuration' (Kretschmann, 1971), cf. Fig. 19).

In a four-wave mixing experiment by Shen and coworkers, employing CARS (Chen et al., 1979) a glass prism was coated by a thin silver film (less than 1 μm thick) and surface plasmons were excited in this film by irradiation with visible light. Two surface plasmon waves with wave vectors $\vec{k}_1$ and $\vec{k}_2$, which travel parallel to the prism surface, were generated by two laser beams ω_1 and ω_2, which were irradiating the prism under the angle of total internal reflection (Fig. 92). The irradiances of the two laser beams (2.5 mJ/cm^2 and 25 mJ/cm^2) were low enough to avoid heating effects in the metal.

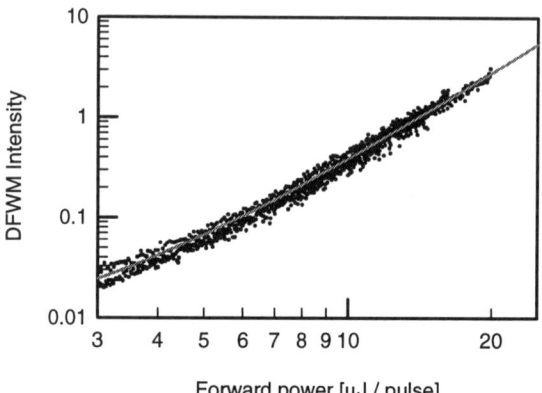

Figure 91 DFWM intensity, induced in the PC geometry by 330 nm light, irradiating a distribution of alkali islands ($r_0=50$ nm, FWHM 50%) grown on mica at 150 K surface temperature. Dependence on laser power P with a fit curve that resembles a P^3-dependence including an additive linear term that represents reflected background light.

At the interface between the silver film and the nonlinear medium (benzene in this case) an anti-Stokes wave with wavevector $\vec{k}_a = 2\vec{k}_1 - \vec{k}_2$ is generated, which is coupled out of the prism. Fig. 88a shows, with the help of an energy level scheme, the generation of the anti-Stokes wave ω_4 via the Raman resonance.

If the frequency differernce $\omega_1 - \omega_2$ coincides with a resonance in the benzene, then the nonlinear signal intensity is strongly increasing (Fig. 92). In order to observe this effect, one has to fulfill at least two conditions. (a) Phase matching between the wavevector K_A of the anti-Stokes surface plasmons and the wavevectors of the irradiating lasers, $(K_A)_\| = 2(k_1)_\| - (k_2)_\|$. The index $\|$ denotes the components along the surface. The phase-matching condition is fulfilled by choosing the correct angle of incidence of the exciting beams with respect to the surface normal. For the experiment shown, laser beam 1 has to irradiate the prism at an angle of 10°. (b) The frequency of the exciting light has to be in a spectral range where surface plasmons can be excited. Since the plasmon resonances in thin films are spectrally broad (a few tens of nanometers), this condition is easily fulfilled.

Surface CARS is of interest especially for the spectroscopy of materials with strong absorption and fluorescence, since the effective interaction length of the laser beams with the medium is limited by the surface plasmons, which are

3.6. HIGHER ORDER WAVE MIXING

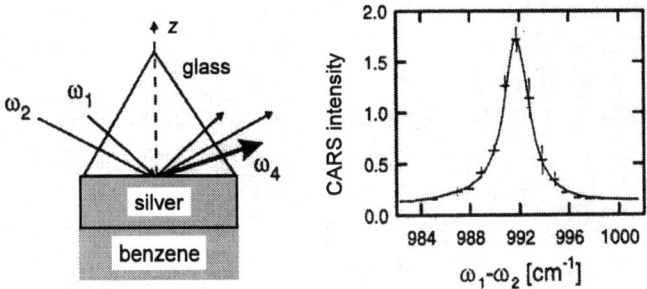

Figure 92 Configuration for the observation of a nonlinear optical signal of third order at a prism surface. The irradiating beams are totally internal reflected and generate anti-Stokes radiation in the benzene, which is strongly enhanced at the spectral position of resonance maxima (right-hand side). Reprinted with permission from (Chen et al., 1979). Copyright 1979 American Physical Society.

excited in the metal, $\lambda_\parallel/2\pi = 1/K_\parallel \approx 10$ μm. Surface sensitivity is given since the evanescent wave has enough field strength to induce a nonlinear effect of third order only in a layer of thickness $\lambda/6\pi$.

A valuable extension of surface-CARS spectroscopy using a prism is to employ waveguides (Stegemann et al., 1983). The waveguide structure usually consists of a silicon substrate with index of refraction n_1, an oxide film of about 1 μm thickness (index of refraction n_2, e.g., ZnO) and a cover of the nonlinear material to be investigated, which has an index of refraction that is smaller than n_2. For $n_2 > n_1$, an electromagnetic wave that is irradiated into the transparent oxide or polymer film will propagate solely along this material since total internal reflection takes place at the upper and lower sides. More details on the propagation conditions of electromagnetic fields in waveguides and the generated modes can be found, for example, in (Yariv, 1985). In order to perform CARS spectroscopy at the nonlinear medium, the exciting laser beams are coupled into the waveguide by an attached prism. A few millimeters downwards they are coupled out of the waveguide together with the generated anti-Stokes signal using another prism (Fig. 93).

The method allows one to perform nonlinear spectroscopy even with small laser power, since the field strength in the spatially restricted waveguide structure is enhanced and the interaction between nonlinear material and laser beams occurs over a wide spatial range (the CARS intensity increases quadratically with interaction length). In that way with pulse energies of 0.1 mJ one might obtain power densities of 200 MW/cm^2 in a 1 μm thick

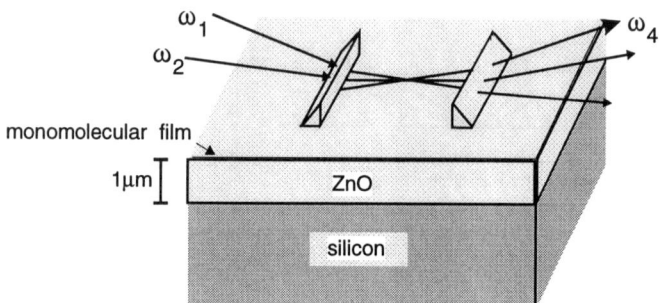

Figure 93 CARS generation in a thin molecular film, adsorbed on a ZnO waveguide. The exciting light frequencies ω_1 and ω_2 are coupled into the waveguide using a prism and the signal ω_4 is coupled out using a second prism.

polystyrene film on silicon. At those power densities along a beam path of 5 mm 5% of the incoming laser energy is used for the generation of a CARS signal in a benzene stretch vibration (Stegemann et al., 1983).

Simultaneously a high spectral resolution is possible, which contrasts conventional infrared or electron energy loss spectroscopies. In that way, for example, different physisorption and chemisorption sites of ethylene on ZnO could be identified via their Raman resonances (Wijekoon et al., 1987). The structure and electronic properties of monomolecular organic films (the building blocks for a molecular architecture, Chapter 1, Section 1.2.3) can be investigated sensitively using this technique too.

It should be mentioned that all of the above applications of four-wave mixing become especially attractive if one uses short or ultrashort laser pulses, which open up the possibility to study dynamic phenomena in adsorbates on surfaces (see Chapter 4, Section 4.1.1).

Finally, the application of higher order nonlinear optical processes such as third-harmonic generation (Berkovic, 1995; Tsang, 1995) or fourth-harmonic generation (Lee et al., 1997) could provide even more detailed interface information. For example, in the case of fourth-harmonic generation Eq. (3.34) has to be replaced by

$$P_j(4\omega) = \sum_{k,l,m,n} \chi^{(4)}_{jklmn} E_k(\omega) E_l(\omega) E_m(\omega) E_n(\omega), \qquad (3.44)$$

i.e., the hyperpolarizability is a tensor of rank 5. Hence it is possible to resolve

3.6. HIGHER ORDER WAVE MIXING

up to five-fold surface symmetries. However, the absolute values of $\chi^{(4)}_{jklmn}$ are small, and thus one needs ultrafast pulses in order to obtain significant signal intensities. The fact that

$$I^{4\omega} \propto \frac{1}{\tau^3} \tag{3.45}$$

makes the use of femtosecond lasers especially attractive.

4
Dynamics and Ultrafast Studies

The absorption of ultrashort laser pulses (pico- or femtoseconds, 10^{-12} or 10^{-15} s) in gaseous molecules, liquids or solids results in extremely high electronic excitation rates, which select among the multitude of possible relaxation channels a few, very fast ones. If, for example, the excitation of a molecule in the gas phase proceeds via a multiphoton process, then losses due to stepwise one-photon processes might be avoided. Electronically excited molecules fluoresce with lifetimes of the order of nanoseconds. An excitation rate that is four orders of magnitude faster (100 fs) then will lead without losses due to fluorescence to direct ionization. This kind of fast ionization is advantageous especially if the molecule has weak bonds, which would result in strong fragmentation upon conventional ionization via electron impact or nanosecond laser pulses. As a result of the femtosecond excitation the energy is strongly localized, which allows nearly fragmentation-free ionization.

Molecules, which are adsorbed on surfaces and experience additional loss channels in the form of adsorbate–adsorbate and adsorbate–substrate relaxations show a similar behavior. Here, femtosecond pulses allow one to obtain fragmentation-free and nonthermal desorption and ionization even in the case of larger molecules up to DNA strands.

As far as structural modifications of metallic or semiconductor surfaces are concerned, purely electronic relaxation processes (for example, electron–electron scattering) proceed on time-scales that are short compared with the pulse length of a typical femtosecond laser. But even in this case at least the coupling to the lattice oscillations via electron–phonon coupling and the subsequent thermal relaxation can be avoided *during* the duration of the laser pulse. More important, new (multiphoton) excitation channels exist especially in the high-power regime where the electric field of the laser pulse might exceed the threshold to optical breakdown and ablated material is transformed on an ultrafast time-scale into a plasma. This then shortens the characteristic time-scales for ablation processes, which are otherwise given by momentum restrictions due to the relatively large mass of the ablated particles.

The absence of direct coupling to the lattice for laser pulses with durations of less than picoseconds reduces significantly unwanted by-reactions such as vibrational coupling in the case of adsorbates or lattice melting in the case of plain surfaces. However, thermal relaxation processes subsequent to the laser pulse usually make *identification* of the initial laser-induced electronic nonequilibrium distribution at a later temporal stage difficult or impossible. Since, in addition, most optoelectronic measurement techniques are slow (picoseconds) compared with the femtosecond time-scale, the full potential of the temporal resolution becomes available solely in correlation or pump/probe experiments.

A simple correlation experiment might be performed by mixing two laser pulses in a nonlinear optical crystal or on a surface to generate frequency-doubled light (Demtröder, 1996). For that purpose the initial laser pulse of frequency ω is split in a Michelson interferometer into two pulses (Fig. 94). One of the pulses is temporally delayed with respect to the other one by changing its optical path. The pulse pair with the temporal delay $\Delta\tau$ is focused onto a nonlinear optical crystal such as BBO, where it is frequency-doubled. The crystal is aligned such that the phase matching condition for frequency doubling is solely fulfilled if two photons from both beams (separated by an angle) hit the crystal.

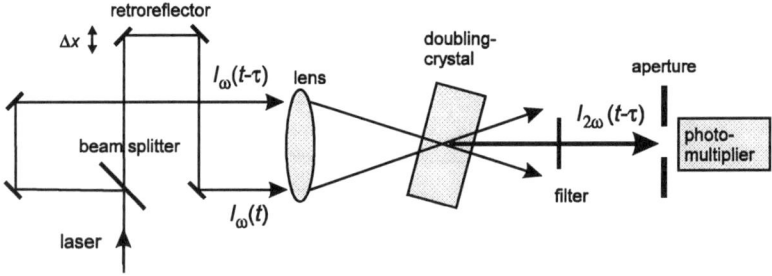

Figure 94 Separation of a femtosecond laser pulse into two pulses with femtosecond pulse delay and overlap of both pulses in a nonlinear optical crystal. By moving one of the retroreflectors by Δx and observing the frequency-doubled signal intensity behind an aperture and a color filter, this set-up serves as an autocorrelator for a background-free evaluation of pulse length.

Behind the color filter light of frequency 2ω is observed as a function of delay time $\Delta\tau$ between the two pulses. Since the intensity of the frequency-doubled light is proportional to the square of the incoming light intensity, the signal intensity is

$$I_{2\omega} \propto \int |[E_0(t)\exp(i(\omega t + \phi)) + E_0(t-\tau)\exp(i(\omega(t-\tau) + \phi(t-\tau)))]^2|^2 \mathrm{d}t. \quad (4.1)$$

Besides the frequency-doubled signal that is generated by each of the two pulses, one will also find a time-correlated signal, which peaks if the delay approaches zero. A typical measurement using 51 fs laser pulses is shown in Fig. 95.

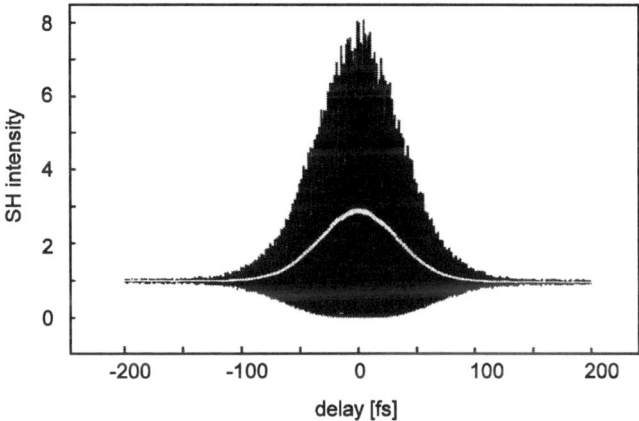

Figure 95 Typical measured interferometric second-order autocorrelation function, obtained by frequency-doubling 51 fs laser pulses in a nonlinear optical crystal (J.-H. Kleinwiele, private communication, 1997). Averaging over the oscillations results in the white curve, which corresponds to a sech^2 temporal pulse shape.

Upon averaging over the interference fringes one finds for the intensity autocorrelation function

$$I_{2\omega} = 1 + \frac{2\int I(t)I(t-\tau)\mathrm{d}t}{\int I^2 \mathrm{d}t}, \quad (4.2)$$

which gives a peak-to-background ratio of 3:1 (Fig. 95). Due to averaging, no phase information is contained and thus no information about the coherence of the pulse.

However, in the case of an instantaneous nonlinear optical response of the crystal one is able to determine in this kind of autocorrelator via variation of the optical path the temporal pulse form of the laser. For Fourier-limited Gaussian pulses this is $\tau_{\text{laser}} = \tau_{\text{ac}}/\sqrt{2}$, where τ_{ac} is the measured pulse half width.[1]

On the other hand, if the pulse width of the laser is known, then the measurement might serve to determine the characteristic time constants for the generation of the nonlinear optical signal. Especially in the case of the measurement of ultrafast processes (characteristic time constants below 100 fs) the propagation of the light pulses through the various optical elements of the set-up has to be carefully taken into account. Such propagation results due to dispersion in a chirp,[2] which has to be compensated before drawing conclusions on the physical origin of measured pulse broadenings or narrowings from the sample under investigation. Here, a useful check of the temporal and spectral quality of the ultrashort pulse is the simultaneous measurement of temporal and frequency spectra. In the case of a bandwidth-limited pulse of sech^2 temporal intensity shape without chirp the product is expected to be $\Delta\omega \times \Delta\tau_{\text{laser}} = 1.978$ (Diels et al., 1985). Deviations from this value point on a chirp, which might be identified more uniquely by measuring the second-order interferometric autocorrelation function instead of the averaged function, Eq. (4.2). The envelope of the constructive interferences has a peak-to-background ratio of 8:1, and this enhanced contrast helps to identify possible chirps in the wings of the pulses.

In the case of pump/probe experiments one has to be equally careful and obtains similar results. In that case usually the pulses have different frequencies. Thus the first pulse can be used to excite the system, while the second one tests the temporal evolution of the excited nonequilibrium state.

The following chapters deal with some selected topics of ultrafast studies on surfaces. Of course, the application of femtosecond pulses proves useful in most analytical applications of laser light on surfaces and consequently their use is mentioned throughout essentially the whole monograph. An instructive overview of time-resolved energy transfer processes at surfaces can be found in (Cavanagh et al., 1994). There, the mechanisms mentioned in Chapter 3, Section 3.2 are also briefly reviewed in a somewhat different context. Most of the problems that are imaginable in the context of the application of femtosecond lasers to real-time monitoring of gas-, liquid- or solid-phase problems are dealt with in the monograph by Diels and Rudolph (Diels and Rudolph, 1996).

---

[1] For the more frequent case of a sech^2 temporal laser pulse shape, the corresponding relation is $\tau_{\text{laser}} = \tau_{\text{ac}}/1.543$ (Diels et al., 1985).
[2] A chirp is a temporal broadening of the pulse. In the case of normal dispersion ($dn/d\lambda < 0$) the high-frequency components are temporally delayed and the low-frequency components are accelerated.

4.1 Electron Relaxation Dynamics

4.1.1 Surfaces and ultrathin films

Besides fundamental interest on the dynamics of systems with confined dimensions the main goals of time-resolved studies of ultrathin films adsorbed on surfaces are potential applications as ultrafast optoelectronic elements. A large body of literature has concentrated on the dynamic properties of metallic films, especially gold or silver films.

The initial absorption of laser light in thin metallic films results in a collective electronic excitation if the films consist of islands, or appropriate excitation conditions are chosen so as to facilitate surface plasmon excitation (see below, Section 4.1.3). In the latter case the spatial decay of non-localized, i.e., propagating, surface plasmons has been determined, resulting in momentum lifetimes of 48±3 fs for a 45 nm thick silver film on a glass prism (Exter and Lagendijk, 1988) and of 20 fs for a 40 nm thick Au film (Kroo et al., 1995). For a 70 nm silver film on a grating structure time-resolved measurements in the ATR geometry (Chapter 1, Section 1.2.2, Fig. 19) have revealed a lifetime of less than 10 fs (Kroo and Szentirmay, 1988). The momentum decay times of coherent multiply scattered surface plasmon polaritons in 35 nm thick gold films have also been determined via time-resolved ATR measurements to be about 56 fs with surface roughness leading to a significant increase in damping rate (Wang et al., 1996). If the films consist of isolated clusters, then because of strong surface scattering the lifetimes of the corresponding localized surface plasmons are expected to be smaller compared with those of nonlocalized plasmons (see below).

The decay of surface plasmons means a loss of coherence in the excitation, but there is still a distribution of highly excited, hot electrons. The dynamics of those electrons has been investigated by a variety of transient techniques taking advantage of ultrashort laser pulses. Besides linear transient reflectivity measurements nonlinear techniques such as transient second harmonic generation (SHG) have proven to be very useful. With laser energies near the interband transition threshold (2.4 eV in gold) the SH signal is expected to be especially sensitive to transient changes in electron temperatures (Hohlfeld et al., 1996; Luce et al., 1997). In contrast to linear thermoreflectivity measurements where this change in reflectivity is of the order of 10^{-3} or less (Sun et al., 1994), changes of the order of a few tens of percent are observed in second-harmonic experiments. Two possible arrangements are shown in Fig. 96.

In general, the change in reflectivity depends on the transient change in the dielectric function of the investigated metal film, which in turn depends on the density of states, which of course is a function of the local electron temperature. As a consequence, the reflectivity of a metal film that has been excited by a pump photon might be increased or decreased, depending on the wavelength of the probe photon being above (decreased absorptivity due to an

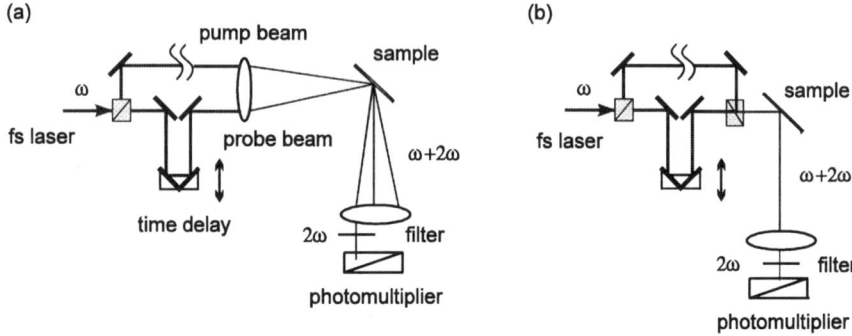

Figure 96 Experimental set-up for the time-resolved measurement of transient linear and nonlinear reflectivities. (a) Pump/probe arrangement. The detector monitors the second harmonic signal at 2ω from a probe pulse, while the pump pulse is temporally delayed by changing the optical path in the μm range. (b) Following temporal delay, pump and probe beams are merged again to obtain a collinear autocorrelation function at the sample.

increase in electron occupancy above the Fermi level) or below the interband transition energy (Sun et al., 1994). From linear pump/probe experiments on thin noble metal films delayed thermalization of the electron gas is concluded with typical time constants of 500 fs, but increasing into the picosecond regime close to the Fermi energy due to state filling effects that block the relaxation (Sun et al., 1994). This electronic thermalization competes with the electron–phonon coupling and makes a quantitative description of the process for metals with strong electron–phonon coupling in terms of the well-known two-temperature model (Anisimov et al., 1974) questionable.

The two-temperature model, which is a simple approach to quantitatively describe the thermal response of electrons, to predict reflectivity changes and to obtain electron–phonon coupling constants, assumes a Fermi–Dirac equilibrium distribution for the electrons and a Bose–Einstein distribution for the phonons and thus allows one to decouple the corresponding differential equations. The temporal and spatial evolution of electron temperature T_e and lattice temperature T_l under the influence of a laser source $g(x,t)$ is described one-dimensionally by

$$C_e \frac{\partial}{\partial t} T_e = K_e \nabla^2 T_e - G(T_e - T_l) + g(x,t) \qquad (4.3)$$

and

4.1. ELECTRON RELAXATION DYNAMICS

$$C_1 \frac{\partial}{\partial t} T_1 = G(T_e - T_1), \tag{4.4}$$

which are two coupled nonlinear diffusion equations with electron–phonon coupling constant G, electronic heat capacity $C_e \propto T_e$ and lattice heat capacity C_1, which is constant above the Debye temperature (Eq. (5.13)). The electronic thermal conductivity K_e for free electrons is

$$K_e = \frac{2 C_e E_F}{3 m_e^* (\nu_{ee} + \nu_{el})} \tag{4.5}$$

with Fermi energy E_F and electron–electron, ν_{ee} (Eq. (4.9)), and electron–phonon collision frequencies ν_{el} proportional to the lattice temperature.

A model including nonthermal electrons has recently been developed (Groeneveld et al., 1995), which works especially well under the conditions of low laser power and lattice temperature and yields slower electron–phonon relaxation times as compared to Eqs. (4.3) and (4.4).

An interesting aspect is the dependence of transient reflectivity on film thickness (Brorson et al., 1987; Hohlfeld et al., 1997b). With pump pulses at 400 nm the optical penetration depth is 12 nm and is thus significantly smaller than the usual film thickness. However, it has been found that the electron temperature is homogeneous, i.e., ballistic electron transport with a velocity of 1000 km/s dominates for films that are thinner than 100 nm. This results in a significant enhancement of transient reflectivity with decreasing film thickness since the energy that is deposited within the optical penetration depth is ballistically distributed over the film. In addition, the picosecond temporal behavior changes from exponential (compare Fig. 106) to linear, in agreement with the prediction of the two-temperature model (Hohlfeld et al., 1997b).

It is expected that for ultrathin films (below 16 nm)[3] the exact value of the relaxation time constant depends on the surface morphology, i.e., the shapes and sizes of the discontinuous distribution of islands. One then enters the regime of 'supported metal clusters', which is discussed below.

The nonlinear optical properties of continuous metal films are not very exciting. More favorable values of nonlinear optical susceptibilities have been found for thin films made of C_{60} molecules. Fullerenes (Kroto et al., 1993), the best-known member of which is the so-called 'soccer-molecule' C_{60} (Fig. 98), have high optical nonlinearities of the order of $\chi^3 \approx 10^{-10}$ esu. This is mainly because they have large numbers of three-dimensionally delocalized π-bindings and a nearly free electron gas, which is quantum confined in a cage with strong boundary conditions. If one adsorbs these carbon soccer-balls on a solid,

---

[3] A measure for 'ultrathin' in terms of 'discontinuous' might be the fact that for nominal thicknesses below 16 nm the Au films are charging if irradiated by low-energy electrons *e.g.*, in a LEED apparatus.

then the nonlinearities might even be enhanced due to the mutual multipole interactions of neighboring molecules.

A way to measure the nonlinear optial response of a thin C_{60} film on an irradiating light wave (i.e., the hyperpolarizability $\chi^{(3)}$) is to use degenerate four-wave mixing (Fig. 97, see also Chapter 3, Section 3.6). For that purpose the initial laser beam (here a 150 fs laser pulse of wavelength 637 nm, generated by a mode-coupled femtosecond laser) is split into two beams (forward and backward pump), which interfere in the nonlinear medium and form a holographic grating. A third beam (probe beam) is coherently scattered at this grating and generates a phase-conjugate signal beam, the intensity of which is measured using a photomultiplier. Since all beams possess the same wavelength (the process is 'degenerate', cf. Fig. 88) the wavelength of the signal beam is also predetermined due to energy conservation rules. The same is the case with the direction of the emitted signal: momentum conservation forces it to be counter-directed to the probe beam. It can be separated from that beam by a beam splitter.

The signal intensity is proportional to the product of the powers of the three irradiating lasers, the square modulus of the nonlinear susceptibility of third order and the square of the interaction length, which in the case of a 10 nm thick C_{60} film is small. However, the large nonlinearity of the films allows one to measure time-dependent signal intensities applying only relatively small irradiances of the order of gigawatts per square centimeter. After adjusting the probe beam onto the optimum temporal overlap of all three partial beams, one delays the backward pump beam with respect to the forward pump beam and obtains as a function of this delay the signal intensity that is plotted in Fig. 98.

Within the pulse width of the laser, which has been determined by an autocorrelator (Fig. 94), the signal intensity decreases strongly. Hence it is probable that this part of the signal is generated by a coherent polarization grating of the π-electrons, which exists as long as the laser beams irradiate the material. In the following the signal decreases with two further exponential time constants of a few hundred femtoseconds and a few picoseconds, respectively. The origin of these signals are gratings of excited electrons in the C_{60} film. The fast component corresponds to the direct decay of the population of the first excited singlet state S_1, while the slow component results from the decay of the long-living triplet state T_1, which has been generated from the singlet state via configuration interaction ('intersystem crossing'). By use of an appropriate laser wavelength the contribution of these long-living components and thus the optical response time of this thin film switch can be varied between femto- and picoseconds.

Another spectroscopic possibility to study ultrafast phenomena in thin films or on surfaces that has gained increasing importance within the last few years is time-resolved image potential state spectroscopy, or more general two-photon photoemission (TPPE) (Schoenlein et al., 1988; Fauster and

4.1. ELECTRON RELAXATION DYNAMICS

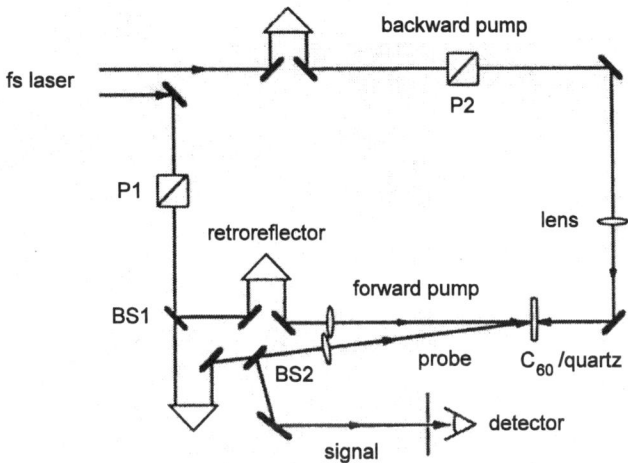

Figure 97 Experimental set-up for the time-resolved measurement of DFWM signals from 10 nm thick C_{60} films, adsorbed on quartz substrates. The retroreflectors are for mutual timing of the beams. The backward pump beam is temporally delayed with respect to the forward pump beam by several hundred femtoseconds to 200 picoseconds. P1 and P2 are polarizers, and BS1 and BS2 are beamsplitters. Reprinted from *Chem.Phys.Lett.*, (Rosker et al., 1992), Copyright 1992, with permission from Elsevier Science.

Steinmann, 1995), c.f. Chapter 3, Section 3.1.5. This enables one not only to observe dynamic changes in surface properties such as temperature or adsorbate growth modes, but also electronic changes in buried interfaces or electron-transfer reactions between the surface and adsorbates. The possibility to generate high population densities of image states makes the preparation of a two-dimensional electron gas feasible with nearly ideal free mobility, but localization in two dimensions due to the adsorption of adsorbates (Lingle, Jr et al., 1994; Ge et al., 1998).

If one applies lasers with pulse lengths below 100 fs, then a variety of elementary steps in surface dynamics open up for real-time studies (Wolf, 1997). These include radiationless energy relaxation processes of electrons close to metallic surfaces (Fann et al., 1992; Lingle, Jr et al., 1996), hot electron dynamics (Schmuttenmaer et al., 1994; Hertel et al., 1996; Petek and Ogawa, 1997; Bauer et al., 1997; Knoesel et al., 1998), polarization dynamics on metal surfaces using interferometric TPPE (Ogawa et al., 1997) or even electron tunneling processes between STM tips and molecules adsorbed on

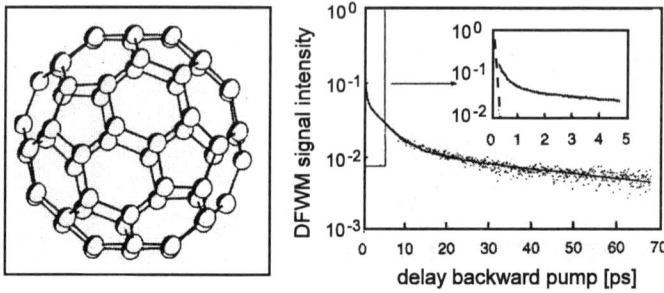

Figure 98 Left-hand side: Structure of C_{60}. Right-hand side: DFWM signal as a function of delay time of the backward pump beam at a wavelength of 637 nm. The solid line corresponds to a threefold exponential decay. The dashed line is due to the measured pulse width of the laser. Reprinted from Chem.Phys.Lett., (Rosker et al., 1992), Copyright 1992, with permission from Elsevier Science.

metal surfaces (Bartels et al., 1998). The strong coherent coupling between irradiated surface and laser pulse (*i.e.*, the manipulation of the phase) can also be used for coherent control (Shapiro and Brumer, 1997) purposes, *e.g.*, of photocurrent (van Driel et al., 1997) or photoexcited electron distributions (Petek et al., 1997).[4] As seen from this exemplary (but far from complete) list of applications, the main areas of research here are metal surfaces since on those surfaces elementary relaxation processes proceed on a femto- or subfemtosecond time-scale. Due to recent developments in laser technology this scale has now become accessible with steadily increasing simplicity.

4.1.2 Embedded nanoparticles

The (nonlinear) optical response of particles with characteristic dimensions of nanometers has been of interest for several decades. Especially in the case of semiconductor nanoparticles this is due to a size-dependent change in the density of states, which in turn gives rise to large optical nonlinearities. It takes little fantasy to imagine that the size-dependent band gap energies can be used to generate optical (laser) diodes with free-fabricated emission spectra.

Again, due to the availability of ultrashort laser pulses, the ultrafast

---

[4] Note, however, that coherent control of nuclear motion on surfaces due to strong homogeneous broadening effects (lifetime quenching) (Jiang et al., 1996) most probably is not possible.

4.1. ELECTRON RELAXATION DYNAMICS

dynamic response of those particles has become of renewed interest (Shah, 1996). Here, as in the case of thin films, one is interested in the exciton dynamics (in semiconductors), the collective and single electron dynamics (in metals) as well as the electron–phonon dynamics. The phonon dynamics of nanoparticles and semiconductor quantum wells has been investigated by Raman scattering. Recent approaches in the field of cluster physics (i.e., particles with a countable number of atoms) to the phononic properties of large clusters (Schröder et al., 1997) might also find some future applications.

As an example of a nonlinear optical investigation, four-wave mixing studies at cadmium–selenide (CdSe) microcrystallites, embedded in glass matrices, might be mentioned. These studies have shown that the time constant for the generation of a four-wave mixing signal increases with cluster size, whereas the large cross section for this process itself (≥ 1 Å^2) is nearly independent on size (Shinojima et al., 1992). For small clusters (radius 1 nm) surface-recombination processes are more important than bulk-recombination since the surface-to-bulk ratio is inversely proportional to the radius. Since it is possible to vary the CdSe cluster size distribution over a wide range, an extremely nonlinear optically active material might be generated from clusters of small radius, the effective time constant of which (given by the charge carrier recombination time) is only about 2 ps.

In the case of metallic nanoparticles most ultrafast studies have concentrated on the dynamic response of particles in glass matrices (*e.g.*, (Inouye et al., 1998)) or in colloidal suspensions. The case of supported particles is discussed in some more detail in the following chapter. As for spectroscopy and ultrafast dynamics of colloidal silver and gold nanoparticles, a recent overview can be found in (Hodak et al., 1998).

4.1.3 Supported metal clusters

Discontinuous metal films are of interest for controlling, for example, the polarization in the cover of a waveguide, since their optical properties depend — similar to the case of metal colloids (Chapter 6, Section 6.2.3) — on the size distribution and average size. The supported metal islands ('clusters'), which form the film, are generated usually by the thermal deposition of metals on insulator substrates (cf. Chapter 1, Section 1.2.3 and Fig. 21). For a given cluster size a maximum in the absorption probability is observed as a function of excitation energy (Figs. 57 and 100). In a classic picture, the clusters act as nanoantennas with size-dependent resonance frequencies, combining receiving (absorbing) and transmitting (scattering) properties. Upon irradiating the nanoantenna with an electromagnetic wave, charges are induced at the surface, which result in restoring forces and thus collective oscillations of the conduction electrons. This 'surface plasmon resonance' (cf. Chapter 3, Section 3.2) coincides with an enhancement of the electromagnetic field strength at the surface of the clusters as illustrated by a Mie calculation in Fig. 99.

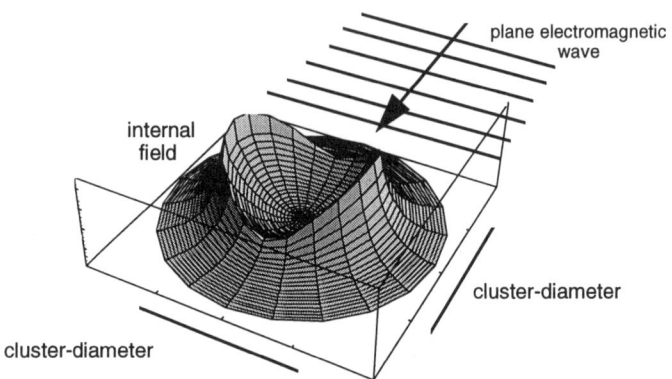

Figure 99 Calculated (classic Mie theory (Mie, 1908)) distribution of the modulus of an electric field around a spherical cluster, induced by a plane, linearly polarized electromagnetic wave of wavelength close to a resonance. Note the field enhancement at the surface of the cluster. Integration over the surface of the cluster results in the field enhancement factors $Q(\omega)$; cf. Eq. (3.32).

The width and spectral position of this resonance change as a function of cluster size. For clusters with a radius smaller than 1 nm (average number of atoms in the cluster about 100) the resonance shifts with decreasing radius to larger wavelengths (Kreibig and Vollmer, 1995).[5] The reason for this shift is that with decreasing size the conduction electrons become less strongly bound to the ionic cores (the 'spill out'(Chapter 3, Section 3.4, Fig. 78) increases), leading to an increase in polarizability of the clusters. Hence this effect depends on the cluster material, namely the size-dependent dielectric function.

With increasing cluster size the electrons are more strongly bound to the cores and the optical properties can be described at least along a limited size range (radii of 1 to 10 nm for metallic clusters such as Na_n) in dipole approximation using classic electrodynamic Mie theory ((Mie, 1908; Bohren and Huffman, 1983)). In this regime the spectral position of the dipole resonance is independent of size.

---

[5] This occurs for metals with quasi-free electrons such as sodium and for excitation energies below the interband transition energy.

4.1. ELECTRON RELAXATION DYNAMICS

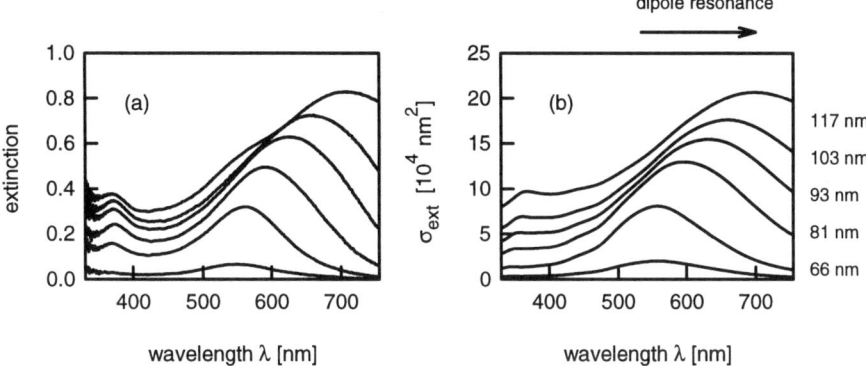

Figure 100 Measured (a) and calculated (b) extinction spectra for sodium clusters adsorbed on mica at a surface temperature of 150 K. Between subsequent curves the amount of adsorbed Na increases by a constant value. The calculated curves assume oblate clusters with $R = a/b = 0.5$ with large semiaxes a of the clusters denoted on the right-hand side; b denotes the small semiaxis, and thus R is a measure of the ellipticity of the clusters. For the lowest two theoretical curves the cluster density is 0.8×10^8 cm^{-2} and 3.2×10^8 cm^{-2}, for the subsequent curves it is 4×10^8 cm^{-2}.

For very large clusters (radii larger than 10 nm)[6] one again observes a red shift of the plasmon resonance. Here, electrodynamic effects such as retardation and excitation of higher order multipole plasmon resonances dominate the spectral position. These latter effects can be reproduced satisfactorily using the (size-independent) optical constants of the bulk material and Mie theory or T-matrix theory (Barber and Hill, 1990), if one is to take the ellipticity of the clusters into account. Fig. 100 compares measured (a) and calculated (b) extinction spectra. The red shift of the dipole resonance is clearly seen. For the calculations one has to assume a size distribution of the clusters, which in analogy to early TEM (transmission electron microscopy) measurements on cold lithium films (Rasigni et al., 1976) and gold decoration measurements on insulator surfaces (Schmeisser and Harsdorff, 1970; Schmeisser, 1974) is usually written as

---

[6] This 'ionic' radius r corresponds to about $N=10^5$ atoms, following $N \approx (r/r_{\mathrm{WS}})^3$; $r_{\mathrm{WS}} = 2.12$ Å is the Wigner–Seitz radius of Na.

$$f_{\pm}(a, a_0) \propto \exp\left[-\frac{(a-a_0)^2}{2\sigma_{\pm}^2}\right], \quad (4.6)$$

with the two widths σ_- and σ_+ related by $\sigma_- = \sqrt{2}\sigma_+$ and the subscripts "+" and "−" respectively denoting cluster semiaxes, a, satisfying the inequalities $a > a_0$ and $a \leq a_0$. As shown in Fig. 101, this asymmetric distribution is characterized by a FWHM of the order of 50% of the mean cluster radius a_0.

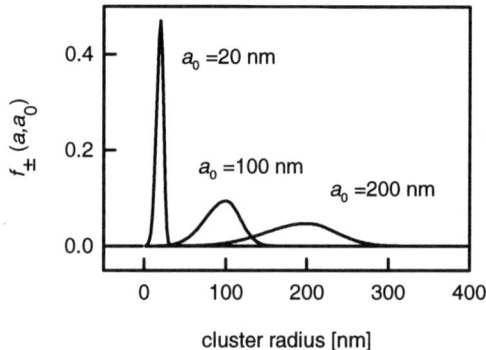

Figure 101 Calculated cluster size distributions (FWHM 50% of a_0), using Eq. (4.6) and for increasing values of a_0. As seen, the distribution outweighs clusters with small radius.

More sophisticated treatments of the optical response of rough cluster films include the interaction of the clusters with their mirror images in the supporting substrates (Royer et al., 1989), the mutual cluster–cluster interactions in terms of a quasi-static dipole–dipole approximation (Singer et al., 1995) and more accurate descriptions of distribution functions of cluster sizes and ellipticities (Balzer et al., 1998c). For example, Fig. 102 shows the distribution of hydroxylized sodium clusters on a mica substrate with the corresponding distribution functions of small and large semiaxes as well as ellipticities parallel and perpendicular to the surface plane shown on the right-hand side. Obviously, the morphology of thermally grown clusters on a surface is very complex, and it has been claimed that the disagreement between classic electrodynamically calculated Mie resonances and measured resonances might even be used as a tool to deduce specific physical and chemical interface properties (Kreibig et al., 1997; Kreibig, 1997).

Detailed characterization of the morphology of rough cluster films (see also Chapter 5, Section 5.1) enables one to deduce the electron relaxation

4.1. ELECTRON RELAXATION DYNAMICS

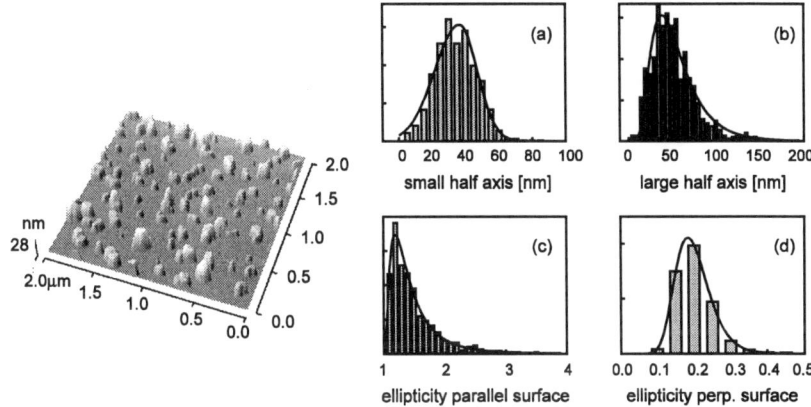

Figure 102 Left-hand side: Scanning force microscopy (SFM) image of NaOH clusters grown as Na clusters in UHV at a surface temperature of $T_S = 300$ K on cleaved mica. The sample size is 2×2 μm^2, the counted number density of clusters is $n = (1.9 \pm 0.2) \times 10^9$ cm^{-2}. Right-hand side: Size distributions of the small (a) and large half axes (b) parallel to the mica surface, distribution of ellipticities parallel (c) and perpendicular (d) to the mica surface. The solid lines are fits to the measured distributions, assuming a double-Gaussian (a) or log-normal distributions (b)–(d). Reprinted from Chem.Phys.Lett., (Balzer et al., 1998c), Copyright 1998, with permission from Elsevier Science.

dynamics as a function of mean cluster size. As shown schematically in Fig. 103, upon resonant laser excitation dynamic processes occurring on the femtosecond (initial electronic relaxation), picosecond (coupling to the lattice) and nanosecond (bond breaking processes) time-scales are expected.

The surface plasmon lifetime is expected to be extremely short (femtoseconds), and it is expected to be size-dependent. For clusters with radii a_0 significantly smaller than the mean free path of the electrons in bulk Na ($\bar{l}$=34 nm) surface scattering is the dominant damping mechanism in addition to Drude damping. Since the ratio of the surface scattering probability (proportional to the area of the cluster) to the number of scattering electrons (proportional to the volume) scales with $1/a_0$, the plasmon lifetime is expected to be (Kreibig and Vollmer, 1995)

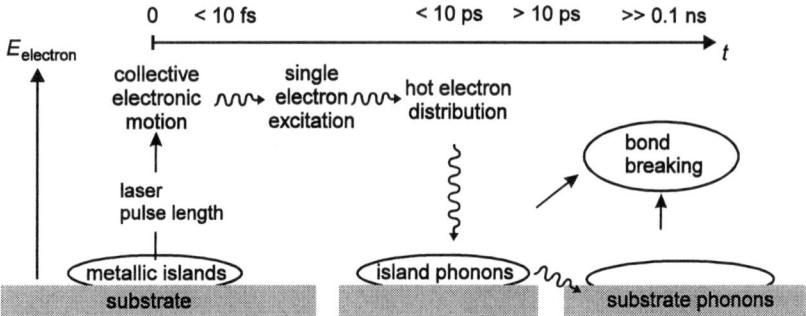

Figure 103 Schematic drawing of optical excitation (at $t=0$) and possible relaxation processes in rough metallic films, consisting of large, surface-bound islands. The ordinate represents the electronic energy E_{electron}, while the abscissa shows typical relaxation times for the decay of collective electronic excitation, single electronic excitation and phononic excitation.

$$\tau_{\text{sp}} = \left(\frac{v_{\text{F}}}{\bar{l}} + \frac{A v_{\text{F}}}{a_0}\right)^{-1}, \qquad (4.7)$$

where the first term refers to Drude damping and the second to surface scattering. Here, v_{F} is the Fermi velocity of the bulk cluster material and A is a size parameter that takes into account electron screening and surface roughness and varies between 0.38 and $4/\pi$ (Kreibig and Vollmer, 1995). For large clusters retardation effects (radiation damping, excitation of higher multipole plasmons) are expected to broaden the plasmon resonances, i.e., shorten the lifetime with increasing a_0 ('extrinsic' or electrodynamic size effects). The total damping rate is then given by

$$\Gamma_{\text{sp}} = \Gamma_{\text{Drude}} + \Gamma_{\text{surface}} + \Gamma_{\text{Mie}}. \qquad (4.8)$$

The additional term to the size dependence of the lifetime of the dipole resonance can be readily calculated via classic Mie theory. It includes the possible effects of interband transitions if the experimentally determined dielectric functions of the bulk are used.

In Fig. 104 the calculated plasmon lifetimes as a function of cluster size ('classic size effect') are shown, including intrinsic and extrinsic damping mechanisms as well as $A=0.45$ (dashed line) and $A=1$ (solid line). The former value has been taken from density functional calculations for Na spheres

4.1. ELECTRON RELAXATION DYNAMICS

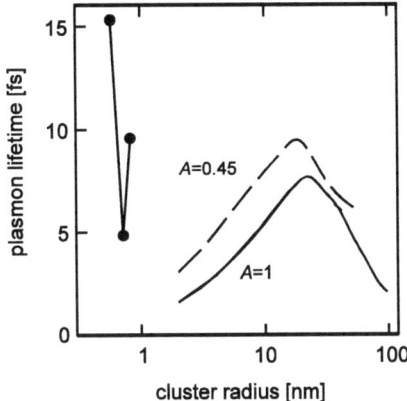

Figure 104 Lifetimes of collective electronic excitations in large sodium clusters. The solid circles are from a quantum mechanical calculation (Yannouleas et al., 1993). On the right-hand side the classic size effect is calculated according to Eq. (4.8) with the A-parameters discussed in the text.

(Apell and Penn, 1983).[7] For very small clusters this calculation certainly is inadequate (Halperin, 1986) and a quantum treatment (Yannouleas et al., 1993; Huang and Lue, 1994) is required. A few discrete values from such a treatment (Yannouleas et al., 1993) are included in Fig. 104.

Experimental information about the absolute value of the decay time constant for the initial collective excitation of surface bound clusters can be obtained in principle both in the frequency domain from the linewidth of the plasmon resonance and directly in the time domain. The former method would result in a surface plasmon lifetime of about 4 fs for sodium clusters of radius 50 nm, adsorbed on mica or, for example, 7 fs for $Na_{n=125}$ clusters adsorbed on BN (Parks and McDonald, 1989). However, the width of the surface plasmon resonance is dominated by various homogeneous and inhomogeneous broadening effects such as the cluster size distribution, mutual interactions of the clusters, chemical interface damping (Hövel et al., 1993), etc. and thus one has to be careful in assigning lifetimes to linewidths. Substantial progress in near-field microscopy allowed one to measure the homogeneous line shape of *single* gold nanoparticles of about 20 nm radius (Klar et al., 1998). From the linewidth a lifetime of about 8 fs has been deduced.

The first direct measurement of the plasmon lifetime of surface-bound clusters in the time domain has been reported for Ag clusters of radius 10 nm,

---

[7] Recent calculations suggest $A=0.58$ (Yannouleas, 1998).

bound to indium tin oxide (Steinmueller-Nethl et al., 1992). In subsequent measurements the lifetime of localized surface plasmons was found to be of the order of 10 fs (Lamprecht et al., 1997) with electron-surface scattering being assumed as the main reason for the damping of the plasmon excitation. In those experiments an array of triangularly shaped silver posts with radii of about 200 nm was used.

The size dependence of the lifetimes of surface plasmons excited in clusters has also been the subject of a recent second harmonic generation study (Klein-Wiele et al., 1998). In contrast to the noble metal studies, here adsorbed sodium clusters were investigated. The set-up is sketched in Fig. 96 and Fig. 105.

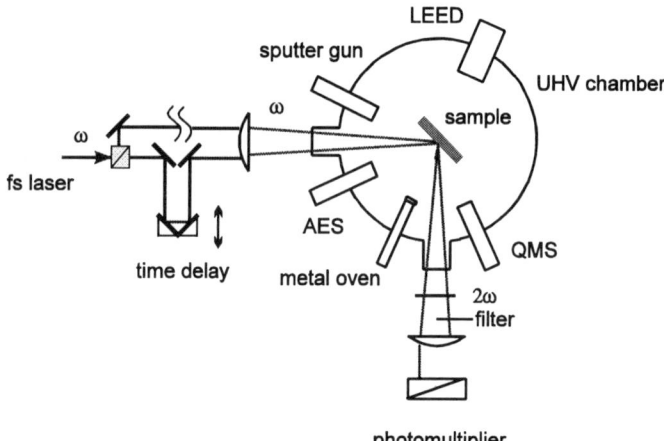

Figure 105 Experimental set-up for studying the ultrafast response of thin metallic films in ultrahigh vacuum. Surface and UHV preparation techniques include a sputter gun, Auger electron spectroscopy (AES), low-energy electron diffraction (LEED) and quadrupole mass spectrometry (QMS). Rough metallic films are prepared on the sample via evaporation from a metal oven.

Since the second harmonic (SH) signal intensity from the adsorbed clusters is governed by the field enhancement (Simon et al., 1975) induced by the surface plasmon excitation in the clusters, the duration of this signal provides a direct measure of the plasmon lifetime.[8] This idea, which was introduced by (Steinmueller-Nethl et al., 1992), assumes that the laser-induced plasma oscillations can be described by a damped harmonic oscillator which is driven

---

[8] There is always a certain probability of SHG due to the nonzero value of $\chi^{(2)}$. However, this signal would be too small to be measured without field enhancement.

4.1. ELECTRON RELAXATION DYNAMICS

by the femtosecond laser field. Then a measured collinear autocorrelation function (Fig.96) from the adsorbed clusters corresponds to the intensity autocorrelation of the convolution of the temporal shape of the laser pulse (given by a sech^2-function; Fig. 95) and the exponential decay of the plasmon excitation. In that way for Na clusters of mean radius 20 nm plasmon lifetimes of the order of 8 fs have been found following excitation near the surface plasmon resonance frequency (Klein-Wiele et al., 1998). A test of the reproducibility of the data was performed by exchanging the Na cluster film for Au thin films, where the SH generation in reflection geometry is instantaneous (Papadogiannis et al., 1997) and where the experimental spectra were found to be unaffected by broadenings.

As a function of mean cluster size the experiment revealed a pronounced maximum for clusters of average radius 22 nm with lifetimes decreasing for both smaller and larger cluster radii as expected from the classic size effect (cf. Fig. 104). For large cluster radii one expects that the simple theory is not able to reproduce the experimental results since it does not take into account cluster–cluster interactions. Those interactions influence the local fields for distances between neighboring clusters smaller than four times their radius (Cruz et al., 1989; Singer et al., 1995). This would already be the case for clusters with $a_0 \geq 34$ nm in Fig. 104.

The decay of the excited surface plasmons results in the distribution of hot electrons. The thermalization time constant of this distribution can be estimated as the electron–electron collision time constant, which can be calculated from Fermi liquid theory to be (Pines and Nozieres, 1966)

$$\tau_{ee} = \frac{128}{\pi^2 \sqrt{3}} \left(\frac{E_F}{E_{laser} - E_F} \right)^2 \frac{1}{\omega_p}. \tag{4.9}$$

With $\hbar\omega_p$=5.6 eV and E_F=3.12 eV for bulk Na, one obtains $\tau_{ee} \approx 10$ fs.[9] Hence it is likely that during a laser pulse of several tens of femtoseconds length local electronic equilibrium is obtained. The subsequent thermalization time of the phonon subsystem, τ_{ph}, is given by the ratio between the phonon mean free path and the speed of sound in bulk sodium. Again, for Na one finds $\lambda_{ph} \approx 1500$ Å at 150 K as deduced from the lattice constant of 4.28 Å and c_s=4340 m/s. Thus $\tau_{ph} \approx 35$ ps, which is three orders of magnitude larger than the thermalization time of the electrons and a factor of 20 larger than the electron–phonon loss time, a measurement of which is described below. Hence the lattice is expected not to be in equilibrium during the whole period

---

[9] This extremely short time constant for interelectronic collisions is due to the high frequencies accompanied by the strong laser excitation. Usually (for thermal excitation) the electron–electron relaxation time is longer as compared to the electron–phonon relaxation time. However, it has been noted recently that even in the case of laser excitation τ_{ee} might increase due to a cascading of the hot electrons (Hertel et al., 1996); see above Section 4.1.1.

of measurement.

The subsequent picosecond dynamics of the highly excited clusters can be investigated by time-resolved pump/probe measurements just as in the case of ultrathin metal films (see above, Section 4.1.1 and Fig. 96). The electronic excitation by a first (pump) laser pulse and the corresponding increase in electron temperature results in unoccupied states below the Fermi edge and enhances the absorbance of the second (probe) pulse. It thus leads to a decrease in SHG from the probe pulse (Fig. 106).

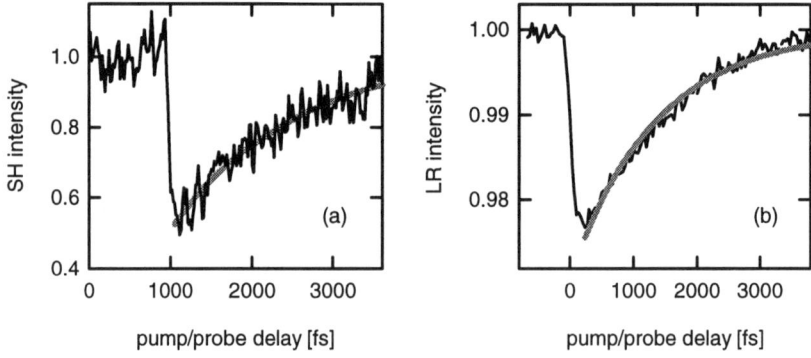

Figure 106 Nonlinear (a) and linear (b) reflectivity changes as a function of pump/probe time delay. (a) Na/LiF, τ=1.1(1) ps. (b) 10 nm Au/mica, τ=1.07(6) ps (J.-H. Klein-Wiele et al., private communication, 1998).

The change in electron density due to the femtosecond pulse and the Fermi edge-smearing associated with the heating of the conduction band electrons affects both the second-order nonlinear susceptibility and the linear dielectric function via the Fresnel factors for ω and 2ω. Both from experiments on polycrystalline Ag and Au surfaces (Hohlfeld et al., 1996) as well as from phenomenological theory (Luce et al., 1997) it is concluded that the electron temperature dependence of $\chi^{(2)}_{zzz}$ determines the picosecond time response. A similar behavior is expected for alkali cluster films.

As the hot electron distribution cools down as a result of collisions with the lattice of the clusters, the SHG from the probe pulse recovers. Thus by detecting the SH signal from the probe beam as a function of pump/probe delay the electron–phonon time constants of the cluster films are directly measured. For Na clusters of radii between 30 nm and 50 nm, adsorbed on lithium fluoride, the values of the temporal half width are $\tau_{ep} \approx 1$ ps (Rubahn, 1997) (Fig. 49), which is similar to the values found on thin films of noble

4.2. RELAXATION OF NUCLEAR MOTION

metals (Elsayed-Ali and Juhasz, 1993).

Finally, a closer inspection of Fig. 106 shows that for long delay times the second harmonic signal (and the linear reflection signal as well) does not reach fully the initial intensity level. This is attributed to phonons returning energy back to the electron gas and thus keeping the electron temperature high over a longer temporal period. The fraction of phonons participating in this effect has been estimated to be of the order of 10^{-3} (Suarez et al., 1995).

4.2 Relaxation of Nuclear Motion

4.2.1 Vibrational state dynamics

The determination of vibrational relaxation rates of adsorbed molecules is possible in principle via the measurement of the linewidths of laser-excited molecules or via direct, time-resolved measurements. In the former case the width of a spectral line is determined via infrared spectroscopy, which is inversely proportional to the lifetime of the investigated state in the case of homogeneous broadening. Energy transfer with the surface obviously affects the effective lifetime (see Chapter 3, Section 3.2), and this is reflected in a change of linewidth and a shift of the line position. This method has the main disadvantage that the homogeneous linewidth is not solely determined by the energy relaxation constant T_1 of the state, but also by the dephasing constant T_2 (Chapter 3, Section 3.1), which might even dominate in some cases. A way to discriminate between both contributions is by using temperature-dependent measurements, since the time constant T_2 depends strongly on temperature due to the temperature dependence of the adsorbate–substrate vibrations.

Direct measurements of relaxation times using pump/probe measurements with short pulse lasers are less affected by these problems. They also should lead further towards direct monitoring of the temporal evolution of individual bond breaking or at least bond-changing events. These then are the basic steps for surface photochemistry.

There are various ways to reach this goal. One might apply a visible ultrashort laser pulse to ultrafast heat the substrate and watch the change in reflectivity of a quasi-cw infrared laser pulse that is tuned to an adsorbate vibration (Culver et al., 1993). The time resolution is provided by difference-frequency mixing infrared and visible pulses. From measurements of that kind it is concluded that the CO stretch vibration interact with the Cu(111) substrate oscillations via the frustrated translational mode (cf. Fig. 55).

If sufficiently short pulses (pico- or femtoseconds) with the correct wavelength in the infrared spectral range are available, then adsorbate–substrate or disturbed adsorbate vibrations can be directly excited. Then transient infrared spectra of adsorbed molecules become available as a function of wavelength and pump/probe delay time. For $CO(v=1)/Pt(111)$ an excited state lifetime of 2.2 ps was found (Cavanagh et al., 1993), quenched via electron–hole pair excitation in the metal substrate.[10]

Even more information on lifetime and dephasing times becomes available if two ultrashort pulses of different frequencies are mixed (difference or sum frequency mixing, SFG, Chapter 3, Section 3.5). An intense infrared laser pulse results in the transient bleaching of the adsorbate in the laser focal spot and the subsequent laser pulse produces an attenuated sum frequency signal from the adsorbate (Chapter 3, Section 3.5). Since SFG is a surface-sensitive method, the temporal changes in the adsorbate layer can be sensitively detected. If one varies the temporal delay between pump and SFG probe pulses, then the successive recovery of the unbleached signal and thus the relaxation lifetime can be determined.

In that way one has determined, for example, the lifetime (corresponding to T_1) of the H–Si(111) stretch vibration (ν=2083.7 cm^{-1}), which is 800 ps (Guyot-Sionnest et al., 1990), as well as its dephasing time (corresponding to T_2), which is 13 ps (Guyot-Sionnest, 1991a). A surface two-phonon bound state ($\nu_{1\to 2}$), which is well localized, has also been observed on the semiconductor system (Guyot-Sionnest, 1991b). In contrast to adsorbates on metals, which possess short vibrational lifetimes due to the possibility of resonant excitations in the metal (for CO/Cu(111) the dephasing time is 2 ps (Owrutsky et al., 1992)), the hydrogen system on silicon is well suited to this kind of measurement due to its relatively long lifetime. This long lifetime is due to (a) the band gap in silicon, which suppresses the coupling of the adsorbate vibration to electronic substrate excitations, and (b) the large energy mismatch between the H–Si stretch vibration and the H–Si bending vibration (ν=637 cm^{-1}) as well as the silicon lattice phonons ($\nu \leq 500$ cm^{-1}). Since, on the other hand, the H–Si bending vibration is quasi-resonant to the highest frequency silicon phonons, its lifetime is short. The measured recovery of the bleached signal is thus dominated by the lifetime of the stretching vibration.

On a silicon surface without steps only a single stretching vibration of adsorbed hydrogen is possible. This is entirely different if the surface has steps and terraces. Depending on the crystallographic cut with respect to the single crystal directions there will be steps with isolated hydrogen atoms and steps with two hydrogen atoms per step atom ('dihydride surfaces'). Fig. 107a shows the SFG-spectrum of such a dihydride surface in the energetic range of the H–Si stretch vibration (Morin et al., 1992a). In addition to the vibrational frequency of hydrogen on the smooth surface at 2084 cm^{-1} there are two step vibrations, C_1 and C_2, which are more strongly bound.

Subsequently, the infrared pump laser is fixed at the energetic position of a vibrational frequency and the corresponding vibration is bleached during the

---

[10] Recently, vibrational energy transfer rates between CO adsorbed on Pt(111) and on Cu(100) have been directly compared using transient infrared spectra (Cavanagh et al., 1995).

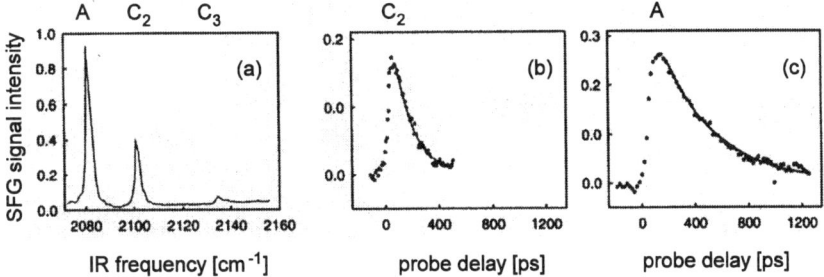

Figure 107 (a) Sum frequency spectrum of the H–Si stretch vibration on a stepped Si(111) dihydride surface. The three maxima are assigned to the vibrational frequency of a hydrogen atom at a terrace (mode A) and the vibrational frequencies of hydrogen atoms at step edges (modes C_2 and C_3). In (b) and (c) the temporal evolutions of the SFG-signal intensities for the vibration at the step edge (mode C_2) and on a terrace place (mode A) are plotted. The exponential decay time constants are 130 ps for the edge-atoms and 420 ps for the terrace-atoms. Reprinted with permission from (Morin et al., 1992a). Copyright 1992 American Institute of Physics.

laser pulse ($\approx$100 ps). The relaxation time of the excited vibration is measured via the recovery of the SFG probe signal as a function of delay time between the pulses (Figs. 107b and 107c). The experimental set-up, including some laser parameters, is shown in Fig. 87. It is observed that the lifetime T_1 of the hydrogen vibration at the step (130 ps) is significantly smaller compared with that of hydrogen atoms bound to terraces (420 ps). Both lifetimes are small compared with the lifetime of the stretch vibration on a smooth surface (800 ps). Thus the energy transfer between terrace and step atoms decreases the lifetime of the vibrations compared with that on a smooth surface. This mechanism can only work if the atoms bound to steps can act as a 'sink' for the atoms bound to terraces, i.e., if they have a short lifetime. The existence of steps alone does not decrease the lifetime of the terrace-bound atoms. The short lifetime of the atoms bound to steps is given by the possibility to relax into energetically close-by modes of the SiH_2, e.g., into the scissors mode, which has a frequency of about 850 cm^{-1}. The most probable physical mechanism for this energy transfer between terrace-bound and step-bound atoms is pairwise dipole–dipole coupling, where the energy transfer rate is determined mainly by the distance between the dipoles ($\propto R^{-6}$) and the square of the dynamic dipole moments (Förster, 1965).

4.2.2 Coherent-phonon oscillations

In Chapter 2, Section 2.4 we discussed the use of laser-induced gratings to study thermal diffusion processes. There, the short-time behavior — mainly due to the generation of acoustic surface waves — was neglected in order to quantitatively determine the thermal constants. However, short-pulse laser-induced transient gratings can of course also be applied to study solely the phononic part of the interaction, namely the generation of *acoustic* surface phonons (Wright and Kawashima, 1992). Essentially, the surface vibrations are excited by two interfering pump pulses and they are interrogated by a time-delayed probe pulse,[11] which is angularly deflected by $\Delta\Theta = 2\psi$ and thus monitors changes in the surface slope ψ. A sketch of a typical experimental set-up is given in Fig. 108.

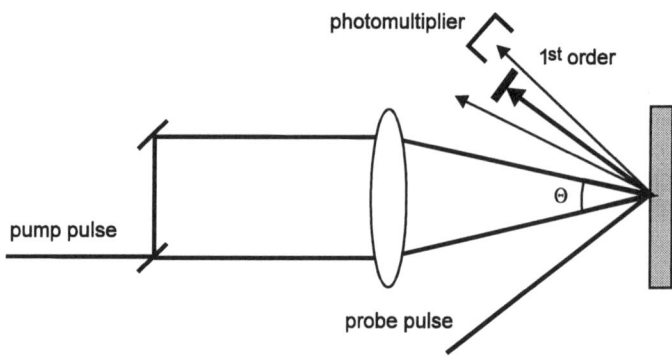

Figure 108 Experimental configuration for the observation of pulsed impulsive stimulated thermal scattering. The pump pulse is split into two beams, which are simultaneously focused on the surface, where they form an interference grating. A probe pulse is scattered off this grating and the first order diffraction efficiency is monitored as a function of pump/probe time delay.

In (Duggal et al., 1992), impulsive stimulated thermal scattering (ISTS) with 100 ps pump and 300 ns probe laser pulses was used to determine the elastic constants of thin-film coatings. The frequency of the 'Rayleigh' acoustic modes of the thin polyimide films is of the order of a few hundred MHz

---

[11] Or by the use of temporal overlap of both pulses and a transient digitizer to obtain the temporal resolution.

4.3. ULTRAFAST PHASE TRANSITIONS

and is determined via Fourier transformation from the pump/probe spectrum with a temporal spacing of the oscillations of about 10 ns. Since it is found that the diffracted signal intensity shows a quadratic decrease with increasing excitation wavevector it is concluded that the physical origin of the deflection is a transient surface ripple structure (c.f. Chapter 6, Section 6.4) induced by the pseudo-Rayleigh waves. The excitation wavevector q is varied by changing the full angle Θ under which the excitation beams interfere:

$$q = \frac{4\pi\sin(\Theta/2)}{\lambda}. \tag{4.10}$$

By the use of transient linear reflectivity changes in a pump/probe configuration and ultrashort laser pulses (50 fs) coherent longitudinal *optical* phonons at the surface of semiconductors were also observed (Cho et al., 1990). They manifest themselves as oscillations in the linear reflectivity ($\Delta R/R \approx 10^{-4}$) on a 100 fs time-scale ($\nu \approx 9$ THz) and result probably from the screening of the surface space-charge field by the laser-induced carriers as well as a stimulated Raman excitation process. The excitation of the longitudinal optical (LO) phonons, which are polarized in the direction of the surface normal, is a coherent process. Hence in terms of 'coherent control' the superposition of coherent phonons and the control of the vibrational amplitude becomes possible (Dekorsy et al., 1993), providing either total extinction or strong enhancement. Interaction of the coherent part with incoherent phonons, resulting from the relaxation of the laser-excited carriers on a picosecond time-scale, as well as decay into coherent acoustic phonons, obviously can also be studied.

Recently, nonlinear reflectivity changes ($\Delta R/R \approx 10^{-1}$) have also been employed to investigate the free-induction decay of coherent longitudinal optical phonons, which have been excited near the center of the Brillouin zone by a pump laser pulse (Chang et al., 1997). Besides the much improved signal-to-noise ratio, additional advantages of this technique as compared with its linear variant are the intrinsic surface sensitivity for centrosymmetric media (cf. Chapter 3, Section 3.4), the possibility to determine the symmetry of the phonon modes, as well as the ability to study buried interfaces. Thus, for GaAs, additional surface modes in the THz range could be identified, which have not been detected previously by electron energy loss spectroscopy.

4.3 Ultrafast Phase Transitions

If one takes into account typical time constants for relaxation processes inside a solid following laser excitation (Fig. 127), one sees easily that it should be possible to induce order–disorder phase transitions in crystalline semiconductors without melting of the crystal lattice by use of femtosecond laser pulses. This process might be called *cold melting* of the crystal lattice or *electronic melting*. The irradiated lattice looses its order on an ultrafast time-scale of less than a picosecond via direct electronic excitation and without previous relaxation of the absorbed energy into lattice oscillations (Shank et al., 1983). Such a well-defined, nonthermal and localized manipulation of crystalline structures is of great interest for applications in surface photochemistry or semiconductor treatment. A necessary condition for this to occur is a high concentration of laser-induced carriers (Stampfli and Bennemann, 1994) of the order of 10^{22} cm^{-3}. Such a concentration can be obtained in GaAs at fluences of about 200 mJ/cm^2 (Sokolowski-Tinten et al., 1995).

In a model experiment for this process, Si(111) has been irradiated in ambient air using two mutually time-delayed, p-polarized, 75 fs pulses from a CPM laser (610 nm) (Tom et al., 1988). The irradiance was so high (200 mJ/cm^2) that the crystal melted in the burn spot of the laser following irradiation with both pulses simultaneously. Fig. 109c shows the resulting sum-frequency signal which provides the temporal resolution of the experiment (180 fs half width). With initial temporal overlap of both pulses the crystal starts to melt and the linear optical properties are modified from that of crystalline to that of liquid silicon. As a result the amount of linearly reflected light is slowly increasing (Fig. 109d) with a time constant of a few hundred femtoseconds, which is a typical electron–phonon interaction time.

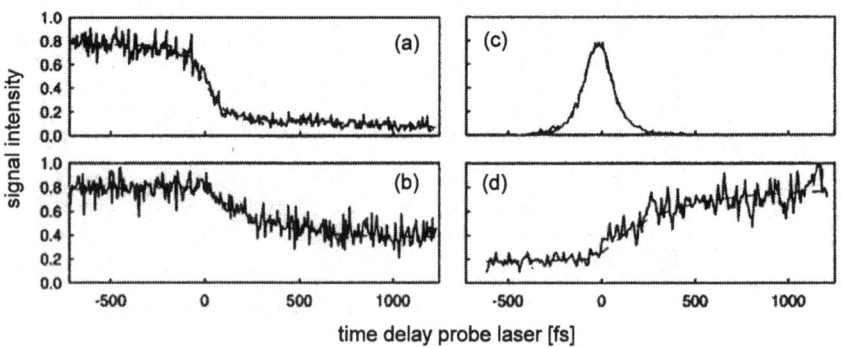

Figure 109 Linear and nonlinear optical response as a function of pump/probe delay time between two 75 fs pulses, which irradiate a Si(111) crystal. (a) S-polarized SH signal; (b) p-polarized SH signal; (c) sum-frequency signal; (d) p-polarized, linearly reflected light. The signals (a), (b) and (c) have been measured simultaneously with three photomultipliers. Reprinted with permission from (Tom et al., 1988). Copyright 1988 American Physical Society.

The frequency-doubled light is changing more drastically and with reversed sign as a function of the beginning of melting (Figs. 109a and 109b). The

4.4. PHOTOCHEMISTRY

s-polarized SH signal decreases with a time constant of 150 fs close to zero, while the p-polarized SH signal decreases with a time constant of 500 fs. It can be shown from the symmetry properties of the second-order nonlinear susceptibility tensor (see also Chapter 3, Section 3.4) that in the chosen laser-crystal geometry the s-polarized light depends only on order-dependent tensor elements, while the p-polarized signal depends on a combination of order-dependent and order-independent elements.

Therefore from Fig. 109 it follows that the crystalline order gets lost within the first 150 fs, which is a time that is too short for the relaxation of the electronic excitation via collisions with the lattice. Additional measurements at a surface with lower Miller indices (Si(100)) furthermore show that the SH signal is generated within the upper 7.5 nm to 13 nm of the silicon-bulk and that it is not a true interface signal.

Within the penetration depth of the laser the silicon atoms are thus extremely fast positioned into a new lattice configuration in which they are — in contrast to the configuration of thermal or equilibrium melting — nearly vibrationally unexcited. The slow decrease of the p-polarized signal intensity shows that the electronically highly excited system relaxes slowly (within a few hundred femtoseconds) into the equilibrium state of liquid silicon.[12]

Similar observations have been made for other semiconductors such as gallium arsenide (GaAs) (Govorkov et al., 1992). In contrast to the centrosymmetrical silicon, GaAs is a noncentrosymmetrical crystal, in which SH generation is allowed. One thus expects that in the molten, metal-like phase the probability for frequency doubling vanishes. It has been observed that within the first 300 fs the uppermost 13 nm of GaAs lose their long-range order following irradiation with 620 nm femtosecond pulses. This results in a change of the nonlinear reflectivity. A possible interpretation is that upon laser irradiation one deals initially with a *centrosymmetrical* semiconductor. In the following the short-range order is also changing, and the semiconductor transforms into a metal-like, molten crystal. This transformation occurs on a time-scale of less than 100 fs (Sokolowski-Tinten et al., 1995).

4.4 Photochemistry

Laser-induced photochemistry at surfaces (Chuang, 1983) is methodologically similar to laser photochemistry in the gas phase: all light sources from the arc lamp to the femtosecond laser can be applied. The existence of a surface imposes several constraints, but also new opportunities for photochemical interactions of adsorbates since the fundamental excitation and relaxation mech-

---

[12] In the case of an insulator the electronic melting induced by femtosecond pulses can result in a transient metallic behavior, which can be identified by a characteristic ripple pattern (Chapter 6, Section 6.4) even without pump/probe techniques (Ashkenasi et al., 1997).

anisms of the excited molecules are vastly different from those encountered in the gas phase. This is caused by the optical properties of the adsorbates, which differ from those of free molecules, and also results from substrate mediated relaxation channels such as electron–hole pair excitation, charge transfer or dipole–dipole couplings. One thus might divide the constraints into peculiarities of the environment of the adsorbate and changed properties of the adsorbate itself.

Environment
Compared with the gas phase, the adsorbate faces reactants in close proximity (corresponding in the gas phase to a pressure of several hundred atmospheres), low or adjustable temperatures, hence suppressing or favoring thermal reactions, a nearly infinite temperature reservoir and thus a nearly infinite reservoir for phononic losses and the possibility of strong field enhancement effects especially at rough surfaces (Nitzan and Brus, 1981b; Nitzan and Brus, 1981a).

The latter point, of course, is one of the bases of the SERS effect (surface-enhanced Raman scattering; see Chapter 3, Section 3.3). Electromagnetic field calculations, for example, show that there is an optimum molecule–surface distance for photodissociation to occur (Leung and George, 1986; Jelski et al., 1988), which possibly could be adjusted by bringing an ultrathin inert spacer layer in between the reactants and the surface (Chapter 1, Section 1.2.3, Fig. 22). Applications of this enhanced photodissociation process to heterogeneous catalysis or improved chemical vapor deposition are easily imaginable.

There is great practical interest also in photochemistry on semiconductor surfaces (Moini et al., 1993), for example due to the possibility of sensitizing semiconductors with large band-gaps by adsorbed dye molecules and thus making them applicable to solar energy conversion or light storage.

In addition to the above points denoting the environmental peculiarities of the adsorbates they also are *aligned* on the surface, i.e., the molecular axis is tilted under a certain angle with respect to the surface normal and there are preferred orientations of the individual atoms in the molecule. For example, CO on Ni(100) adsorbs with the carbon atom situated on the surface and the oxygen nearly upright above. This preferred orientation makes surface-aligned photochemistry (SAP) possible. In order to disturb the adsorbed molecules as little as possible, one might use dielectrics with crystalline order such as lithium fluoride (LiF) and initiate the photochemical processes with an UV excimer laser at a frequency where these dielectric substrates are nearly transparent. This idea has been exploited successfully in a large number of studies, starting with aligned photochemistry of CH_3Br on LiF(001) (Bourdon et al., 1984; Harrison et al., 1988a), via H_2S (Harrison et al., 1988b) and HBr (Bourdon et al., 1991) up to $(NO)_2$ adsorbed on LiF(001) (Jackson et al., 1995). The main experimental workhorse has been angularly resolved time-of-flight studies of the photodissociation products. Besides simple desorption, various dissociation but also reaction channels within the adsorbate layers could be detected and the photodynamics to some extent revealed. More recently the studies have been extended to charge-transfer processes between coadsorbates (alkyl halides) on metallic surfaces (Dixon-Warren et al., 1993).

Modified adsorbate properties
Among the adsorbate properties that differ from that encountered in the gas phase are a strong electronic coupling to the substrate in the case of metals and thus largely changed bond energies, the existence of additional modes of internal vibration such as frustrated translation and rotation (Fig. 55) and a weakening of internal bonds due to the adsorbate–substrate bonding.

Most of the above aspects for particles adsorbed on metal surfaces are discussed in detail in (Dai and Ho, 1995). A recent overview on desorption, dissociation and reactions of adsorbed molecules that have been irradiated by visible or UV light is given in (Zhu, 1994). This covers different excitation mechanisms such as direct photoexcitation of the adsorbate, which is strongly affected by the electromagnetic field intensity at the surface (see above), or charge-transfer mediated adsorbate excitation (and dissociation) mechanisms such as found for molecular oxygen on Pd(111) (Hasselbrink, 1994). The latter mechanism proceeds via photo-induced electron–hole pair creation in the metal, migration of the hot electrons to the surface (assuming that the penetration depth of the light is comparable to the mean free path of the excited electrons) and resonant tunneling into an affinity level of the adsorbate a few electronvolts above the Fermi energy (Fig. 110).

Besides providing information on the energetics of photochemistry on surfaces, laser methods are also well suited for obtaining dynamic information and thus unraveling details of the molecule–surface interaction potential. The laser methodology (Rettner et al., 1996) is often based on the classic approach, which is to employ scattering of thermal energy atomic and molecular beams (Barker and Auerbach, 1984). Via laser irradiation, state-selective information about the outcome of the surface encounter is provided, i.e., translational and internal final state distributions of the reactants, electron- and nuclear-motion coupling factors, as well as spatial alignment or orientation of the angular momentum vectors (Zimmermann and Ho, 1995). One might also study the scattering of nonthermal, vibrationally excited species (Gostein et al., 1995; Jongma et al., 1997) or metastable particles (Conrad et al., 1982; Böttcher et al., 1994). In the latter case the local density of states of the outermost surface layer is sensitively tested, whereas the former studies reveal, for example, vibrational deactivation probabilities.[13] For some model systems (H_2/Cu(111) or NO/Pt(111)) rather exhaustive studies have been

---

[13] In general, on dielectric surfaces those are at least two orders of magnitude smaller compared with electronic deactivation probabilities.

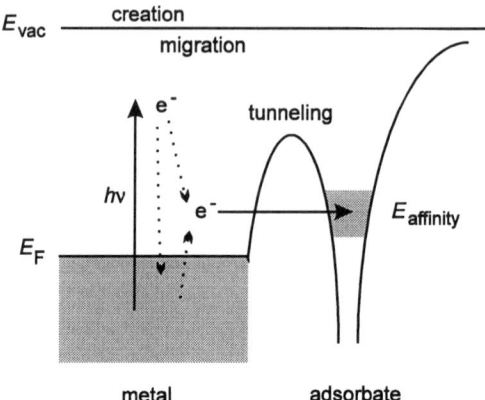

Figure 110 Possible elementary steps for substrate hot electron mediated photodissociation of adsorbates from metal surfaces.

performed in the past, leading to a profound understanding of energy transfer and trapping dynamics. For example, direct gas–surface abstraction reactions (Eley–Rideal mechanism) as well as reactions of fully accommodated species (Langmuir–Hinschelwood mechanism) have been observed. By exploiting modern laser methods, those studies might well be extended to more complicated systems (including polyatomic species (Francisco et al., 1996)) in the near future.

In addition to obtaining stereodynamic information from polarized investigations of the desorbing particles (Jacobs et al., 1987) one might get detailed insight into the surface reaction dynamics if the reactants are aligned prior to their interaction with the surface, *e.g.*, via strong electric fields. Recently, the stereoreactivity of the dissociative adsorption of deuterium on Cu(111) has been investigated that way (Hou et al., 1997), revealing that the dissociation has a much higher probability for broadside as compared with end-on collisions.

The use of short and ultrashort laser pulses should allow one to follow the break and the formation of individual bonds on the surface (Gadzuk, 1995). Here again, as in general in the field of surface photochemistry, the main experimental emphasis has put on the simplest reaction, namely the break of the bond between substrate and adsorbate, i.e., the femtosecond laser-induced desorption from a metal surface (Ho, 1996).

Upon femtosecond laser irradiation high desorption efficiencies have been reported with extremely nonlinear dependencies of desorption rate Y on laser fluence F_L ($Y \propto F_L^n$, $n=6$ (Misewich et al., 1994), in general $n=3-7$). In contrast, irradiation with nanosecond pulses usually results in a linear fluence dependence for such a 'direct' desorption process. Time-resolved correlation measurements have shown that the response time is picoseconds (Budde et al., 1991) or, in the case of CO/Cu(111), less than half a picosecond (Prybyla et al., 1992), providing a hint that the laser-excited electron–hole pairs in the substrate are strongly coupled to the adsorbate vibrations. This assumption has been confirmed by state-resolved measurements, which reveal an adsorbate vibrational temperature that is significantly higher as compared with the substrate temperature (Budde et al., 1993).

The impingement rate of photons induced by a femtosecond pulse is very high, resulting in a high excitation rate as compared with the relaxation rate of the excited state. Hence the usual DIET mechanism for desorption (Chapter 2, Section 2.2) evolves into a 'DIMET' process (desorption induced by multiple electronic transitions) (Misewich et al., 1992). The transition between DIET and DIMET just depends on the excitation rate, and thus one is able to observe this in a plot of desorption yield vs. fluence by an abrupt change from a linear dependence to a dependence with a high power exponent. Recently, this argument has been driven even further (Ho, 1996) by invoking similarities between femtosecond laser-induced desorption and electron-induced desorption, stimulated in the near field of a scanning tunneling microscope. In the latter case the transfer rate of individual atoms from the surface to the tip of tunneling microscope depends nonlinearly on the tunneling current. It is tempting to conclude that the same physics might hold at extremely small lengths and extremely short time-scales.

4.5 Picosecond Electron Diffraction

In previous chapters we have discussed methods to study — in real time — vibrational and electronic energy transfer on surfaces by the use of ultrafast laser spectroscopic techniques which might be called 'conventional' in the sense that they also could be (and have been (Zewail, 1994)) applied to problems appearing in the gas or liquid phase. Some indirect information on structural changes of surfaces could be deduced from the sensitivity of second harmonic generation on surface symmetry (Section 4.3). However, it would be rewarding to improve the temporal resolution of conventional surface structural probes such as electron diffraction and in that way monitor microscopic changes of surface morphology in real time. This would then ultimately include watching the vibrational motion of surface atoms, the hindered rotational motion of adsorbed molecules (Stipe et al., 1998) or the evolution of elementary steps of surface reactions.

Real space methods involving ultrashort laser pulses and scanning tunneling microscopy such as 'correlated optical reactivity and STM, CORSTM'

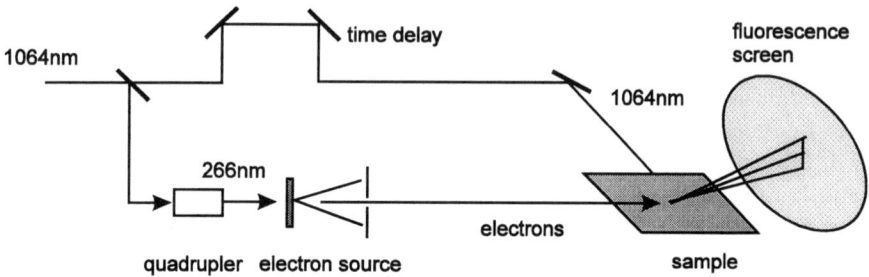

Figure 111 Experimental set-up for the time-resolved measurement of surface structural changes via reflection electron diffraction.

(Feldstein et al., 1996; Feldstein and Scherer, 1998) are just evolving,[14] while scattering methods have a long history.

An intriguing idea for picosecond electron diffraction is to employ a streak camera (Mourou and Williamson, 1982), which upon irradiating the photocathode with an ultrashort laser pulse generates a photoelectron replica with a temporal width of picoseconds and spatial resolution of several tens of micrometers (Aeschlimann et al., 1995). By synchronizing the electron pulse with a heating laser pulse the temporal evolution of heat on a surface can be monitored by means of the transient surface Debye–Waller effect (i.e., intensity changes of reflection high-energy electron diffraction (RHEED) peaks) (Elsayed-Ali and Herman, 1990; Aeschlimann et al., 1995). A typical set-up is sketched in Fig. 111.

First time-resolved structural studies have been performed on model surfaces (Pb(110), Pb(111) and Pb(100)), where the phenomenon of surface melting has been investigated extensively in the past. In contrast to bulk melting the surface melts by forming a disordered layer, several atomic layers thick, at temperatures significantly below the bulk melting point (Frenken and van der Veen, 1985). Heating the surface by a short-pulse laser and watching the transient structural changes by RHEED has revealed that the surface disordering occurs on a time-scale below 180 ps (compare Fig. 109), that the surface disorder temperature in the case of Pb(110) is the same as in the

---

[14] If one is interested in macroscopic structural information on laser-induced surface morphology changes, then conventional light microscopy might be combined with ultrashort pump/probe techniques. For some recent progress see (von der Linde et al., 1997).

4.5. PICOSECOND ELECTRON DIFFRACTION

static case even if the heating rate is of the order of 10^{11} K/s and that the process is reversible, i.e., crystalline order regrowths (Herman and Elsayed-Ali, 1992b). In contrast, on the more close-packed Pb(111) surface the high heating and cooling rates allowed superheating up to 120 K before melting started (Herman and Elsayed-Ali, 1992a); i.e, the disorder temperature was seen to be significantly above the bulk melting point. Finally, in the case of Pb(100) one observes incomplete surface melting with finite disordered layer thickness and residual order up to the bulk melting temperature and above (Herman et al., 1993). These studies are especially important since there is a strong interplay between surface melting and surface roughening. Reversible or irreversible roughening is of main importance for such diverse phenomena as growth of thin films or annealing of sputtered surfaces.

Even more detailed information can be expected if one applies picosecond RHEED and laser heating to different growth phases of adsorbates on surfaces, where, for example, the typical RHEED growth oscillations deliver additional structural information on the ultrathin films. Also, very recently cross-correlation measurements combining femtosecond laser pulses (λ=800 nm) with picosecond X-ray pulses have provided information about ultrafast structural disordering processes, at least in the bulk of an InSb crystal (Larsson et al., 1998). Extension of these measurements to the study of surface processes might become feasible in the near future.

Finally it is noted that the picosecond RHEED method in turn now has also been applied to obtain data for ultrafast changes in the structural properties of gas phase particles (Williamson et al., 1997).

5

Fundamental Laser Surface Treatment

5.1 Heating and Melting

Assume a laser beam hits a substrate under an angle α with respect to the surface normal (Fig. 112). The beam will be partially reflected, partially transmitted and partially absorbed.[1]

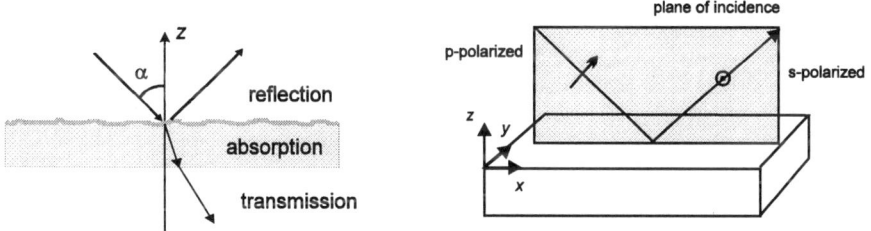

Figure 112 Reflection, absorption and transmission of a laser beam that hits a substrate. On the right-hand side the definition of p- and s-polarized light is visualized.

The reflectance is given as the ratio between reflected and incident intensity (Fresnel's formulas (Born and Wolf, 1975)). In the case of a light beam entering under normal incidence ($\alpha = 0°$) from vacuum (index of refraction $= 1$) we find

---

[1] The sum of reflection and absorption cross section is usually called 'extinction cross section'.

$$R = \left|\frac{E_r}{E_i}\right|^2 = \left|\frac{\tilde{n}-1}{\tilde{n}+1}\right|^2, \qquad (5.1)$$

with E_r meaning the reflected and E_i the incident electromagnetic field strength and $\tilde{n} = n' + in''$ the complex index of refraction. If the irradiated medium is glass ($n \approx 1.5$), then one finds for $\alpha = 0°$ a reflectivity of 0.04. Thus at normal incidence on a glass plate 4% of the incoming light intensity will be reflected from the frontside. Depending on the orientation of the electric field vector of the incoming, linearly polarized light with respect to the plane of incidence one usually denotes light that is polarized perpendicular to this plane as 's-polarized' light and light that is polarized parallel as 'p-polarized' light. In the former case the reflectivity coefficient E_r^s/E_i^s is given by $-\sin(\alpha - \beta)/\sin(\alpha + \beta)$ and in the latter case E_r^p/E_i^p is given by $\tan(\alpha - \beta)/\tan(\alpha + \beta)$, where β is the angle that the refracted light forms with respect to normal incidence.

The complex index of refraction $\tilde{n}$ is the square root of the complex dielectric function ϵ, which fully dictates the response of the material to the irradiating electromagnetic field. In the case of homogeneous absorption one finds the incoming light intensity I to be attenuated with traversed substrate thickness dz as

$$\frac{dI}{dz} = -\alpha I. \qquad (5.2)$$

Hence the linear absorption coefficient α describes the exponential decay of intensity of the electromagnetic wave in z-direction,

$$I = I_0 exp(-\alpha z). \qquad (5.3)$$

Since

$$I \propto \exp(2i(kz - \omega t)) \qquad (5.4)$$

for plane electromagnetic waves, propagating in z-direction with wavevector k and angular frequency ω and

$$k = \frac{2\pi}{\lambda} = \frac{\omega \tilde{n}}{c}, \qquad (5.5)$$

one obtains for the absorption coefficient

$$\alpha = \frac{2\omega n''}{c} = \frac{4\pi n''}{\lambda}. \qquad (5.6)$$

Here, $c/\tilde{n} = \omega/k$ is the phase velocity of light in the substrate.

The complex index of refraction depends on the wavelength of the light (*dispersion*), and thus the absorption coefficient α also depends on the wavelength. Resonances in the dispersion curve $\tilde{n}(\lambda)$ result in resonances

5.1. HEATING AND MELTING

of α and thus R. These resonances are accessible in the case of nonmetals by direct infrared excitation of lattice vibrations or via optical transitions between valence and conducting bands ('interband transitions'), which are separated by a band gap (Fig. 113). Due to the small photon momentum the transitions occur mainly at the Γ-point without momentum transfer to the lattice of the solid. The 'direct' band gap amounts to 1.47 eV for GaAs and might be varied largely via doping of the semiconductor or (to a lesser extent) via temperature changes. For silicon the direct interband transition requires a photon energy of 3.4 eV. A phonon-assisted, indirect transition can be excited already at a photon energy of 1.12 eV.

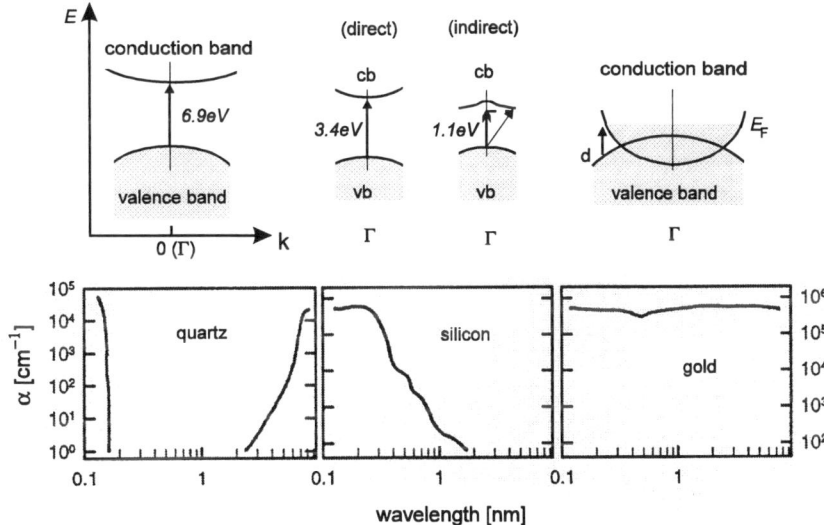

Figure 113 Schematic band structure of electronic transitions in insulators (quartz), semiconductors (silicon) and metals (gold). The corresponding absorption coefficients (reprinted with permission from (Allmen, 1995); copyright 1986 Springer-Verlag) are also shown as a function of wavelength.
The solid arrows characterize optical transitions (perpendicular in energy/momentum space, i.e., at the γ-point), whereas the thin arrow stands for a phonon-assisted, indirect transition along minimum energy differences between the participating bands. For gold the transition from the 5d-band to the Fermi edge E_F (long-wavelength absorption edge) is denoted.

For insulators the corresponding transitions are situated in the ultraviolet spectral range; for quartz (fused silica) at 7.8 eV. In that case the excitation of defects (*e.g.*, color centers) becomes crucial. A typical 'color center' (Fowler, 1968) in the case of a K^+Cl^- crystal is a missing chlorine atom, which is exchanged by an electron that can be optically excited. The excited electron interacts with the bulk phonons, resulting in a red shift and a strong broadening of the emission lines. Hence upon irradiation with photons the crystal appears to be 'colored'. This effect becomes even more pronounced if additional foreign ions (*e.g.*, Na^+) are inserted into the crystal ('F_A center') and is one of the main reasons for the destruction of optical components by intense laser radiation. In quartz a green afterglow can be observed following irradiation with intense excimer laser light ($\lambda \leq 351$ nm). In lithium fluoride (a typical window material for UV radiation with less than 5% absorption at 222 nm over 2 mm thickness) the F-band absorption is maximum at 5.1 eV, which is close to the wavelength of the KrF-excimer laser. The excitation is followed by a relaxation of the center due to strong vibronic coupling with the lattice, which leads to a relaxed state about 100 to 200 meV below the conduction band. This state is de-excited by radiation in the blue (450 nm) with a time constant of μs.

At the position of the color center the absorption cross section strongly increases, the probability for multiphoton processes is enhanced compared with a perfect insulator, and the resulting strong heating leads to material destruction ('laser damage'). Also, the relaxation of the lattice around the center is the first stage in the generation of a photoacoustic shock in the crystal.

For a two-photon absorption process Eq. (5.2) has to be completed by an additional term:

$$\frac{dI}{dz} = -\alpha I - \beta I^2, \qquad (5.7)$$

where β is the two-photon absorption coefficient [cm/W]. Calorimetrically measured values of linear and nonlinear absorption coefficients for several nominally transparent UV window materials are compiled in Fig. 114. The laser source was an excimer laser with a pulse length of 25 ns.

In *metals* free electrons in the conduction band are excited, the characteristic frequency of which is the bulk plasmon frequency[2]

$$\omega_p = \sqrt{\frac{N_e e^2}{m_e \epsilon_0}}, \qquad (5.8)$$

where e denotes the electrons charge and m_e its mass. The plasmon frequency increases with increasing density N_e of the conduction band electrons: it is the generated space charge that drives the oscillation.

---

[2] The quantum mechanical analog of the plasma oscillations is called 'plasmon'.

5.1. HEATING AND MELTING

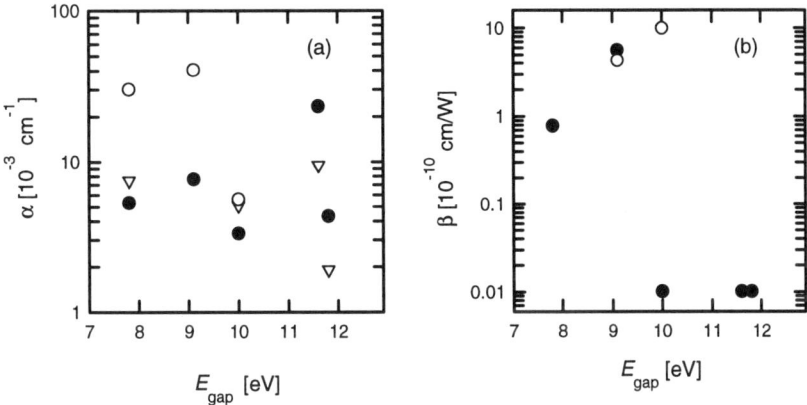

Figure 114 Measured linear (a) and nonlinear (two-photon, (b)) absorption coefficients (Eva and Mann, 1996) for several optical lens materials as a function of band gap energy and for three photon energies: 3.5 eV (λ=351 nm, $\triangledown$); 5 eV (λ=248 nm, •); 6.4 eV (λ=193 nm, ○). The band gap energies of 7.8 eV, 9.1 eV, 10 eV, 11.6 eV and 11.8 eV correspond to fused silica, BaF$_2$, CaF$_2$, LiF and MgF$_2$, respectively.

Note that light of frequency $\omega \geq \omega_p$ is not reflected by the metal since the real part of the index of refraction vanishes at ω_p (cf. Eq. (3.14)). Hence the plasmon frequency ω_p is the limiting frequency above which the metal becomes transparent. For smaller frequencies, on the other hand, both α and R can adopt large values (the latter one close to unity).

Another important limiting frequency (ω_e) results from the collision time of excited electrons with the lattice atoms, τ_e. Typical values of τ_e between 10^{-12} s and 10^{-14} s correspond to wavelengths in the range $\lambda_e \approx 3 - 300$ μm. If the frequency of the exciting light source is small compared with ω_p, but large compared to ω_e (i.e., for wavelengths below 3 μm), then one might write in Drude approximation (free electron metal, Eq. (3.14)) (Abelés, 1972)

$$\alpha \simeq \frac{2\omega_p}{c}. \tag{5.9}$$

As a consequence, the absorption coefficient for ideal metals in the visible and near IR spectral range is independent of wavelength.[3] For silver with a calculated plasmon energy of 8.8 eV (λ_p = 141 nm) one obtains α = 9×10^5 cm^{-1}. The condition $\omega \leq \omega_p$ for large absorptivity is fulfilled for nearly all metals, especially for irradiation with Nd:YAG (λ=1.064 μm) and CO_2 lasers (λ=10.6 μm).

However, those estimates are valid only for bulk material. At the surface the plasmon frequency is smaller by a factor of $\sqrt{2}$ compared with the bulk plasmon frequency (Chapter 1, Section 1.2.2). If the thickness of the investigated material is comparable to the penetration depth of the light intensity, which is given by $c/2\omega n'' = 1/\alpha$ (hence of the order of a few ten nm), i.e., in the case of thin metal films or clusters, then the *surface plasmon frequency* is the limiting frequency.

With increasing temperature (for example as a consequence of absorbed laser power) the Fermi edge is smeared out, the conductivity of the metal is decreasing and the absorbance $\tilde{A} = 1 - R$ is increasing. As a result more laser power is absorbed, the temperature increases further, $\tilde{A}$ increases, and so on. If one takes into account processes that release heat (especially radiation, a process the importance of which is increasing with the fourth power of temperature), then one finds that the absorptivity increases linearly with temperature below the melting temperature of the metal. The absolute value of absorptivity of course depends on the wavelength of the irradiating light. Above the phase transition to melting $\tilde{A}$ increases rapidly and is independent of wavelength. Hence the boiling point is reached quickly and part of the material begins to evaporate. The resulting cloud of vapor and plasma absorbs most of the laser light; in this regime of 'anomalous absorption' the absorbance depends strongly on irradiance and increases only weakly with increasing substrate temperature.

In the course of laser–matter interaction usually more heat is generated than carried away. This heat is strongly localized. Therefore even materials with high thermal conductivity (*e.g.*, copper) or low absorptivity (*e.g.*, diamond) can be annealed inside the focal spot of a laser without strongly affecting the surrounding material.

The subsequent redistribution of the motion of the excited particles via elastic collisions and of the energy via inelastic collisions between particles or particles and lattice results in a statistical equilibration of energy over the entire heat bath of the solid (except that energy that is lost by non-thermal processes, for instance desorption of particles). The time-scale for elastic collisions is short compared with the length of typical laser pulses (nanoseconds). For inelastic collisions the relevant time-scales strongly depend on the irradiated substance and can vary between hundreds of femtoseconds (metals) and microseconds (insulators). The induced heat finally is released

---

[3] In the visible spectral range some metals have resonances in the absorption curve due to interband transitions, i.e., due to their deviation from the free electron behavior.

5.1. HEATING AND MELTING

from the localized focal point of the laser with a time constant that is usually significantly larger compared even with the duration of a nanosecond laser pulse.

If the primary relaxation process proceeds via the generation of heat, then the release of the heat flow q/A (with respect to the area A) might proceed via *materials diffusion, radiation, convection* or *heat transfer*.

Materials diffusion

The characteristic time for materials diffusion is usually long (Chapter 2, Section 2.4) compared with the temporal period for absorption of energy, which is mainly given by the laser pulse length. Therefore materials diffusion plays only a minor role. Laser treatment induces, at least in the course of the pulse duration, 'islands' with properties that largely differ from those of the surroundings.

Radiation

Heat flux q per area A of the heated substrate (temperature T) to the surroundings (temperature T_0) is for a partially absorbing ('grey') body according to the Stefan–Boltzmann law given by

$$\frac{q}{A} = \epsilon^* \sigma_B (T^4 - T_0^4), \tag{5.10}$$

where σ_B, the Stefan–Boltzmann constant, is $\sigma_B = 5.6697 \times 10^{-8}$ W/m²K⁴ and $\epsilon^* = P/P_{bb}$ means the emissivity of the substrate[4] with P the emitted power and P_{bb} the emitted power of an ideal black body ($\alpha=1$) of the same temperature (Özişik, 1985).

Heat transfer via radiation becomes important especially at high temperatures since the absolute temperature enters to fourth power.

Convection

Under certain circumstances the release of heat via convection might become an important relaxation channel. Those circumstances include very high temperatures, where a fluid phase emerges, materials treatment which involves removing ablated material with a beam of gas or materials treatment in air. The energy flux is proportional to the temperature difference between solid, T_s, and liquid or gas, T_{fl},

---

[4] Note that the emissivity depends on the angle of the radiation with respect to normal incidence. For a metal the generation of an image charge suppresses radiation from a 'parallel dipole' which would emit along the surface normal, and thus the emitted radiation is directed mainly along the surface plane. For an insulator no such restriction exists and thus the emitted radiation obeys a cosine law along the surface normal due to geometrical reasons.

$$\frac{q}{A} = h(T_s - T_{fl}), \tag{5.11}$$

where h is the heat transport coefficient for conduction. In the case of surface experiments performed in vacuum ($p_0 \leq 10$ mbar) below the melting temperature convection has no significant influence.

Heat transfer via conduction

In most cases this is the primary energy transfer mechanism. The heat flux is proportional to the temperature gradient $\frac{\partial T}{\partial n}$, which is normal to the plane along which the flux proceeds and is directed from high to low temperatures (Fourier's law) :

$$\frac{q}{A} = -K\frac{\partial T}{\partial n}. \tag{5.12}$$

The thermal conductivity K in general depends on the temperature of the substrate. A useful compendium of values is provided in (Touloukian and Ho, 1973). Below the Debye temperature[5] Θ_{Debye} it is determined for metals by electron transport. Metals with a low number of defects and at low temperatures will show better conductivity since the resistance by lattice vibrations becomes smaller. Above the Debye temperature K stays approximately constant since thermal and electrical conductivity depend directly on each other (Wiedemann–Franz law) :

$$K(T) = \text{const.} \cdot \sigma(T) \cdot T \tag{5.13}$$

and the electrical conductivity σ is proportional to Θ_{Debye}/T. The proportionality constant is about $3(k_B/e)^2$. For insulators the phonon–phonon scattering rate dominates the thermal conductivity. This rate decreases exponentially with increasing temperature,[6] corresponding to the exponential decrease in the mean free path of the phonons, which eventually becomes a $1/T$ dependence. In semiconductors a $1/T$ dependence is also observed in many cases. For example, in the case of silicon one finds for temperatures above 1000 K (Nissim et al., 1980):

$$K(T) = \frac{299\ K}{T - 99\ K} \quad [\text{Wcm}^{-1}\text{K}^{-1}]. \tag{5.14}$$

For a thorough calculation of heat flow through the material an energy rate equation has to be constructed which determines the distribution of

---

[5] The Debye temperature defines the 'softness' of a solid. In solids with low Debye temperatures phonons might be excited already at low temperatures, hence the solid is 'soft'. For example, the Debye temperature of the 'soft metal' sodium is 150 K, that of the rigid insulator lithium fluoride is 700 K.

[6] At very low temperatures the T^3 dependence of the specific heat c_v leads to a subsequent decrease of the thermal conductivity.

5.1. HEATING AND MELTING

temperatures within the investigated area as a function of time and space. For a known distribution of temperatures the temperature gradients might be evaluated which determine the rate of heat transfer.

Let us assume a homogeneous material and temperature independent constants K and specific heat c_p. Then the temporal change of internal energy U within a test volume $\Delta x \Delta y \Delta z$ is determined by the difference between incoming and outgoing amounts of heat, $q_{x1} - q_{x2}$, as well as the transformation of energy inside the volume, induced for example by the laser irradiation (cf. Fig. 115).

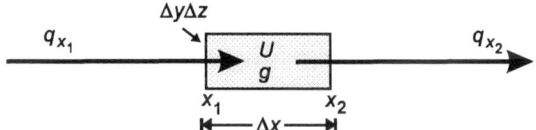

Figure 115 Incoming and outgoing flux of heat into and from a test volume with internal energy U, which is heated by a laser heat source g.

In the x-direction the energy rate equation reads

$$q_{x1} - q_{x2} + g \cdot \Delta x \Delta y \Delta z = \frac{\partial U}{\partial t}, \qquad (5.15)$$

with $g = g(x, y, z; t)$ being the external heat source density [W/m³]. For a Gaussian laser beam, which is absorbed in a homogeneous medium, g might be written as

$$g = I(t)(1 - R)\alpha \exp(-\alpha z) \qquad (5.16)$$

with the intensity distribution

$$I(t) = I_0(t) \exp\left(-\frac{r^2}{w^2}\right). \qquad (5.17)$$

The Gaussian radius w determines that point where the intensity has been lowered to $1/e$ of the initial intensity, and $r = \sqrt{x^2 + y^2}$. The temporal change in internal energy is via the heat capacity

$$C = \rho \cdot c_p \cdot V \qquad (5.18)$$

(with density ρ of the material) proportional to the temporal change in temperature:

$$\frac{\partial U}{\partial t} = \rho c_p (\Delta x \Delta y \Delta z) \frac{\partial T}{\partial t}. \tag{5.19}$$

The incoming amount of heat in the x-direction is given by Fourier's law:

$$q_{x1} = -K \Delta y \Delta z \left(\frac{\partial T}{\partial x} \right)_{x1} \tag{5.20}$$

and q_{x2}, respectively. A Taylor expansion of the temperature gradient around x_1 and x_2 results following insertion in Eq.5.15 in the fundamental differential equation for heat conduction

$$\nabla^2 T - \frac{1}{\kappa} \frac{\partial T}{\partial t} = -\frac{g}{K} \tag{5.21}$$

with the thermal diffusivity κ of the material,

$$\kappa = \frac{K}{\rho c_p} \quad [\text{m}^2/\text{s}]. \tag{5.22}$$

The thermal diffusivity defines a characteristic length (Jost, 1960)

$$L = 2\sqrt{\kappa t} \tag{5.23}$$

(thermal diffusion length) over which temperature differences are equalized via heat conduction within time t. The corresponding time constant usually is much larger compared with typical electronic relaxation times. Possible exceptions are electronic multiple scattering processes, which might have comparable time constants.

The ratio between diffusion length L and optical absorption depth α^{-1} determines the temperature profile in the substrate in the z-direction, along which the laser irradiates the substrate. For $\alpha^{-1} \ll L$ (a surface source) the absorbed laser power $(1-R)P_L t_p = P_a t_p$ is used to heat a layer of thickness L. Since the change of internal energy of the substrate, ΔU, results directly from the absorbed laser power $P_a t_p$ and is proportional to the change of temperature ΔT along the absorbed volume V, one immediately obtains for the average increase in temperature in the heated layer

$$\Delta T \approx \frac{P_a t_p}{c_p \rho V}. \tag{5.24}$$

For laser beams with Gaussian radii much larger than the diffusion length the volume that the heat wave experiences during the laser pulse is determined by the radius w of the laser:

$$V \simeq \pi w^2 L. \tag{5.25}$$

5.1. HEATING AND MELTING

In the opposite case a half sphere with the radius of the diffusion length is heated:

$$V \simeq \frac{2}{3}\pi L^3. \tag{5.26}$$

The time that the heated material needs for cooling equals approximately the heating time t_p.

For $\alpha^{-1} \gg L$ light absorption results in an exponentially decaying temperature profile with characteristic length α^{-1}. Hence the increase in temperature is

$$\Delta T \approx \frac{P_a t_p \alpha \exp(-\alpha z)}{\pi c_p \rho w^2} \tag{5.27}$$

and the cooling time is (heat conduction into the material along the absorption depth, thus $\alpha^{-1} = L$) about α^{-2}/κ.

In order to obtain analytical solutions of the heat equation (without radiation and convection) we assume in addition to K, c_p and κ being independent of temperature

- the source term to be given by a continuous laser with Gaussian intensity profile,
- the sample to be laterally infinitely extended; and
- $\Delta z \gg L$.

These assumptions are usually justified for excitations in metals and interband excitations in insulators.

At the surface ($z=0$) and within the laser focus ($r=0$) the surface temperature increases during laser irradiation by

$$\Delta T(0,0,t) - \frac{I^*}{\sqrt{\pi}}\arctan\left(\frac{L}{w}\right). \tag{5.28}$$

Here, we define

$$I^* = \frac{\tilde{A} I_0}{\pi w K} \tag{5.29}$$

as the ratio between the absorbed intensity $\tilde{A} I_0$ and the product of Gaussian radius and thermal conductivity. In the case of equilibrium between irradiating laser intensity and heat release (for $t \to \infty$) one finds, since $\arctan \to \pi/2$,

$$\Delta T(0,0,\infty) = I^*\sqrt{\frac{\pi}{4}} \simeq 0.9 \cdot I^*. \tag{5.30}$$

Ninety percent of this equilibrium value is reached after a time of $t \approx 10 w^2/\kappa$.

The radial temperature dependence at the surface ($z=0$) is determined by the electromagnetic field strength in the Gaussian laser radius, which decays exponentially in r :

$$\Delta T(r, 0, \infty) = I^* \sqrt{\frac{\pi}{4}} \exp\left(-\frac{r^2}{2w^2}\right) J_0 \left(\frac{r^2}{2w^2}\right), \qquad (5.31)$$

where J_0 represents the Bessel function of zeroth order.

Fig. 116 shows the numerically calculated, normalized dependence of laser-induced substrate heating as a function of depth z (left-hand side) and time (right-hand side).

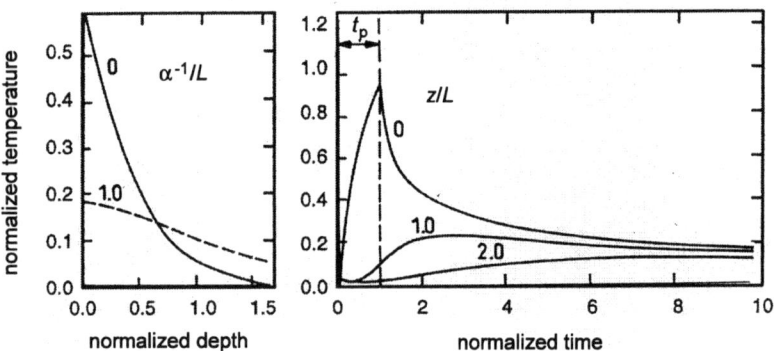

Figure 116 Calculated and normalized temperature TK/I of a laser-irradiated sample as a function of depth z/L (normalized to the diffusion length, left-hand side) and time t/t_p (normalized to the laser pulse length t_p) for different absorption depths α^{-1}/L and substrate depths z/L. Reprinted with permission from (Allmen, 1995). Copyright 1986 Springer-Verlag.

For the predictions to be independent of material parameters and absorbed laser intensity the substrate temperature T has been normalized to the ratio between thermal conductivity K and laser irradiance I. The depth z has been normalized to the diffusion length L and the time t to the pulse length t_p. On the left-hand side two depth-profiles of the temperature are shown for the case of surface absorption ($\alpha^{-1}=0$) and for the case of an absorption depth that equals twice the diffusion length. In the case of the deeper absorption the temperature finally reached is significantly lower and the curve is very shallow. Its shape follows mainly the absorption profile. The right-hand side of Fig. 116 shows the temporal evolution for different normalized depths following irradiation with a laser pulse of length t_p. At the surface the temperature is increasing until the laser irradation stops. In the following it decreases instantaneously. With increasing depth inside the

5.1. HEATING AND MELTING

substrate the decrease in temperature is delayed and the maxima are less strongly pronounced.

If the losses via heat conduction are not too strong, then the surface temperature will reach the melting temperature after a time t_s, which might be calculated by the use of Eq. (5.28). For mica this temperature is $T_{\text{melt}} \simeq 1300$ K. If one irradiates the surface with a cw laser of power 10 W and radius 100 μm (25 kW/cm^2), then this temperature is reached after a few ten microseconds.

Subsequently the melting zone, which has been generated at the surface, travels with a velocity

$$v_S \propto \frac{A \cdot I_0}{Q_s} \tag{5.32}$$

into the sample. Here, Q_s denotes the heat of melting. If one does not blow out the molten material with a gas beam (which is done usually in commercial materials treatment machines), then the surface temperature reaches the boiling point of the material and an evaporation front is generated, which travels into the sample too. The velocity of this layer

$$v_D \propto \frac{A \cdot I_0}{(Q_s + Q_v)} \tag{5.33}$$

rapidly reaches, with increasing laser intensity, its limiting value, which is given by the velocity of sound of the material. Here, Q_v is the heat of evaporation. For most materials this limiting velocity (≈ 10 km/s) is correlated with a saturation irradiance of 10^8 W/cm^2.

Melting and evaporation processes usually cannot be described by analytical solutions of the heat transfer equation but require numerical calculations. However, we note that numerical solutions even for temperatures below the melting point can become important if (i) in addition to heat conduction radiation and convection processes are to be taken into account, (ii) the dimensions of the sample are small compared with characteristic thermal length scales or (iii) the material parameters exhibit significant dependences on temperature within the range of temperatures relevant for the investigated process. As seen from Eqs. (5.10) – (5.12), heat conduction and convection depend linearly on temperature differences, whereas the radiation shows a dependence to the fourth power. If one takes into account that K, κ and ϵ depend on temperature and spatial position and that the irradiated material is infinitely extended, then one obtains a complex system of nonlinear, coupled differential equations instead of Eq. (5.21).

The most widespread method for the numerical solution of this problem is the method of finite differences (Jaluria and Torrance, 1986). Here, one discretizes the sample into small, mutually independent areas and calculates the heat exchange between these areas. The different heat transfer processes are, to first order, taken into account additively in order to determine the final temperature. After obtaining values of temperature for all areas, the differences in temperature can be used to calculate the heat fluxes between the areas and thus new temperatures. This procedure is iterated until the preselected temporal range has been covered.

At the beginning the surface is divided by a rough mesh in order to estimate how far the heat is transferred over the total investigated temporal range. The partial differentiations ∂z and ∂t in Eq. (5.21) are exchanged by the differential quotients and values are calculated. In the following a finer mesh is used only within the areas of largest gradients.

If during the numerical calculation the surface temperature T reaches the melting temperature T_s, then this can be taken into account approximately by the introduction of an effective specific heat,

$$c_p(T) = c_{p0} + \frac{\Delta H}{\Delta T_s}. \tag{5.34}$$

ΔH stands for the latent melting heat and $\Delta T_s = T - T_s$. If one reaches the evaporation temperature at a certain point of the mesh, the temperature is set to be equal to that of the neighboring point. In the case of evaporation part of the laser power is absorbed by the resulting plasma. The corresponding losses can be taken into account by the use of an effective irradiance

$$I_{x,y} = I_0 \exp(-\beta \Delta z). \tag{5.35}$$

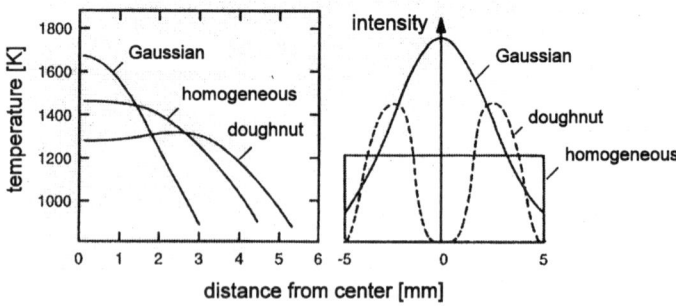

Figure 117 Calculated surface temperatures for different laser beam profiles of equal average power 3 kW, with $\bar{A}$=0.8 and beam diameter 10 mm. On the right-hand side the corresponding intensity distributions are plotted. Reprinted with permission from (Steen, 1991). Copyright 1991 Springer-Verlag.

Figure 117 shows the result of a numerical simulation of the influence of

5.1. HEATING AND MELTING

the laser mode profile on the distribution of surface temperatures (Steen, 1991). The calculations have been performed as a function of distance from the center of the laser beam. For a Gaussian mode ($TEM_{0,0}$) the temperature decreases nearly linearly as a function of distance from the center of the beam. Note that without including heat release one would expect an exponential distance dependence. The reason for this behavior is that for regions of higher temperature the heat release is significantly stronger compared with the regions of low temperature. For the square profile a significantly wider range of uniformly high temperatures is obtained. For laser-material treatment this is obviously the most advantageous case. However, interestingly the doughnut mode also results in a wide range of nearly constant temperatures. The doughnut mode is a combination of $TEM_{1,0}$ and $TEM_{0,1}$ modes, which in the temporal average forms a ring-shaped distribution of laser intensity. Such a mode can be accidentally generated if one tries to enforce a $TEM_{0,0}$ mode in the resonator by introducing an aperture with a slightly too large diameter. We note that with respect to materials treatment the Gaussian fundamental mode provides an unfortuituous distribution of laser energy.

A direct comparison of the results of numerical simulations with experimental findings is most easily performed for a simple system. Here, we show results for a thin mica plate (thickness 400 μm), which is irradiated in vacuum by an Ar^+ laser (absorbed irradiance 40 W/cm^2). The irradiance is low enough that melting is avoided. Since convection can be neglected in vacuum, only conduction and radiation losses have to be taken into account in the calculation.

Figure 118 is a top-view of the crystal with laser irradiation at (x=11 mm, y=13 mm). The solid lines are calculated isothermes of laser-induced increase in temperature ΔT in increments of 2 K (innermost line at ΔT=20 K, next line ΔT=18 K, etc.). In Fig. 118a results for a spherical piece of mica are shown, which has been irradiated in the center. In Fig. 118b similar results for a quadratic crystal are shown, which has been irradiated far from the center. Obviously, irradiation close to the border of the crystal results in significantly different distributions of surface temperature as compared with irradiation at the center.

In order to compare the calculated with the true distribution of temperatures on the surface, the surface temperature has to be determined in vacuum. The most straightforward way to do this is to apply a thermocouple and to measure the contact voltage, which depends on temperature. However, both the absolute increase in temperature as well as the temporal behavior of the thermocouple are affected by the contact pressure between thermocouple and surface, which is not known very well. A possible way to avoid those problems is to adsorb a thin, conducting layer on the surface, the resistance of which can be determined spatially resolved as a function of surface temperature (Zenobi et al., 1988; Schreck et al., 1993). Similarly, one might use standard laser calorimetric techniques (Ahrens et al., 1973; Bubenzer

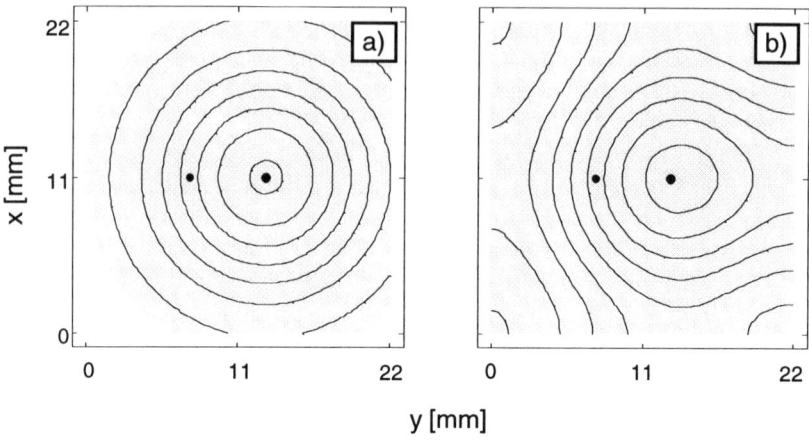

Figure 118 Two samples of mica (spherical (a) and quadratic (b)), irradiated by an Ar$^+$ laser at (x=11 mm, y=13 mm). Calculated isotherms are shown. The temporal evolution of temperature at the point which is located at the left-hand side of the laser irradiation point (quadratic sample) is shown in Fig. 119.

and Koidl, 1984; Sahoo and Apparao, 1992). Precise results can also be obtained by adsorbing dye containing polymers on the surface, which possess a temperature-dependent fluorescence yield (Romana et al., 1989). If the film consists of islands, which are not connected to each other, then the thermal properties of the probe film can be neglected and the determination of the change in temperature-dependent, characteristic properties such as desorption rate (Balzer and Rubahn, 1994) or atomic beam scattering intensity (Balzer et al., 1998b) results in precise values of surface temperature. Finally, all-optical techniques such as photothermal (Jackson et al., 1981; Welsch and Ristau, 1995) or photoacoustic beam deflection (Fig.142) might also lead to accurate values of surface temperature changes.

Fig. 119 shows measured and calculated temporal evolution of the heat wave, which proceeds along the mica crystal following laser irradiation. The laser has irradiated the mica sample for 10 s and has been blocked thereafter (vertical line). At a distance of 5.5 mm from the irradiation point it takes a few seconds until the maximum of the temperature rise appears, followed by a slow temperature decrease. Calculation (solid grey line) and measurement agree well upon usage of the correct values of K and κ.

The maximum increase in temperature ΔT_{max} is expected to decrease

5.1. HEATING AND MELTING

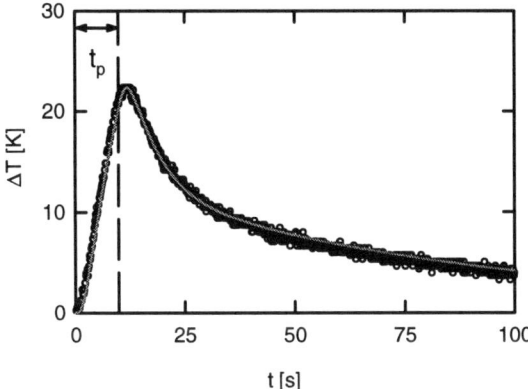

Figure 119 Comparison between measured (symbols) and calculated (grey line) evolution of temperature increase of a point on a mica surface (room temperature) following irradiation by an Ar$^+$ laser of pulse length t_p at a distance of 5.5 mm.

with increasing distance between laser and temperature sensor. This effect is demonstrated in Fig. 120. There the position of the laser has been moved in the direction of the crystal border while the position of the temperature sensor has been fixed (r = distance between laser and temperature sensor). The finite size of the mica crystal results at a laser position near the border (r_{border}=15 mm) in back pressure of heat. Therefore the heating increases in the opposite direction as compared with an infinitely extended crystal. Both calculation and experiment show this effect. The final decrease of temperature change below the value for the infinitely extended crystal (cf. Fig. 120) results from the fact that a fraction of the laser beam no longer hits the crystal and thus takes not part in the heating process.

The reliability of calculated values of the laser-induced temperature rise can be nicely confirmed via a truly nondestructive method, namely helium atom scattering (HAS) (Balzer et al., 1998b) (Fig. 28). The surface temperature dependence of the intensity I_{HAS} of a specularly reflected helium atom beam is governed by the Debye–Waller exponent,

$$I_{\text{HAS}} \propto \exp\left(-\frac{3\hbar^2 \mathbf{k}^2 T_s}{k_B \Theta_D^2 m_s}\right) \quad (5.36)$$

with scattering vector $\vec{k} = \vec{k}_f - \vec{k}_i$, Debye temperature Θ_D and the mass of

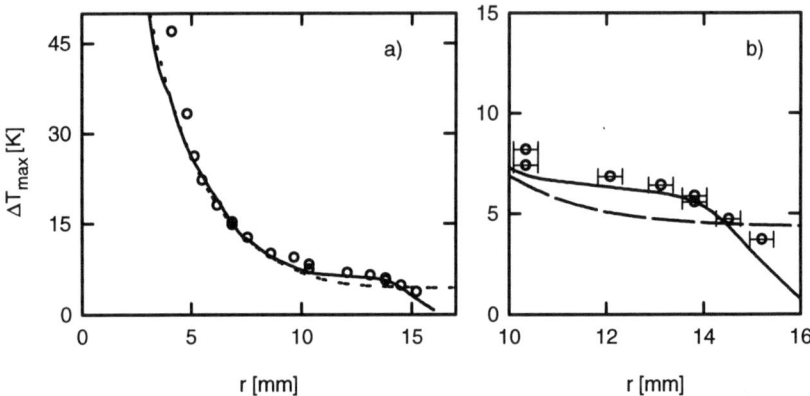

Figure 120 (a) Influence of the finite size of a mica crystal on the maximum laser-induced temperature increase. The dashed curve would result for an infinitely extended crystal. (b) Enlargement of the data obtained in the edge region of the sample. Reprinted from *Chem.Phys.Lett.*, (Balzer and Rubahn, 1995b), Copyright 1995, with permission from Elsevier Science.

the substrate atom m_s. Determining the effective Debye temperature from the measured dependence of specular intensity on substrate temperature by thermally heating the substrate (Fig. 121a) enables one to calculate laser-induced temperature changes (Fig. 121c) from measured laser-induced changes of specular reflectivity (Fig. 121b). The straight line in (Fig. 121c) is from a calculation using the above-mentioned analytical formulas and the thermal properties of the substrate. The good agreement between experiment and theory indicates that for the presented example laser heating proceeds exclusively via the substrate.

Industrial applications of laser heating (Bergmann et al., 1995) elucidate especially the possibiliy of obtaining at well localized spatial points, high temperatures (due to the high coupling efficiencies and large energy densities) and high cooling rates. This allows one to improve the tribological properties of the materials, the fatigue strength and the corrosion resistance. If the heating is limited to below the phase transition temperature, then the material might be *formed* well defined. Above this temperature *hardening* occurs, e.g., in steels with high carbon contents. Further applications include *LCVD* (laser chemical vapor deposition) or *magnetic domain control, MDC.*

5.1. HEATING AND MELTING

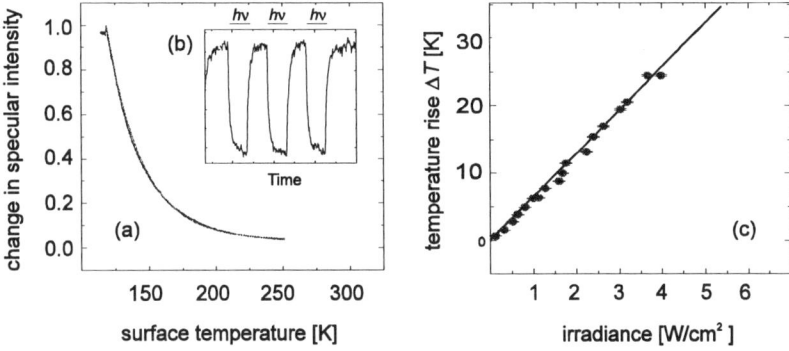

Figure 121 (a) Surface temperature dependence of the specularly scattered He intensity from an alkali-covered insulator surface. The fit line corresponds to a Debye temperature of 110 K. (b) Decrease of specularly scattered He intensity as an Ar^+ laser (1 W/cm^2) irradiates the surface. (c) Comparison of laser-induced change in surface temperature as deduced from the HAS data (filled circles) with theoretical predictions (solid line).

In laser-CVD reactive gases are blown at the surface, which is heated at a well-defined location with a laser. Since the reaction rate depends exponentially on temperature according to an Arrhenius law ($k(T) = k_0 \exp(-E_A/k_BT)$ with E_A the activation energy), reaction products are deposited at the point of laser irradiation. If one moves the focus of the laser away from the surface, three-dimensional structures might be generated ('stereolithography', cf. Chapter 6, Section 6.6).

In MDC large magnetic domains are divided into smaller units. Since the current-losses in transformers are induced by the movement of domain borders in AC-fields, these kinds of losses are reduced by laser-treatment.

From an historical point of view the primary goal of laser materials treatment was material hardening in order to minimize aging phenomena. An improved fatigue life as compared with induction hardening has indeed been reported (Mazumder, 1983). The hardening is obtained by thermal phase changes in the crystal structure of the material; see below. Following heating of the material (in most cases steel) the phase transitions that occur under thermal equilibrium conditions are avoided during the cooling phase and a new structure with improved hardness is formed. Thus the usefulness of laser treatment relies again on the extremely high cooling rates.

Fig. 122 shows a typical apparatus. The laser beam (usually from a high power CO_2 laser) is directed via a system of optical elements for beam

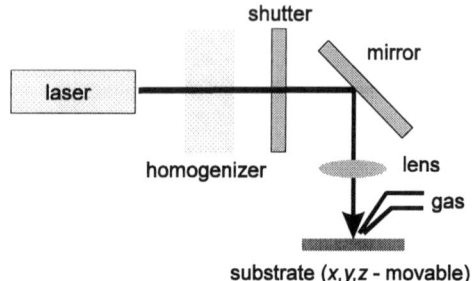

Figure 122 Set-up for industrial laser-heat treatment of materials.

homogenization under inert gas onto the sample, which is mounted on a computer-controlled working table. Reflection losses are minimized by either roughening of the substrate, oxidizing it or coating it with an absorbing layer. Another possibility is to irradiate the material at the Brewster angle ($\alpha \approx 8°$ for metals) with light that is polarized parallel with respect to the plane of incidence. If p-polarized light hits a metal, then dipole radiation perpendicular to the direction of oscillation of the electromagnetic field vector is induced. If the angle of incidence is chosen such that the direction of the reflected beam is perpendicular to the beam that is diffracted into the material, then in the direction of the refracted beam no light is reflected and all light is absorbed. This is the case for $\tan \alpha = \tilde{n}$. This ideal case cannot be obtained for a real system. However, another increase in the amount of absorbed energy can be obtained by irradiating the material against the direction of movement. In that case the Poynting vector of adsorbed energy is directed *opposite* to the direction of movement.

In order to start the phase transformation, *e.g.*, in a steel with a high amount of carbon ($\geq 0.25\%$), the laser-induced increase in temperature has to overcome the transformation temperature, and also has to be restricted to below the melting temperature. Heating of the steel to 1000 K (the equilibrium eutectoid or 'critical' temperature T_c) transforms the pearlite colonies (the α-structure which consists of a granular structure of ferrite and zementite Fe_3C) to austenite (the tetragonal γ-structure). Upon further heating carbon migrates from the austenite into the surrounding ferrite and generates new, carbon-rich austenite with an fcc structure.

If one stops the laser irradiation, cooling takes place with a cooling rate of

5.1. HEATING AND MELTING

usually more than 10^4 K/s. At this high cooling rate[7] the austenite transforms into martensite, but the back-diffusion of carbon is impossible. Martensite is a structure with high internal stress and stable dislocations, i.e., a structure that is 'hard' and brittle compared with the initial ferrite/pearlite mixture. The depth of hardening depends on the rate of thermal carbon diffusion and thus on the heating time, whereas the homogeneity of the resulting martensite depends on the cooling rates and the size of the initial ferrite domains. The hardness[8] increases about linearly with increasing percentage of carbon until a limiting value of 0.7% to 0.9% has been reached. Fig. 123 shows typical values of untreated and laser-treated steels (20 CrMo). As seen, the laser treatment increases the hardness of the steel up to a factor of five.

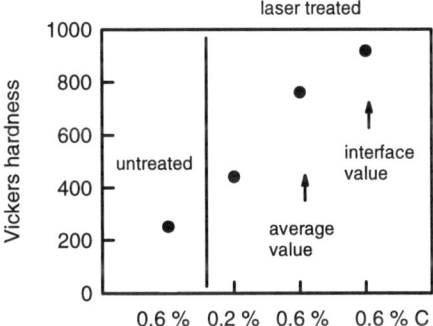

Figure 123 Hardness of untreated (0.6% carbon content) as well as laser-treated steels (values from (Steen, 1991)) for 0.2% carbon content and 0.6% carbon content. In the latter case a typical value ('average') and the maximum value at the interface between laser-treated and untreated material ('interface') are given.

In order to obtain homogeneous layers by laser treatment a thorough estimate of the evolution of the expected hardening processes is advantageous. For that purpose an exact knowledge of the phase diagram and information about the chemical nature of the material as well as a calculation of the thermal heat evolution are necessary. The latter calculation becomes possible one-dimensionally (in the x-direction) and analytically if the region which

---

[7] For example, a cooling rate of 10^2 K/s would lead to the formation of bainite instead of martensite, a much less 'hard' structure.
[8] The Vickers hardness is defined as H_V [N/mm^2] $= 0.189 \times F/D^2$ with F [N] the force that is exerted on the investigated material by a proof weight (a diamond tip, cut at an angle of 136°) and D the average indentation diagonal of the diamond tip on the surface. Typical forces are of the order of 500 N.

is to be hardened is shallow and lies parallel to the surface plane. Usually this is not given due to the inhomogeneous laser beam profile and thus a two-dimensional calculation in the (x,z)-direction is necessary in order to predict the true temperatures. Assuming a homogeneous material density, temperature independent values of K and κ, negligible convection and radiation losses and an infinitely narrow, only in the y-direction extended source, one might deduce 'masterplots' of possible transformation depths vs beam parameters (Ashby and Easterling, 1984). These plots provide depths of hardened steel, normalized to the laser beam radius w, as a function of scan velocity times laser beam radius, which has been normalized to the diffusivity, for different values of absorbed, dimensionless beam energy $q^* = AP_L/w \cdot k_B(T_{trans} - T_0)$. Here, T_{trans} means the transformation temperature of steel. A more simplified approach, which considers hardening to be accomplished at the maximum depth at which the laser-induced temperature reaches T_c, results in even simpler, albeit experimentally reproducible formulas (Bradley, 1988).

More exact calculations of course again ask for numerical methods, *e.g.*, the method of finite differences. The same is the case if the surface temperature exceeds the melting temperature. Then a melt trough is generated (Fig. 124), the cooling rate of which can be calculated only numerically.

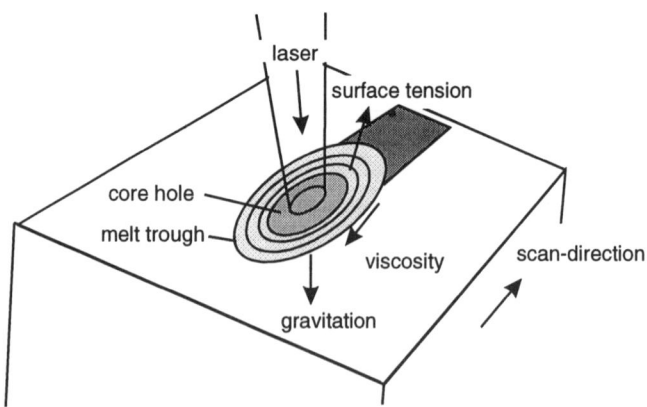

Figure 124 Laser-induced melt trough with forces that act on it. The sample is scanned below the fixed laser. Reprinted with permission from (Steen, 1991). Copyright 1991 Springer-Verlag.

5.1. HEATING AND MELTING

In Fig. 124 the forces are plotted that act on the melt trough. Vapor and photon pressure are generated directly by the laser beam and force the melt to flow in the scan direction. The viscosity of the melt, which increases in the direction of the colder regions, and the gravitational force act in the same direction. The surface tension tends to pull material in the direction of the colder regions. The melt trough is elongated in the scan direction; on the opposite side a 'prow wave' is generated. Obviously, the time-dependent sum of all those forces due to irregularities of the sample material and amplitude fluctuations of the laser light results in vorticies and strong convection phenomena.

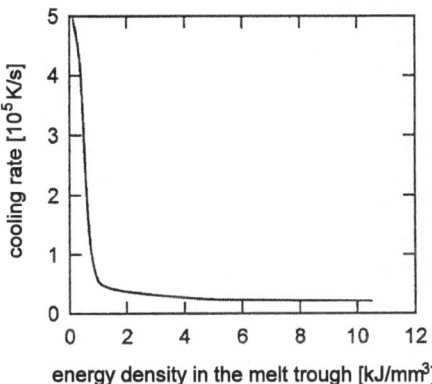

Figure 125 Cooling rate of the center of the melt trough as a function of laser-induced energy density. Reprinted with permission from (Steen, 1991). Copyright 1991 Springer-Verlag.

Fig. 125 shows the calculated effective cooling rate of the center of the melt trough as a function of laser-induced energy density. Only a small range exists, where the cooling rate is high enough that an effective material hardening can be obtained. If the energy density is too high, then the phase transitions take too much time and the material solidifies only slowly; in addition the molten region is much larger than absolutely necessary. This is detrimental to an effective hardening.

Industrial applications of laser melting are, in addition to *hardening*, mainly *surface structuring* and *sealing, surface alloying and nitriding, melt quenching, welding* (Duley, 1998), *laser liquid-phase epitaxy* (LPE) and *laser-CVD*.

Surface-structuring (cf. Chapter 6, Sections 6.2, 6.4, 6.5, 6.7) means, for example, the well-defined generation of holes for optical (compact disc) recording (Holtslag, 1989) or for the fabrication of meshes in stainless steel. The latter goal can be reached via a pulsed Nd:YAG laser, which might generate in the single-shot mode a stainless steel sieve with 100 μm diameter holes. In order to melt iron an energy per volume of 20 J/mm^3 is necessary. If one does not blow out the melt with a gas jet but aims to directly evaporate the material, then 180 J/mm^3 is necessary (evaporation temperature 2735°C). The maximum working speed for this evaporation process is given by the restriction that the resulting holes are supposed to be spherical. This means that the relative velocity between material and laser has to be small compared with the hole diameter per laser pulse length. In the case of a 100 μm hole and a laser pulse length of 100 μs maximum velocities of 1 m/s can be obtained. The minimum obtainable hole diameter is of the order of 5 μm for 0.01 mm thick steel foil and 30 μm for 1 mm thick steel plates. The advantage of laser-generated sieves as compared with mechanically drilled or etched sieves is their smoother structure and their longer lifetime due to the spatially limited region where phase-change processes did occur; hence the stainless steel is not rusting at the edges of the holes.

Surface alloying results in changed thermal, electrical and chemical properties and thus serves mainly to increase the corrosion and wear firmness. Exposure of an aluminum or iron surface to 308 nm excimer laser radiation in a nitrogen atmosphere, for example, results in the generation of a few micrometers thick surface melt and ionization of the nitrogen. The ions penetrate the aluminum liquid and form an homogeneous aluminum nitride layer ('excimer-laser nitriding') with significantly enhanced corrosion resistance. The importance of material transport during irradiation for the nitrogen uptake (and thus the achieved hardness) has been noted recently (Illgner et al., 1998). Lateral material transport is facilitated mainly by the laser-induced pressure p_{tot} at the surface of the melt trough (Fig. 124), which acts as a 'piston' to squeeze material out of the high-pressure region into the cold border, where it condenses. This leads to a ridge around the melt trough in the case of no lateral movement of the substrate. The material's velocity (Illgner et al., 1998)

$$v_{\text{mat}} = \sqrt{\frac{2p_{\text{tot}}}{\rho_0}} \qquad (5.37)$$

relies, besides the density ρ_0 of the liquid, only on the total surface pressure $p_{\text{tot}} = p_{\text{rec}} + p_{\text{plasma}}$. Here, the two most relevant contributions are the recoil pressure of the evaporated material (Bäuerle, 1996b) (assuming that the melting is accompanied by partial evaporation)

$$p_{\text{rec}} \approx \frac{\tilde{A} I_0 \sqrt{\langle v^2 \rangle}}{\Delta H_v + \langle v^2 \rangle} \qquad (5.38)$$

5.1. HEATING AND MELTING

with absorbed irradiance $\tilde{A}I_0$, heat of vaporization ΔH_v and average velocity of species leaving the surface, $<v^2>$, and the plasma pressure, derived from the model of 'Laser Supported Detonation', which is relevant for irradiances above 10^{11} W/cm², (Raizer, 1970)

$$p_{\text{plasma}} \approx 0.3 \cdot I_0^{2/3} \rho_0^{1/3} \tag{5.39}$$

for monatomic species[9] with an initial density ρ_0. Typical plasma pressures for irradiances of 10^{12} W/cm² are of the order of 10^7 Pa, resulting in materials velocities of the order of 100 m/s and removal of layers of the order of a few tens of nanometers per pulse.

Via *melt quenching* one can generate, for example, in gold–titanium films on sapphire substrates metallic glasses due to the extremely high cooling rate following pulsed laser irradiation ($\approx 10^{12}$ K/s for picosecond pulses). The laser liquid-phase epitaxy (LFE) (see Chapter 6, Section 6.1) restores crystalline order in ion-implanted substrates. Since crystal growth occurs directly from the melt and not from an amorphous phase, the number density of defect sites is significantly reduced. Compared with conventional liquid-phase epitaxy, LFE results in a fabrication speed which is about a factor of ten faster.

The most important problems for laser-induced melt hardening compared with conventional transformation hardening is a destruction of the initial coating which served to increase the absorptivity. Also, inside the laser focus a plasma is generated (Section 5.2), which increases the absorption of laser light up to a certain density, but also attenuates the transmission of light to the substrate.

The latter problem can easily be solved by blowing out the generated plasma with a rare gas beam. The destruction of the coating is not significant since the reflectivity of most materials decreases with increasing temperature. Hence following removal of the coating the material itself absorbs enough laser light. The reason for this increase in absorptivity with increasing temperature is the increasing number density of phonons, which couple with the laser-excited electrons.

Due to the high irradiances self-induced optical coupling processes occur, such as:

• Self-focusing. The index of refraction of the medium contains, besides a part which is independent of laser irradiation (linear part), another part which depends on laser intensity (nonlinear part): $n = n_0 + n_2 \cdot I$. If n_2 is positive, then the index of refraction of the medium is highest in the middle of the Gaussian laser beam profile. This spatial variation of the index of refraction acts as a lens which focuses the laser into the medium. With an increase in focusing the change of index of refraction increases, resulting in a further increase in

---

[9] For diatomic species the constant 0.3 has to be exchanged for 0.22.

focusing and so on. A counteracting effect is the widening of the beam due to diffraction off defect sites.
- Thermal or optical generation of free charge carriers in semiconductors, which result in an increase in absorptivity.
- Electron avalanches in insulators, which also result in enhanced absorptivity.
- Generation of macroscopic dips in the substrate, which channel the laser beam.
- Corrugation, melting or crystallization at the surface, which results in a decrease in reflectivity.

These phenomena begin at 'hot spots' in the laser beam, i.e., at positions of highest power density. In order to obtain homogeneous heating of the substrate over the whole focus area of the laser, *homogenization* of the beam profile is necessary. The simplest approach to this is a defocusing lens. More uniform profiles are obtained by a movement of the beam in the surface plane via a set of mirrors, multiply reflecting and expanding kaleidoscopes or by the use of special optics such as axicons or toroidal mirrors. An axicon ('axis image'; a glass cone) (McLeod, 1954; Sochacki et al., 1993), for example, images a point source on its axis of revolution to a range of points along its axis; i.e., it generates a focus line instead of a focus spot. An axicon telescope would be in focus for close-by targets and simultaneously to targets at infinite distance.

If one aims to apply the laser to mask projection (cf. Chapter 6, Section 6.2.2) more homogeneous illumination profiles can be obtained by sending the laser light through arrays of typically 5×5 or 10×10 individual lenses. The lenses expand selected parts of the incoming beam profile to illuminate the whole mask and the array serves to overlay subsequently all the components (Mann and Hopfmüller, 1994).

A principal problem for every method of homogenization is of course that laser materials treatment occurs also on and in the used optical components, resulting in destruction of optical elements inside the hot spots via, for example, color center generation ('laser radiation damage' (Wood, 1986; Stoneham, 1994)). The mechanism of laser damage[10] depends very much on the optical material and, especially, on the question of whether the initial absorption process is linear or nonlinear. Linear absorption might be caused by defect states in the surface region, which grow with increasing number of laser pulses until the accumulated energy is high enough to allow, for example, melting. In the case of nonlinear absorption the increase in energy of conduction band electrons results in an increasing impact ionization rate, i.e., an avalanche of ionized carrier density, which eventually causes damage.

---

[10] A discussion of laser damage phenomena occurring in a variety of different materials can be found in the Proceedings of the 'Boulder Laser Damage Symposia', published by NIST (Gaithersburg) 1969 – 1993.

5.2. PLASMA GENERATION

Also, as suggested by recent pump/probe measurements (Chase, 1994), cracks at the surface might increase the local fields, thus leading to enhanced avalanche ionization. Thus in contrast to volume damage, which is dominated by intrinsic, multiphoton absorption, surface damage often is dominated by defect absorption.

It is thus not surprising that imperfections in the coating of the optical elements (variations in stoichiometry or thickness of the individual layers) can give rise to serious problems. Nonuniformities of the thermal conductivity, induced by thickness variations, result in inhomogeneous heating and thus cracks and defect generation. Those defect sites then have an enhanced absorptivity since the optimized optical properties of the dielectric elements are changed, resulting in even stronger local heating, etc. Typical damage thresholds are of the order of a few J/cm^2. Above 10 J/cm^2 nearly all transmitting optical elements degrade if more than a single pulse is applied, even for wavelengths larger than 200 nm. For high-reflective interference multilayer coatings the damage threshold is of the order of 20 to 25 J/cm^2 at 248 nm.

Recently, improved coatings have emerged for the use with high power excimer lasers, the improvement being the result of 'laser conditioning' (laser heating and recrystallization of the surface structure; see Chapter 6, Section 6.1) (Kozlowski and Thomas, 1994) or the use of dedicated coating materials such as fluorides as well as contamination-free coating deposition techniques (Kaiser, 1996).

5.2 Plasma Generation

If the irradiance exceeds some threshold irradiance I_{thr}, then multiphoton and penning ionization or thermionic emission processes result in a dense cloud of electrons and charge carrierswithin the laser focus. In that case the number of collisions between charge carriers is larger than the number of collisions between lattice and charge carriers: the charge carriers show collective behavior, i.e., they form a *plasma*. The initial extension of this plasma is much shorter than the wavelength of the irradiating light. In the case of ultrashort pulses the electromagnetic (ponderomotoric) forces engage an oscillatory motion of the plasma film, thus leading to sidebands of the reflected light. This higher harmonic generation at surfaces (von der Linde and Rzazewski, 1996; Plaja et al., 1998) needs femtosecond pulses with high contrast ratio and irradiances of about 10^{17} W/cm^2, but in contrast to higher harmonic generation in the gas phase it does not suffer from phase matching requirements. Higher harmonics up to order 35 have been observed, meaning a wavelength of 20 nm.

Following irradiation in most cases the plasma cloud leaves the surface. Experiments with lasers of different pulse lengths (10 ns to 1 ms) and different wavelengths (248 nm to 10.6 μm) have demonstrated that the product of laser irradiance [W/cm^2] and square root of pulse length has to be higher than 4×10^4 Ws$^{1/2}$/cm^2 in order to form a plasma (Sappey and Nogar, 1994). In Fig. 126 a picture of such a plasma is shown, which has been obtained by a CCD camera following irradiation of a mica surface with a KrF excimer laser (λ=248 nm).

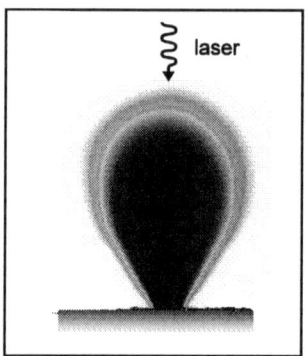

Figure 126 Laser-induced ablation plasma, induced by irradiating a mica surface in ultra high vacuum with an excimer laser (λ=248 nm, τ=20 ns) at a fluence of 3.5 J/cm^2. The picture was obtained using a CCD camera.

The extension and shape of the plasma cloud depend on the fluence. Usually, the laser-generated particles undergo mutual collisions if the density is high enough (if more than half of a monolayer has been desorbed) and thus form a 'Knudsen layer' via gas-dynamic expansion. These collisions result in a narrowing of the angular distribution with increasing fluence (Kelly et al., 1991; Kelly et al., 1992), very similar to the behavior of a nozzle beam expansion which was described in Chapter 1, Section 1.2.

Laser wavelength, pulse length, irradiated material (isolator, metal or semiconductor) and surrounding medium, inside which the plasma has been ignited (*e.g.*, vacuum or air), determine which mechanism dominates and whether a plasma is generated or not.[11] The above factors also influence the primary excitation and decay channels. Basic discussions and review articles can be found in (Fogarassy and Lazare, 1992; Miller, 1994; Miller and Geohegan, 1994; Metev and Veiko, 1994).

A useful equation for an estimate of the material rate m [kg/(s· cm^2)]

---

[11] Ablation might occur significantly below the threshold irradiance that is discussed in this chapter; see Chapter 6, Section 6.2. The ablated products are not necessarily highly excited.

5.2. PLASMA GENERATION

that has been ablated inside the microplasma as a function of the absorbed irradiance I_a [W/cm^2] for a given wavelength λ [μm] is (Sappey and Nogar, 1994)

$$m = 110 \cdot \lambda^{-4/3} \cdot (I_a/10^{14})^{1/3}. \tag{5.40}$$

Within the laser-induced plasma extreme UV (10–100 eV) or soft X-ray radiation (100 eV–2 keV) is generated for irradiances above 10^{11} W/cm^2 as a result of inner shell excitation processes. Since this radiation is very intense and spatially well focused, it is well suited as an irradiation source for X-ray lithography. At wavelengths around 10 nm structures can be written efficiently into polymers with a resolution of far better than 100 nm (Kubiak et al., 1994). In that case the laser plasma source serves as a surrogate for synchrotron radiation. If one increases the irradiance in the range of 10^{18} W/cm^2 and if one uses pulses with femtosecond length, then — as noted above — higher harmonics and high energetic X-ray pulses can be generated, which are of interest as potential sources for X-ray lasers. It has been demonstrated that defined morphological changes of the surface in order to obtain electromagnetic field enhancement effects significantly improves the efficiency and brilliance of the X-ray source (Falcone et al., 1994). This effect is related to field enhancement effects observed in nonlinear optics.

Note, however, that with increasing electron density the plasma frequency also increases (Eq. (5.8)). If the plasma frequency is higher than the laser frequency, then the laser radiation will be effectively screened and the ablation process stops. This screening effect too is material-specific and cannot be described (and avoided) by the use of a general recipe.

6
Advanced Treatment

The availability of intense laser sources with a wide range of possible wavelengths from the ultraviolet to the infrared spectral range as well as pulse lengths as short as a few femtoseconds has contributed significantly to the recent, fast development of microstructuring and microtechnologies (Rai-Choudhury, 1997). Extended discussions can be found in the literature, especially in (Büttgenbach, 1991) (conventional micromechanics) and (Metev and Veiko, 1994) (laser-based methods). Laser-assisted microstructuring was discussed also in Chapter 5, Section 5.1 of the present monograph.

6.1 Laser Annealing

Irradiation of a metal or semiconductor surface with a laser of high power density results in a spatially and temporally localized electron plasma (cf. Chapter 5, Section 5.2), which relaxes via electron–phonon and phonon–phonon coupling to the heat bath of the lattice and might result in melting and evaporation of the substrate. The extreme high heating and cooling rates (10^{15} K/s) can be used to recrystallize amorphous surfaces of semiconductors ('pulsed laser annealing', PLA, $e.g.$, (Kittl et al., 1993)). This technique is, from the viewpoint of technical applications (micro-electronics), of special interest for semiconductors of groups IV and III/V ($e.g$, Si or GaAs), which have been made amorphous in the course of an ion-implantation process. Ion implantation serves to dope the semiconductor with electron acceptors or donors, respectively. Using PLA-assisted methods, single-crystalline surfaces can be obtained which have dopant concentrations of up to 10^{21} cm^{-3}, since in contrast to thermal annealing the high heating rates avoid diffusion of the dopants into the bulk.

The basic physics (Akhmanov et al., 1989) is of primarily thermal nature if pulses with lengths above picoseconds are applied: following excitation of a hot electron plasma the crystal-lattice is heated within picoseconds via electron–phonon coupling. This results in a local melting of the crystal along the path of the penetrating laser light. A transition from the solid to the liquid phase occurs if the average deflection of the lattice atoms from their equilibrium position amounts to a significant fraction x of the size of the unit cell, a^2 ('Lindemann-criterion').[1] This fraction x is significant if its value is of the order of 0.25. The melting temperature that has to be reached by laser irradiation is about (Akhmanov et al., 1989)

$$T_{\text{melt}} = \frac{x}{9\hbar^2} m k_B T_{\text{Debye}}^2 a^2, \qquad (6.1)$$

with atomic mass m and Debye temperature T_{Debye}. In the course of the following, fast cooling (a receding of the melting front towards the surface) epitaxial crystal growth occurs, resulting finally in a single crystalline surface.

In Fig. 127 recombination processes are shown which occur in the course of the generation of a hot electron plasma in the semiconductor following irradiation by an intense laser pulse (irradiance MW/cm^2 to TW/cm^2). Depending on the material, the laser beam penetrates the semiconductor a few tens up to thousands of nanometers. The photon energy is assumed to be larger than the band gap, meaning that electrons can be excited as free carriers from the valence into the conduction band.

The generation rate Y_c for carriers is proportional to the absorbed irradiance, i.e., it decreases exponentially with increasing penetration depth into the material, z, and increases with decreasing length of the laser pulses, τ_p:

$$Y_c \approx \frac{\alpha(1-R)}{h\nu\tau_p} I_L \exp(-\alpha z) = Y_c^0 \exp(-\alpha z). \qquad (6.2)$$

For a fluence $I_L = 100$ mJ/cm^2 from a laser with wavelength 532 nm and pulse length $\tau_p=15$ ns one obtains in the case of silicon (R=0.37 at 300 K, α=1.25·10^4 cm^{-1}) $Y_c^0 = 1.4 \times 10^{29}$ cm^{-3}/s.

A carrier density of $n_c = 10^{19}$ cm^{-3} separates the regime of 'strong excitation' from the regime of 'weak excitation', since the electron–electron collision rate (10^{14} s^{-1} for $n_c = 10^{19}$ cm^{-3}) is higher than the electron–phonon collision rate above this density. The electron–electron collision rate increases in proportion to the carrier density, while the electron–phonon collision rate is nearly independent of carrier density.

In the regime of *strong excitation* intraband relaxation via electron–electron collisions and radiationless Auger recombination processes dominate. In the latter processes an electron and a hole recombine and the released energy is transferred to a third charge carrier. Since three carriers are involved, the total relaxation rate amounts to

---

[1] This is a simplified approach since one has to take into account that the upper few atomic layers melt at a lower temperature as compared with the bulk ('surface melting', cf. Chapter 4, Section 4.5). If the penetration depth of the laser is much larger than about 3 nm, then this effect can be neglected.

6.1. LASER ANNEALING

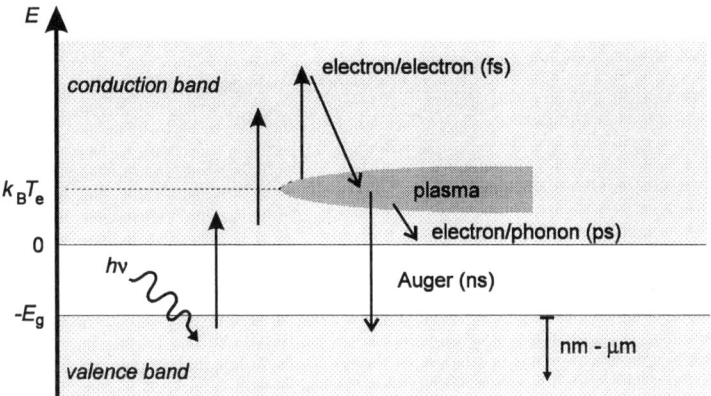

Figure 127 Interband excitation of a semiconductor by a laser pulse and following relaxation processes. Typical time constants are denoted. The penetration depth of the laser light is between 10 nm and some μm, depending on wavelength, angle of incidence and irradiated material.

$$R_c = C \cdot n_c^3. \qquad (6.3)$$

On the other hand, since

$$R_c = n_c/\tau_r \qquad (6.4)$$

the relaxation time is given by

$$\tau_r = 1/(C \cdot n_c^2). \qquad (6.5)$$

Hence it decreases with increasing carrier density. The limit is about 6 ps in the case of silicon ($n_c \approx 10^{21}$ cm^{-3}), since at such high plasma densities the mutual screening effects of the charges limit the maximum possible carrier density. With $C = 4 \times 10^{-31}$ cm^6/s (Akhmanov et al., 1990) (silicon) one obtains for the above discussed example a recombination time of $\tau_r \approx$ 0.5 ns. This is significantly smaller than the pulse length of the laser. Thus a stationary carrier concentration is to be expected.

The fast intraband relaxation results on a femtosecond time-scale in a hot, 'thermalized' plasma with electronic equilibrium temperatures of a few up to ten thousand Kelvin. At the same time the lattice temperature remains close to the initial value, i.e., at room temperature.

In the regime of *weak excitation* with small carrier densities the excited carriers relax mainly into the lattice. Hence transversal and longitudinal optical phonons are excited on characteristic time-scales of several hundred femtoseconds. The direct excitation of acoustic phonons is improbable, but the optical phonons can decay into two energetically lower lying acoustic branches. Here, the characteristic time constants for phonon–phonon scattering events are of the order of a few to a few tens of picoseconds.

If one takes these time constants into account, then for nanosecond laser excitation a relaxation into the heat bath of the lattice occurs during the laser pulse. This kind of 'instantaneous thermalization' justifies a calculation of laser-induced temperature changes via conventional thermodynamic methods (see Chapter 2, Section 2.1). If one excites with ultrashort pulses from femtosecond lasers, then the theoretical description becomes significantly more complex. One has to take into account the collective and nonlinear effects within the generated plasma. The resulting changes in the index of refraction of the absorbing medium may lead to self-focusing, to the generation of optic harmonics (energy loss process) and to spatial and temporal gratings ('surface periodic structures', SPS [2]; Section 6.4), which trigger coherent and incoherent nonlinear optic processes. In all cases absorption and recombination processes depend on the strength of excitation and cannot be included as constant parameters into the calculation.

If the carrier temperature exceeds 10^4 K, then diffusion of the carriers no longer can be neglected compared with the relaxation in optic phonons. This occurs in silicon with a diffusion constant of 100 cm^2 s^{-1} and results within the relaxation time in a smearing out of the carrier distribution over a distance of up to 10^{-4} cm and thus to a significant reduction of the heating rate for the lattice.

6.2 Laser Ablation

Usually one defines the laser-induced removal of substrate material as laser *ablation*, whereas laser *desorption* refers to the removal of adsorbed materials. Compared with conventional photolithographic techniques, laser ablation methods have the huge advantage of being 'self-developing'. Therefore no wet chemistry is necessary in order to remove the irradiated material. In this section selected ablation processes are described, which are induced by ultraviolet, visible and infrared laser radiation. Emphasis is put on polymeric, insulating and metallic materials. Of course, semiconductors can be treated

---

[2] In the simplest case the SPS are induced by interferences between the incoming laser beam and the beam, which has been reflected at surface roughnesses. At the positions of maximum field strength the material liquifies or a plasma is generated.

6.2. LASER ABLATION

equally well[3] and they represent also a very important field of applications (Haglund,Jr and Itoh, 1994; Haglund,Jr, 1996). A more exhaustive discussion of recent developments in the field of laser ablation can be found in (Bäuerle, 1996b).

6.2.1 Polymers

The well-defined ablation of polymers, induced by irradiation with intense UV light, is of great technical interest for a variety of different areas of application such as microelectronics, the packing industry, medicine (Srinivasan, 1986), fundamental biological investigations, etc. In all cases the applications are based on the generation of well-defined, clean structures with dimensions in the micrometer regime. First observations of UV laser-induced ablation date back to 1982 (Srinivasan and Mayne-Banton, 1982), when a PET film (polyethylene terephthalate) ('mylar', polyester) (see Fig. 134) of 250 μm thickness was irradiated with UV light (193 nm) from an excimer laser of pulse length 14 ns and fluence of several hundred mJ/cm^2. Electron microscopy pictures of the irradiated areas (Fig. 128) showed sharp edges between irradiated and nonirradiated areas and no hint of thermal damage to the surroundings. Obviously there was no significant heat transfer into the substrate outside the laser burn spot.

A possible explanation for this phenomenon is that in the 'melt pot' an overheated liquid is generated, which (analogous to rocket exhaust gases) vaporizes instantaneously. If the released energy E_{ex} is converted completely into kinetic energy of the evaporating molecules, then the fragments should have a velocity of

$$v = \sqrt{\frac{2E_{\text{ex}}}{m}}. \qquad (6.6)$$

Here, m is the mass of the molecular fragments and

$$E_{\text{ex}} = E_{\text{ph}} - E_{\text{B}} \qquad (6.7)$$

with photon energy E_{ph} and binding energy E_{B}. For ablation of PMMA at 193 nm (E_{ph}=6.4 eV) one obtains with E_{B}=2.7 eV and m=8.3×10^{-25} kg a velocity of the MMA fragments of 1200 m/s. Hence the ablation process should result in fast fragments with an angular distribution that is aligned normal to the edge hole. However, measurements of the real velocity distribution (Fig. 133) suggest that the energy conversion is not complete.

The measured edge depth amounts to 120 nm per pulse at a fluence

---

[3] An interesting multiwavelength approach to the treatment of semiconductors, which promises to provide a solution to the problem of remaining debris around ablation edges, is discussed in (Zhang et al., 1998).

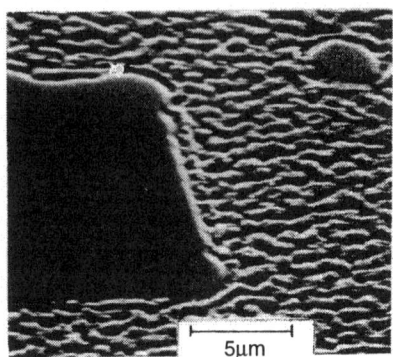

Figure 128 Scanning electron microscopy (SEM) image of a PET film which has been irradiated with an excimer laser at 193 nm and with 110 mJ/cm². The rippled areas have been irradiated by the laser, the dark areas not. Note the sharp edges between irradiated and nonirradiated areas. Reprinted with permission from (Srinivasan and Mayne-Banton, 1982). Copyright 1982 American Institute of Physics.

of 370 mJ/cm². A threshold value of laser fluence, F_L^s, was observed, below which the material was only thermally ablated. Nonthermal ablation with 193 nm light was observed also for other polymers such as PMMA (polymethylen methacrylate), PI (polyimide, 'kapton'), polycarbonate, PTFE (polytetrafluoroethylene, 'Teflon'), etc. although with different values of threshold fluence, ranging between 0.01 and 1 J/cm² for polymers and 0.5 and 2 J/cm² for insulators.

Early investigations of the *wavelength dependence* of the ablation threshold showed that absorption often begins with a critical *absorbed* threshold energy density. The threshold F_L^s is lowered with increasing photon energy; however, since simultaneously the effective absorption coefficient α_{eff} increases, the product $\alpha_{\text{eff}} \cdot F_L^s \approx 5\pm 2 \times 10^4$ J/cm³ (for PI (Brannon et al., 1985)) appeared to remain approximately constant. Of course, this is only the case if thermal diffusion is negligible (i.e., for a surface source), the remaining material is not chemically modified and the absorption obeys Eq.3.4 (Beer's law). In order to account for experimentally observed deviations from this simple rule, more sophisticated theories have been developed, including explicitly thermal diffusion (Furzikov, 1990), screening by the laser-generated plume and bleaching of the chromophores (Schmidt et al., 1998) or a receding heated

6.2. LASER ABLATION

surface layer. Inclusion of thermal diffusion,[4] for example, results in a new scaling of $\sqrt{\alpha_{\text{eff}}} \cdot F_L^S \approx \text{const.}$ (Furzikov, 1990), which is valid for weak absorbers (exponentially depth-dependent heat profile).

In what follows we will discuss some basic ideas of those models.

Photothermal model

A *photothermal* explanation of the observations implies that the absorbed photon energy is transferred directly into excitations of lattice vibrations. A 'hot spot' is generated, which results in thermally activated fragmentation of the material. Then the critical threshold energy fluence corresponds to a critical temperature, which differs for materials with different values of absorbance. This temperature has to be high close to the ablation threshold (≥ 1300 K). However, the thermal fragmentation rate of polymers in this temperature range is not very well known. Hence one can a priori only assume that the irradiated polymers decay within nanoseconds.

An experiment that underlines the importance of this fast thermal decay channel has been discussed in (Dijkkamp et al., 1987). There a thin polymer layer has been adsorbed on a silicon substrate and has been irradiated with 248 nm light. Although the absorption coefficient of silicon is much higher compared with that of the polymer the typical instantaneous ablation process has been observed. It has thus been concluded that the laser power has been absorbed in silicon and has been thermally transferred into the polymer. Numerical simulations of the heat flow resulted in a temperature of 1500 – 1600 K at the interface between silicon and polymer.

The most simple thermal *model* for UV laser ablation (Cain et al., 1992; Cain, 1993) assumes a thermal fragmentation of first order of the polymer; this means that the number n of single bondings in the monomer is decreasing with the rate constant $k(T)$ following absorption of UV photons:

$$\frac{dn}{dt} = -k(T) \cdot n. \tag{6.8}$$

The rate constant depends on temperature $T = T(r,t)$, which is a function of position and time. The thermal rate constant might be described via an Arrhenius-type expression:

$$k(T) = \nu_0 \cdot \exp\left(-\frac{E_A}{k_B T}\right), \tag{6.9}$$

with activation energy E_A and pre-exponential fequency factor ν_0. At time t_e, at which an etch hole has been generated, the fraction of residual bondings is

$$n_f = \exp(-k(T) \cdot t_e). \tag{6.10}$$

---

[4] Thermal diffusion affects also the etch depth as a function of laser fluence (Eq. (6.13)), which grows faster than expected for the case of negligible diffusion.

This fraction might be estimated from the molecular weight of the ablated fragments. For the strong UV absorber PI one finds mainly small fragments following excitation at 248 nm, i.e., $n_f=0.5$. For the weak absorber PMMA, on the other hand, intact monomers dominate, i.e., $n_f=0.99$.

If the radiation has been absorbed following Beer's law, i.e., if linear one-photon absorption with effective absorption coefficient α_{eff} dominates, then the distribution of temperatures in the substrate at the end of the laser pulse can be estimated from

$$T(z) = \frac{\alpha_{\text{eff}} \cdot F_{\text{L}}}{\pi c_p \rho w^2} \cdot \exp(-\alpha_{\text{eff}} z) \qquad (6.11)$$

with $F_{\text{L0}} = P_a \cdot t_p$ (see Eq. (5.27)). Here, one has neglected heat conduction. Implementing this distribution into the expression for the rate constant (Eq. (6.9)), and this into the expression for n_f, Eq. (6.10), one finds

$$\ln\left(\frac{\nu_0 t_e}{\ln(1/n_f)}\right) = \frac{\pi c_p \rho w^2 E_A}{\alpha_{\text{eff}} F_{\text{L}} k_B} \cdot \exp(\alpha_{\text{eff}} z_e), \qquad (6.12)$$

or equivalently for the depth of the etch hole z_e ('etch depth'):

$$z_e = \alpha_{\text{eff}}^{-1} \ln\left(\frac{F_{\text{L}}}{F_{\text{L}}^{\text{S}}}\right) \qquad (6.13)$$

with

$$F_{\text{L}}^{\text{S}} = \alpha_{\text{eff}}^{-1} \frac{\pi c_p \rho w^2 E_A}{k_B \ln\left(\frac{k_0 t_e}{\ln[1/n_f]}\right)}. \qquad (6.14)$$

This logarithmic dependence of etch depth on the laser fluence close to the threshold fluence F_{L}^{S} has been observed experimentally too. Note that this dependence is not a peculiarity of the thermal model, but follows readily from assuming a one-photon absorption process with ablation threshold. Getting over the threshold is necessary since the laser fluence is absorbed in the material with the absorption coefficient α_{eff}[5] and is thus exponentially attenuated with increasing depth z.

The numerical calculation of the etch depth is performed in two steps:

1. The total absorbed energy is calculated as a function of time from the density of excitable molecules ('chromophores') n_{chr} and the absorption cross section σ_1 via $\alpha_{\text{eff}} = \sigma_1 \cdot n_{\text{chr}}$ by assuming that initially the excited state is unoccupied. Typical values are $n_{\text{chr}} = 8.7 \times 10^{21}$ cm^{-3}, $\sigma_1 = 6 \times 10^{-19}$ cm^2, $n_f = 0.5$ for PI at 193 nm and $n_{\text{chr}} = 7.1 \times 10^{21}$ cm^{-3}, $\sigma_1 = 5 \times 10^{-17}$ cm^2, $n_f = 0.99$ for PMMA at 193 nm (Cain et al., 1992). The calculated value of

---

[5] In general, α_{eff} deviates significantly from the temperature-dependent low-intensity absorption coefficient that has been introduced in Eq. (5.2) and has contributions from multiphoton processes and radiation-induced defects.

6.2. LASER ABLATION

internal energy of the polymer is recalculated into a temperature via division by the heat capacity.

2. The thermal decay of the polymer and the diffusion of heat in the substrate are calculated by solving the heat conduction equation under the boundary condition that the polymer is ablated if $n_f \geq n$.

Figure 129 shows etch curves that have been calculated that way for a PI film and different wavelengths. Experimentally the hole depths for irradiation with a given irradiance are determined by irradiating a film of known thickness long enough to drill a hole through the entire film. Counting the number of necessary pulses then allows one to evaluate the etch rate per pulse. One might also determine the pulse depth after a certain number of pulses using a profilometer or from electron microscopy. All those methods have the disadvantage that it is not possible to discriminate between the first and the following laser pulses. This might result in misleading conclusions since for certain polymers and wavelengths incubation pulses greatly change the subsequent ablation behavior. In that cases the first pulse modifies the material constants significantly, thus enabling ablation with the following pulses. In addition, usually the ablation rate changes with changing number of pulses (saturation, bleaching effects, etc.).

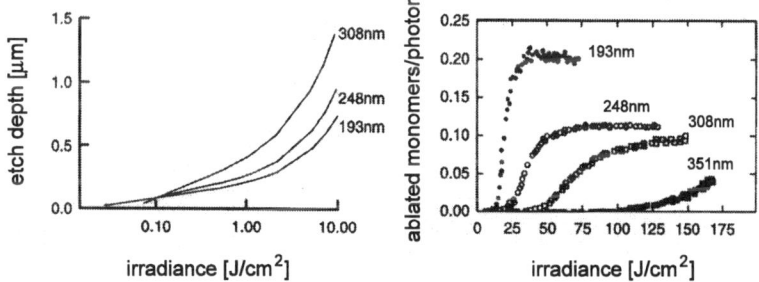

Figure 129 Left-hand side: Calculated etch curves for PI at different UV wavelengths. Reprinted with permission from (Cain et al., 1992). Copyright 1992 American Institute of Physics. Right-hand side: measured curves. Reprinted with permission from (Küper et al., 1993). Copyright 1993 Springer-Verlag.)

A way to determine the ablation rate per laser pulse is to use a crystal microbalance, which essentially consists of a quartz crystal, which is stimulated to resonant high-frequency oscillations (typically 6 MHz) (Lu and Czanderna, 1984). The absolute value of the resonance frequency depends sensitively on the mass of the crystal. Mass changes, induced for example by adsorbing or desorbing material, result in a frequency change, which can be measured with high accuracy. Since changes of film thickness in the subnanometer regime result in frequency changes of kHz, subnanometer depth resolution is easily achievable.

For the determination of the mass of UV-laser ablated material from a polymer film it is evaporated onto the quartz crystal (or glued using a thin layer of pyraline in the case of PI), irradiated and the mass change is measured per pulse. The measured ablation rate per pulse (recalculated into ablated monomers per incoming photon) is plotted on the right-hand side of Fig. 129 for irradiation with 193 nm, 248 nm, 308 nm and 351 nm light.

Obviously, with decreasing wavelength smaller irradiances are necessary to obtain the same ablation yield. In the case of long wavelength pulses (351 nm) in the regime of small irradiances a nearly exponential increase of ablation rate without significant threshold is observed. In contrast, the ablation at 193 nm shows a steep increase with strong threshold behavior. All curves saturate with increasing irradiance due to the opening of competing channels, for example the absorption of laser light in the plume or thermal processes. Note that the phenomena induced by 193 nm light show a significantly different behavior as compared with the phenomena observed following irradiation with longer wavelengths: in the limit of high irradiances five photons are sufficient to ablate a monomer, whereas at 248 nm (and at the other wavelengths too) more than ten photons are necessary. If one takes into account that the energy difference per photon between 193 nm and 248 nm is 1.4 eV, whereas that between 248 nm and 308 nm is 1 eV, then one is even more tempted to assume that the significantly different behavior induced by the 193 nm light is caused by more than the stronger heating of the surface at the higher photon energy.

This conclusion is strengthened by the results from the thermal model calculation (Fig. 129 left-hand side). For all three wavelengths the model predicts an exponential increase of etch depth with increasing laser fluence. The experimentally observed, much stronger increase for 193 nm is not accounted for.[6]

Two possible sources of error inherent to the crystal balance method have to be taken into account: (i) The resonance frequency of the microbalance depends sensitively on temperature. Laser-induced changes in crystal temperature result in apparent mass increases or decreases. These spurious effects show a complicated, difficult to compensate temporal behavior, which is dictated by the temporal evolution in the course of the laser pulse and the thermal constants of the crystal. (ii) The etch rate might also result from the loss of gases, which have been stored in a layer below the surface itself. Hence the initially observed mass loss is not automatically

---

[6] However, it has been shown that all the curves can be fit reasonably well by a more elaborate interpolation formula, taking into account screening of the laser intensity in the ablation plume (Bäuerle, 1996b).

6.2. LASER ABLATION

correlated with the evolution of an etch hole.

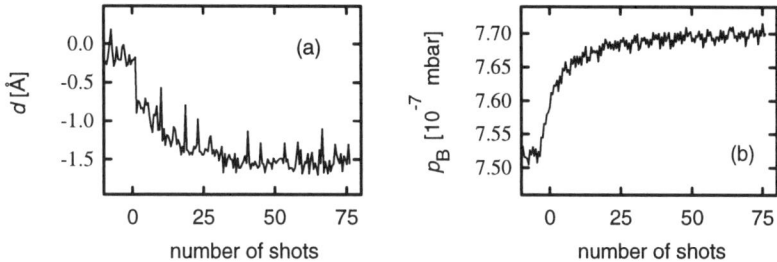

Figure 130 (a) Ablated mass thickness from a 270 nm Au film following irradiation by an excimer laser (248 nm, 8 mJ/cm^2), measured via a quartz microbalance and plotted as a function of the number of laser shots. (b) Simultaneously measured increase in total background pressure.

As an example, Fig. 130 shows the ablated mass thickness from a Au film, following irradiation by an KrF excimer laser. Note that with increasing number of shots the ablated mass is *not* linearly increasing as would be expected for a true ablation process. Instead, the mass thickness reaches a saturation value. As shown by the simultaneously measured change in background pressure (Fig. 130b), the laser serves mainly to desorb adsorbed background gas from the surface. After a while a steady state with the simultaneous adsorption rate from the background gas has been reached. This figure thus demonstrates that one has to be especially cautious in interpreting the results from measurements with a quartz microbalance. If one, for example, were to increase the laser fluence in order to obtain ablation from the film surface, then heating effects would tend to severely complicate the interpretation of the microbalance data, as discussed above.

The calculations on the basis of the photothermal model allow one to deduce information on the temporal evolution of the ablation process, which of course depends on the irradiated material. In the case of a strong absorber such as PI the degradation occurs on a time-scale of less than picoseconds (Fig. 131). The surface itself becomes hot much slower. Thermal diffusion transfers energy *during* the nanosecond laser pulse out of the focal spot; thus the etch depth quickly acquires (within picoseconds) a constant value and does not increase during the pulse. In the case of a weak absorber such as PMMA the degradation is delayed and the etch depth increases on the time-scale of picoseconds.

The expected increase in temperature at the surface is, for PMMA, significantly smaller (ΔT=826 K for 200 mJ/cm^2 193 nm light) as compared with PI (3915 K) if one does not take into account the degradation of the polymer. If one takes the etch process into account, then the expected temperature increase is about the same for PMMA, but reduces to $\Delta T = 595$ K for PI. From the absorbed energy the equivalent temperature (PI: 10^5 K, PMMA 5800 K) and from that the average thermal velocity

$$v = \sqrt{\frac{8k_B T}{\pi m}} \qquad (6.15)$$

of the ablated particles (using an average mass of the fragments of $m = 8.3 \times 10^{-27}$ kg (PI), $m = 8.3 \times 10^{-25}$ kg (PMMA)) might be calculated. One finds that the PI fragments are ablated with 22 times the velocity of sound, whereas the PMMA fragments possess only Mach 0.5. The supersonic velocity of the PI fragments results for ablation in air in a characteristic crack that can easily be heard.

Ablation temperatures

For further exploration of the physical mechanism of the UV ablation process the experimental determination of the *temperature* at the polymer surface is important. A way to determine such a temperature is the measurement of the rotational and vibrational state distribution of the ablated products. Then one only has to assume that the products are ablated in thermal equilibrium with the laser-induced *hot spot* at the surface. Unfortunately, this indirect determination of surface temperatures results in very inconsistent data.

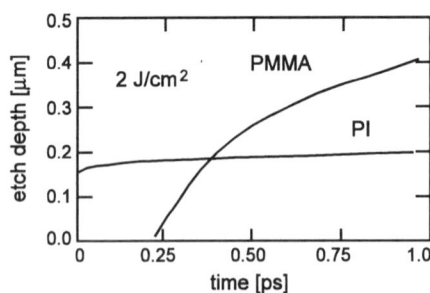

Figure 131 Calculated temporal evolution of the etch depth for a strong absorber (PI) and a weak absorber (PMMA). Reprinted with permission from (Cain et al., 1992). Copyright 1992 American Institute of Physics.

In Table 2 measured temperatures for different fragments are listed. As seen, rotational and vibrational temperatures as well as the temperatures of different ablation products differ greatly. More detailed measurements have

6.2. LASER ABLATION

shown that the temperatures change also as a function of distance to the surface. If one uses the rotational temperature as a measure of the surface temperature, then one has to take into account that this temperature is strongly influenced by mutual collisions of the ablated particles in the course of their removal from the surface due to the large probability for rotational energy transfer. Hence one would have to perform measurements close to the surface, which is a difficult task. The vibrational temperature, on the other hand, might be totally different from the surface temperature since many products are ablated in the electronically excited state. Then they decay into the electronic ground state within nanoseconds, resulting in a vibrational energy distribution which is dictated mainly by the optical transition moments between ground and excited states (Fig. 58) and is thus far away from thermal equilibrium with the surface heat bath.

Table 2 Measured Boltzmann temperatures of ablated fragments for different materials and ablation wavelengths λ_d [nm]. T_{rot} means the rotational temperature, T_{vib} the vibrational temperature (both in Kelvin) and P_L the fluence [mJ/cm^2]. The values are from: a: (Davis et al., 1985); b: (Srinivasan et al., 1986); c: (Goodwin and Otis, 1989); d: (Koren, 1988).

Material	λ_d	Fragment	T_{rot}	T_{vib}	P_L	Reference
PMMA	193	CH	3200 ± 200	3200 ± 200	≈500	a
PMMA	193	C$_2$	1000 ± 200	-	≈500	b
PI	193	CO	700 + 50	900 ± 150	6	c
PI	193	CO	1150 ± 50	3400 ± 300	150	c
PI	248	HCN	2250 ± 150	-	400	d

Measurements of translational temperatures of ablated fragments via, for example, time-of-flight techniques (see below, Fig. 133) (Dyer and Srinivasan, 1989) results in velocities of the ablated fragments from PET of nearly 5 km/s at low irradiation (below 100 mJ/cm^2), but strongly depending on the laser fluence. Laser beam deflection measurements (Ventzek et al., 1991) (Fig. 142) on PET result at 0.1 J/cm^2 irradiation with 248 nm light in 1.9 km/s, but at 9 J/cm^2 in 16 km/s. From that data in principle the temperature of the substrate can be calculated. This would result for 1 J/cm^2 and benzene as ablated particle in about 3000 K (Ventzek et al., 1991). However, besides the fact that the beam deflection method does not provide direct information on the mass of the ablated particles other effects such as collisions of the ablating particles or plume absorption (Schmidt et al., 1998) might significantly influence the kinetic energy distributions.

More direct values of surface temperatures might be determined by a measurement using, for example, a pyroelectric crystal (Mihailov and Duley, 1991) or by detecting the change in resistance of an adsorbed thin film (Brunco et al., 1992). In the case of kapton, irradiated at 308 nm, a temperature in the range between 1560 K and 1730 K has been determined (Mihailov and Duley, 1991). The temperature, determined for irradiation with 248 nm, is 1660±100 K (Brunco et al., 1992) and thus is within this range of values. For the determination of this latter temperature a 140 nm thick NiSi layer has been adsorbed on a quartz substrate, followed by a 50 nm thick silicon-oxide layer and a 100(200) nm thick PI layer. The resistance of the NiSi layer increases linearly with temperature. Hence following irradiation of the PI layer with a laser the interface temperature just below the PI layer can be determined via the change in resistance. The temperature at the PI surface is then calculated via a one-dimensional numerical simulation of heat flow through the whole adsorbed structure. In that way the temperature-dependent values for K and κ are also estimated. Integration of the temperature profile then results in the effectively absorbed irradiance, which is plotted vs the applied irradiance in Fig. 132.

A significant ablation threshold of 36 mJ/cm^2 is observed, below which the absorbed irradiance increases linearly with increasing applied irradiance. The deviation from a straight line with slope unity reveals directly the reflection losses, which are 12% in the present case. Above the threshold the absorbed photon flux is no longer used for an increase in the surface temperature but for ablation of polymeric material. Consequently, the absorbed flux remains constant. This observation is consistent with the assumption that a critical absorbed threshold energy density does exist. At the value of this energy density the surface temperature is 1660 K.

However, this way of determining the surface temperature also is indirect. It depends sensitively on the correct description of the evolution of temperatures through the multilayer system. A more appropriate method would be to determine the temperature *on* the surface in the close neighborhood of the laser spot, *e.g.*, via monitoring the evaporation of an adsorbed film if the binding energies of the adsorbates are well known (cf. Figs. 119 and 121) or via observing characteristic changes in the lattice structure.

As denoted above, one also might not expect to be able to determine the surface temperature directly from the kinetic energies of the ablated particles. However, the *mechanism* of ablation (thermal vs photochemical) might be reflected in the velocity distributions. A measurement of the time-of-flight (TOF) distribution of the MMA (methyl methacrylate) fragments following irradiation of a PMMA sample with 193 nm light of 300 mJ/cm^2 is shown in

6.2. LASER ABLATION

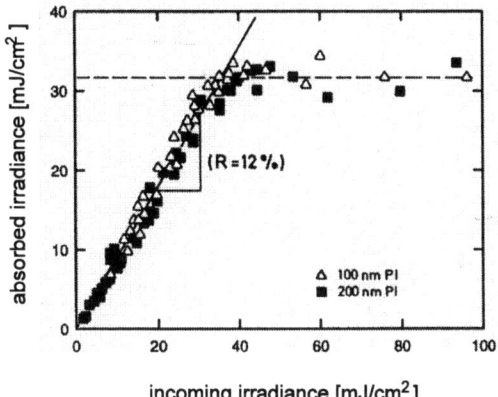

Figure 132 Absorbed irradiance vs applied irradiance of 248 nm light onto PI films of various thicknesses. Reprinted with permission from (Brunco et al., 1992). Copyright 1992 American Institute of Physics.

Fig. 133.

The ablated fragments are ionized after a flight path of 9.2 cm via electron bombardement, are mass selected in a quadrupole mass filter and are detected using an open electron multiplier. The measured time delay between laser pulse and generation of electrons in the multiplier (taking into account the drift time within the mass filter) and use of the known flight path results in the velocity of the ablated particles. The flux of slow particles (broad maximum in Fig. 133), which has been ablated from the surface and has a number density n, obeys a Maxwell–Boltzmann velocity distribution in polar coordinates θ and ϕ (Comsa and David, 1985),

$$d^3 f(v) \propto nv^3 \exp\left(-\frac{mv^2}{2k_B T}\right) \cos\theta \sin\theta \mathrm{d}\phi \mathrm{d}\theta \mathrm{d}v \tag{6.16}$$

and thus can be described using an effective temperature T. For an irradiance of 80 mJ/cm^2 one finds $T=1200$ K, which corresponds to a mean kinetic energy

$$E_{\mathrm{kin}} = 2k_B T \tag{6.17}$$

of 230 meV.[7] This translational energy increases exponentially with increasing

---

[7] The $2k_B T$-dependence too results from the fact that a *flux* of particles is ablated from the surface.

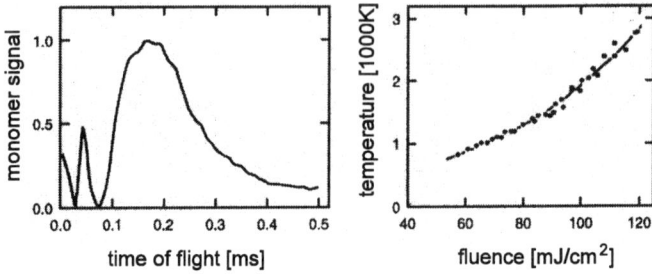

Figure 133 Left-hand side: Time-of-flight distribution of MMA fragments following excimer laser (193 nm) irradiation of a PMMA sample (300 mJ/cm^2). The maximum at flight time zero results from ions that are created directly by the laser. Right-hand side: Maxwell–Boltzmann temperatures of the slow fragments as a function of laser irradiance. Reprinted with permission from (Danielzik et al., 1986). Copyright 1986 American Institute of Physics.

irradiance. If one takes into account that the irradiating photon energy is 6.4 eV and the binding energy of MMA fragments in PMMA is 2.7 eV, then it becomes obvious that a large amount of irradiated energy has been lost in substrate excitations or that the adsorbate is highly internally excited.

These observations agree with a *photothermal* explanation of the ablation process. Only at high irradiances is a second, fast peak in the TOF spectrum observed (Fig. 133 left-hand side, $E_{kin} \approx 10$ eV), the origin of which might be nonthermal. This fast peak characterizes the transition from a Knudsen to a nozzle beam expansion. The collisions between the ablated particles then lead to an increase in translational energy at the cost of energy in internal degrees of freedom (see Chapter 1, Section 1.2.4).

Product distributions

Although the photothermal model correctly reproduces the etch rates and the principal difference between a weak and a strong absorber, it cannot explain why 193 nm ablation differs in principle from 248 nm ablation. One of the obvious weaknesses of the model is that the initial rate equation is based on a single activation energy E_A, i.e., the breaking of the 'key' bond in the monomer with resulting total degradation. The molecular structure of the polymer and its chemical response on the laser pulse are not adequately taken into account. For instance, the theory is not able to predict — except via the global parameter n_f — the internal state distribution of the ablated fragments. A measurement of this *product distribution* could provide hints as to whether the basic process is of photothermal or photochemical nature.

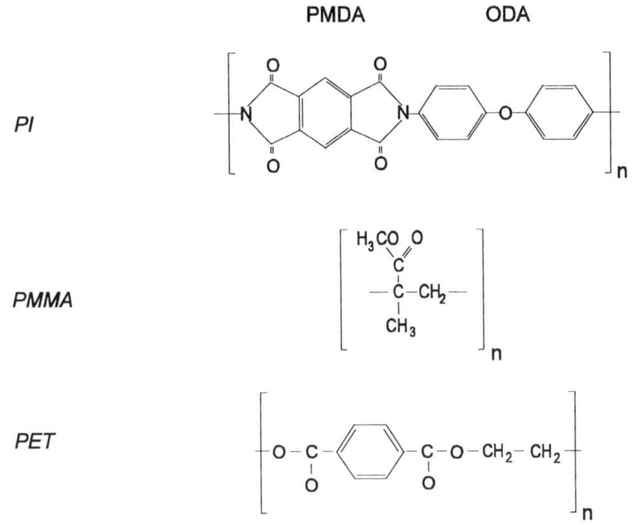

6.2. LASER ABLATION

Figure 134 Chemical structures of PI, PMMA and PET. PMDA means pyromellitic dianhydride and ODA oxy dianiline, these being the components from which the polymer is synthesized.

From the chemical structures of PI, PMMA and PET (Fig. 134) one might guess which products result from the laser-induced dissociation process. Optical excitation of PI is induced by the PMDA group in the wavelength range from 300 to 330 nm. In the case of PMMA it proceeds via the COOCH$_3$ group. A possible degradation mechanism, which results in the total destruction of the benzol rings in PI, is plotted in Fig. 135 (Srinivasan, 1993a).[8] On average 0.5 UV photons at 193 nm are necessary to disrupt the central PMDA group from the residual polymer. At least five further photons are necessary to destroy the resulting three benzol rings.

The energetic most favorable path to the destruction of a benzol ring is

---

[8] Competing processes to the direct degradation are multiphonon excitation (*e.g.*, via the benzol rings), singlet–triplet conversion (resulting in electronically excited triplet states), electronic de-excitation via internal conversion (IC) between the excited and ground states of the same symmetry, fluorescence, phosphorescence and collisions. The cross sections for all those processes have to be known for a quantitative prediction of the fragmention pattern.

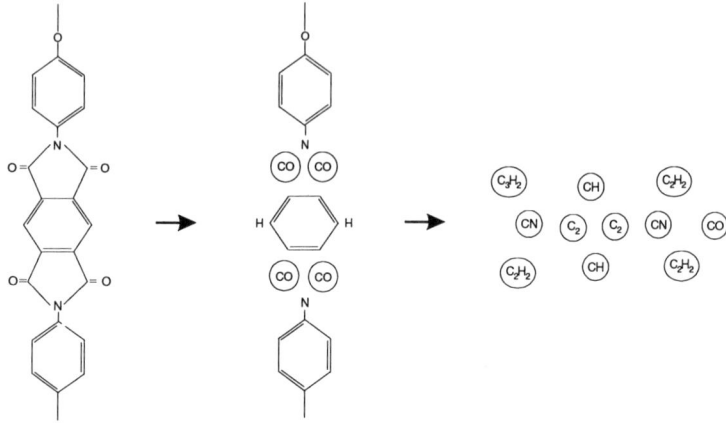

Figure 135 Possible laser-induced degradation of PI. Reprinted with permission from (Srinivasan, 1993a). Copyright 1993 Springer-Verlag.

$$C_6H_6 \to C_4H_2 + C_2H_2 + H_2; \qquad (6.18)$$

here, the necessary energy is 5.9 eV. The following step is

$$C_4H_2 \to C_2H_2 + C_2, \qquad (6.19)$$

which requires an energy of 6.5 eV. Hence, even for the first step a 193 nm photon is necessary to allow direct fragmentation.

The resulting product distributions and their temporal evolution into the surrounding medium might be observed 'in situ' by irradiation with a second laser pulse. Fig. 136 shows the experimental set-up: the ablation is induced by an excimer (193 nm, 248 nm) or CO_2 laser pulse (9.17 μm). Following an adjustable delay time between nano- and microseconds a dye laser pulse (λ=596 nm) irradiates the substrate surface either parallel to the surface or under an angle of 45°.

The irradiation of the surface with the dye laser pulse at 45° with respect to the excimer pulse enables one to observe transient changes, i.e., changes that would not be visible in the course of a subsequent investigation of the irradiated surface with, for example, an electron microscope. Fig. 137 shows as an example the modification of a PI surface following irradiation with 248 nm light of different fluences 60 ns (a) and several seconds (b) following irradiation. Obviously, the blackening of the surface (i.e., the enhanced absorptivity for the dye laser light) is only a transient process, at least for

6.2. LASER ABLATION

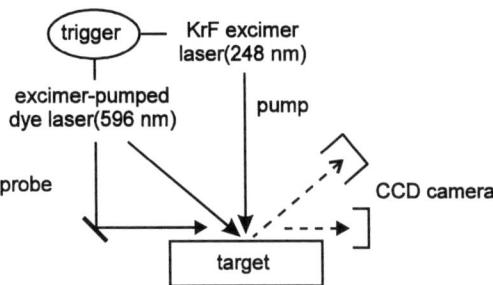

Figure 136 Experimental set-up for the direct observation of ablation fragments. Irradiation with the probe pulse along the surface results in images of the ablated products (Figs. 138 and 139), while irradiation at 45° allows one to image the irradiated surface (Fig. 137).

low irradiances.

The temporal evolution of the product distributions perpendicular to the PI surface can be observed via irradiation with the dye laser pulse parallel to the surface (Fig. 138). In the case of ablation with an IR pulse (upper part) one observes the ejection of large, pyrolitic (i.e., thermally) generated fragments such as C_6H_5CN or C_6H_5OH (dark cloud). Following UV ablation mainly gaseous fragments are detected, which merely refract the dye laser light.

In the case of a weak absorber such as PMMA the product distribution looks significantly different (Fig. 139). Immediately after starting the irradiation process with 248 nm light a shock wave evolves ($v \approx 600$ m/s), followed by a contact front, which is engaged by the ablated material. In contrast to excitation with 9.17 μm light, where strongly refractive material (MMA monomers) becomes visible inside the contact front, here the ablated parts are mainly C_2 and smaller polymer fragments. The thermal destruction of PMMA (even without a laser) proceeds via an 'unzipping' reaction, in which case the monomers are stripped as a whole; the (photochemical) UV excitation leads via electron transfer in the carboxylic group to the unzipping of CH_3, CO and other small fragments.

Following ablation of the polymer fractions strongly refractive material is ejected ($v \approx 150$ m/s). This material is initially of needle-type consistency, transforms (after 6 μs) into droplets and forms a beam, which narrows in the following until it has nozzle beam character. Apparently the surface melts after about 6 μs and then solidifies from the edges. Since the melting temperature of the ablated polymer fractions is smaller than 50°C, the surface temperature

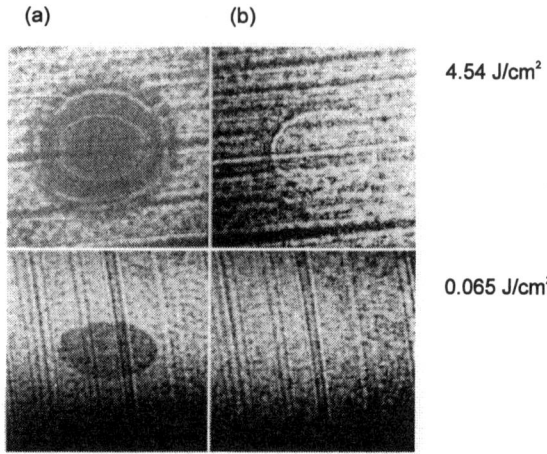

Figure 137 Irradiated surface of a PI film about 60 ns after (a) and a long time after irradiation with an UV laser (b). Reprinted with permission from (Srinivasan et al., 1990). Copyright 1990 American Institute of Physics.

is expected to be at the same order of magnitude during the melting process; immediately after laser irradiation, of course, it might have been much higher.

Time resolved investigations

A more thorough investigation of ablation dynamics makes it necessary to apply laser pulses with sub-nanosecond length. However, the evolution of electronic excitation induced by a femtosecond pulse could be *essentially* different from that induced by nanosecond pulses (see also Chapter 4, Section 4.1.3). This then results in a different ablation behavior. For example, it has been found that PTFE shows thermal degradation effects following excitation at 248 nm with 16 ns pulses, whereas 300 fs pulses result in a clean etch hole (Küper and Stuke, 1989). This behavior is not observed in PMMA, where even nanosecond pulses result in a clean etching pattern.

A rationale for this difference can be found in the absorption behavior: PTFE absorption requires a two-photon process. The threshold is a factor of three or four higher compared with PMMA and the corresponding higher power density might be coupled to the material only via femtosecond pulses ($P_\text{L} \geq 100$ GW/cm^2). In PMMA, in contrast, one needs 'incubation pulses' for ablation, since one initially has to produce strongly absorbing unsaturated fragments (for example the disrupted COOCH$_3$ group). The dissociation energy of the weak bonds in PMMA is 2.7 eV, in PTFE 4.2 eV.

6.2. LASER ABLATION

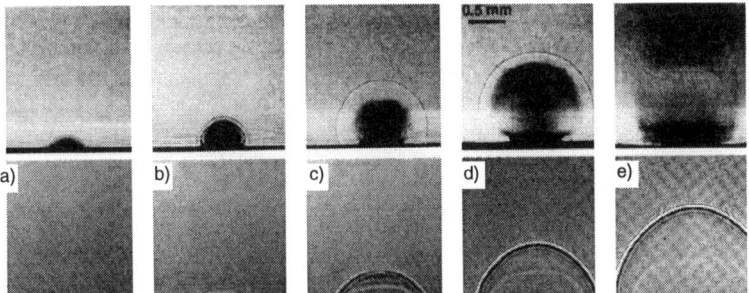

Figure 138 Material ejection following IR laser (upper row) and UV laser (lower row) irradiation of PI. The consecutive pictures a) to e) represent in the upper row the following temporal delays between ablating and illuminating laser pulses: (a) 150 ns, (b) 300 ns, (c) 500 ns, (d) 800 ns, (e) 2600 ns. And in the lower row: (a) 50 ns, (b) 100 ns, (c) 200 ns, (d) 400 ns, (e) 700 ns. Upper row: Reprinted with permission from (Srinivasan, 1993a). Copyright 1993 Springer-Verlag. Lower row: Reprinted with permission from (Srinivasan et al., 1990). Copyright 1990 American Institute of Physics.

If one uses two femtosecond pulses from a single laser, which are delayed temporally by use of different spatial paths, and if one measures the correlation function between the pulses (cf. Fig. 94), then the differences between strong and weak absorbers become even more prominent (Fig. 140). The weak absorbers (PTFE or PMMA) show an increase in ablation rate as soon as both femtosecond pulses are temporally overlapped. Evidently, in this case one is dealing with a two-photon process. In the case of a strong absorber (PI) the initially high ablation rate decreases as soon as both pulses irradiate the sample at the same time. Here, the absorption of the second pulse in the material is enhanced by irradiation with the first pulse, leading to a decrease in the penetration depth of the light and thus to a smaller ablation rate. The half width of the correlation function thus reflects the lifetime of the intermediate state that has been excited by the first laser pulse.

The initial stage of the ablation process too might be investigated with the help of correlation measurements by applying subsequently two lasers pulses of different energy densities. If in between the first and the second pulse material has been ablated, then the ablation will be stronger if the more intense pulse hits the sample before the weaker does. Otherwise it would have been attenuated by the material that has been ablated by the weaker laser. This is the case for nanosecond delayed pulses, but not if the delay between the

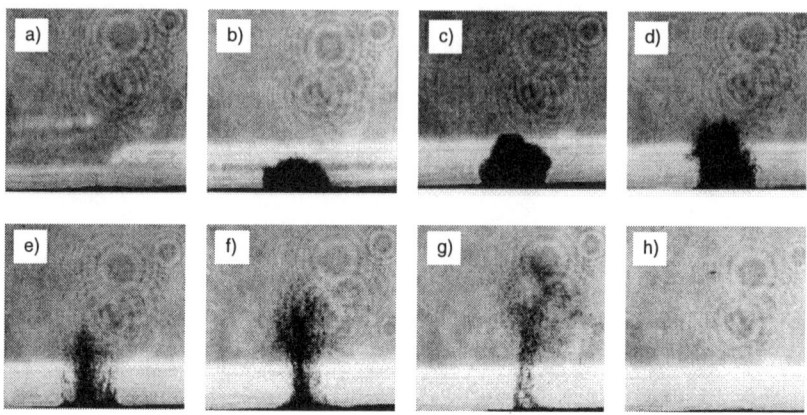

Figure 139 Material ejection following UV laser (248 nm) irradiation of PMMA. The time delays between ablation and illumination pulses are: (a) 750 ns, (b) 1000 ns, (c) 3000 ns, (d) 6100 ns, (e) 9700 ns, (f) 15000 ns, (g) 20700 ns, (h) 1 s. Reprinted with permission from (Srinivasan, 1993b). Copyright 1993 American Institute of Physics.

pulses is smaller than 200 ps. An obvious conclusion from that observation is that the ablation of material (for PI and PMMA) occurs between 200 ps and several nanoseconds after the begin of irradiation with the laser.

Photochemical model

Thus far the discussed observations support the assumption that UV ablation with wavelengths smaller than 248 nm at least contains a nonthermal component. If one were to take into account only the observed etch curves, then one could believe that a thermal process is responsible for ablation at 248 nm. However, the etch curves can also be described by the following *model*, which does not possess an explicit thermal component but assumes that the absorbed laser fluence results in direct bond breaking ('photochemical model') (Pettit and Sauerbrey, 1993). The model also takes into account multiphoton absorption and saturation processes (Fig. 141).

Let there be a threshold value for the absorbed laser power, which defines the maximum etch depth. Hence one can calculate the etch depth via the attenuation of light in the material. Let us further assume that the laser-excited polymers relax only via stimulated emission of radiation. Then one finds as a solution of a rate-equation approach for the change in photon density

6.2. LASER ABLATION

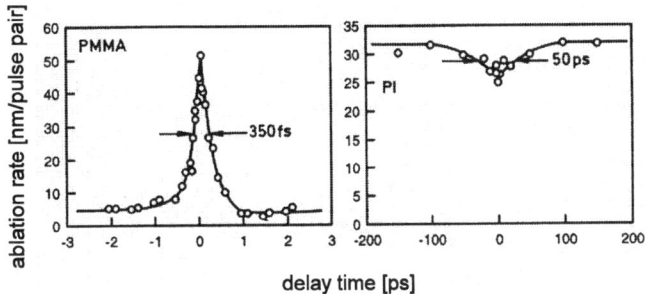

Figure 140 Ablation rate as a function of delay time between two laser pulses of 500 fs length for irradiation of PMMA (left-hand side) and PI (right-hand side). Reprinted with permission from (Preuss et al., 1993). Copyright 1993 American Institute of Physics.

S with depth z

$$\frac{dS}{dz} = -\frac{n_{chr}}{2}(1 - \exp(-2\sigma_1 S)) \quad (6.20)$$

in the case of one-photon absorption with an absorption cross section σ_1 and a chromophore density N_{chr}. For small photon densities, $2\sigma_1 S \ll 1$, this equation leads to Beer's law:

$$\frac{dS}{dz} \approx -n_{chr}\sigma_1 S, \quad (6.21)$$

$$n_{chr}\sigma_1 = \alpha_{eff}. \quad (6.22)$$

For large photon densities, $2\sigma_1 S \gg 1$, saturation occurs, i.e., an equilibration of population in excited and ground states, induced by stimulated emission:

$$\frac{dS}{dz} \approx -\frac{n_{chr}}{2}. \quad (6.23)$$

Upon integration, Eq. (6.20) predicts for the etch depth:

$$z_e = \frac{2}{n_{chr}}(S_0 - S^s) + \frac{1}{n_{chr}\sigma_1}\ln\left(\frac{1 - \exp(-2\sigma_1 S_0)}{1 - \exp(-2\sigma_1 S^s)}\right), \quad (6.24)$$

with the threshold photon density

$$S^{\text{s}} = \frac{P_{\text{L}}^S}{h\nu}. \qquad (6.25)$$

In the case of n-photon absorption (e.g., for PTFE) the integration from S_0 to S^{s} depends on the temporal and spatial characteristics of the laser pulses, K_n,

$$z_{\text{e}} = \frac{2}{n \cdot n_{\text{chr}}} \int_{S^{\text{s}}}^{S_0} \frac{\text{d}S}{1 - \exp(-2\sigma_n K_n S^n)}. \qquad (6.26)$$

Figure 141 shows, as an example of the quality of this model, the calculated etch curve for 248 nm ablation of PI in comparison with measured values. Obviously, close to the threshold and also up to high laser irradiances the agreement is very good although no explicitly thermal mechanism has been invoked.

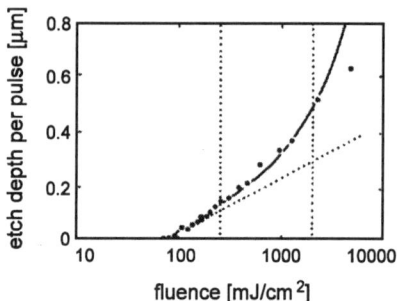

Figure 141 Calculated (solid line) and measured (dots) etch rate as a function of laser fluence for 248 nm ablation of PI. The dashed line corresponds to the linear absorption behavior (Eq. (6.21)). The perpendicular lines separate with increasing fluence the regimes of linear absorption, absorber saturation and cloud attenuation. Reprinted with permission from (Pettit and Sauerbrey, 1993). Copyright 1998 Springer-Verlag.

For low fluences linear absorption dominates, leading to a logarithmic increase in etch depth on fluence. For intermediate fluences absorber saturation dominates. Here the etch depth per pulse increases even more strongly compared with the linear absorption regime. Finally the screening of the laser pulse by the ablated material decreases the slope of increasing etch depth, as seen by the deviation between final measured point and calculated curve in Fig. 141. This so-called 'cloud attenuation' or plume absorption might play an important role even for lower irradiances and for polymers with large as well as small absorption coefficients (Schmidt et al., 1998). Obviously, it

6.2. LASER ABLATION

cannot be accounted for by the model discussed above.

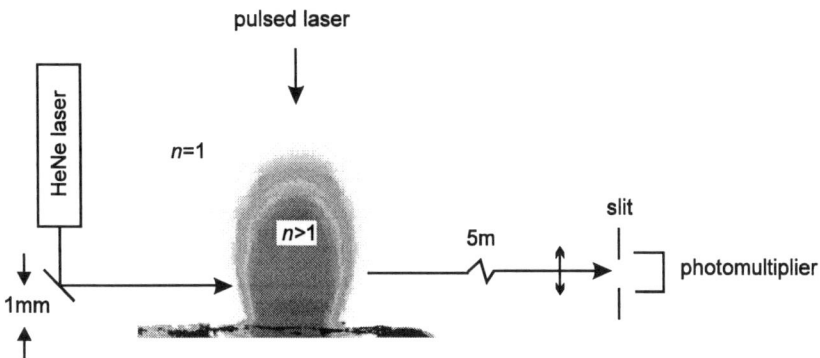

Figure 142 Photoacoustic deflection technique ('mirage') for the determination of damage thresholds. Since the method relies on the propagation of acoustic waves, it only works for samples surrounded by ambient gas.

The following remarks are in order.

1. A thermal interpretation of the ablation process implies that the value of the ablation threshold depends on the detector sensitivity. In many cases it proves useful to discriminate a *macroscopic* ablation threshold (evaporation of material, resulting in an ablation hole with measurable depth) from a *microscopic* ablation threshold (melting processes and a change in the optical properties of the material).[9] As can be seen from Fig. 153, the laser-irradiated material undergoes significant morphological changes even below the ablation threshold (see also Fig. 150). The microscopic site-sensitivity of the bond-breaking efficiency has been investigated for metallic (*e.g.*, Pt(111) (Frohn et al., 1994)), semiconductor (*e.g.*, InP(100) (Moison and Bensoussan, 1982), GaP(110) (Hattori et al., 1992)) and insulating (*e.g.*, CaF_2(111) (Gogoll et al., 1996)) surfaces. In the case of semiconductor surfaces especially the 7×7 reconstructed Si(111) surface has found attention (Tanimura and Kanasaki, 1998).

A sensitive technique for the determination of single-pulse damage thresholds uses either changes in the reflection of a HeNe probe beam from the

---

[9] An useful account of the transition from laser-induced desorption to surface damage is given in (Matthias and Dreyfus, 1989).

laser-irradiated surface (Schwarzenbach et al., 1984; Woelker et al., 1991) or an additional deflection, induced by shock waves in the surrounding gas, which result from the ablated material (Petzoldt et al., 1988). Upon irradiation, an acoustic wave is generated, followed by the expansion of a plasma or a dense gas cloud. Hence the index of refraction (n') is changed, in the case of ejection of neutral gas ($n' > 1$) or plasma ($n' < 1$). A HeNe laser, traversing parallel to the surface, is deflected by this new index of refraction ('mirage' effect (Green et al., 1977)), and the deflection is measured with a position-sensitive photodiode or behind a slit with a photomultiplier (Fig. 142). This method results in valuable information about shock and cooling waves in laser-induced plumes (Koren, 1987) (Chapter 5, Section 5.2), but can also successfully be applied as a nondestructive technique for the evaluation of thermal properties of thin films (McGahan and Cole, 1992; Machlab et al., 1993).

The deflection of the HeNe probe beam is proportional to the gradient of the refractive index and thus the time-integrated deflection measurement provides values of the energy of the acoustic pulse (Petzoldt et al., 1988). Moreover, measurements of the photoacoustic energy as a function of diameter of the damage spot show an approximately cubic dependence, indicating that the volume of the ablated material determines the acoustic energy and thus the magnitude of the probe beam deflection. Using this method, accurate damage thresholds for UV window materials such as LiF or CaF_2 (Petzoldt et al., 1988), for oxide thin films such as TiO_2 or HfO_2 (Reichling et al., 1994), for polymer (Matthias et al., 1995) and metal thin films (Siegel et al., 1997; Güdde et al., 1998) have been determined. Different slopes in the rise of the deflection amplitude as a function of incident fluence clearly demonstrated that the mirage method is able to discriminate between microscopic (melting-induced) and macroscopic (ablation) damage thresholds (Matthias et al., 1995).

Thin metal films are discussed in more detail in Section 6.2.3. As for the oxide films (Section 6.2.2) an interesting (but up to now not fully understood) finding is that the damage threshold increases exponentially with band gap energies from 0.03 J/cm^2 (TiO_2, E_g=4 eV) up to 1 J/cm^2 (HfO_2, E_g=6.5 eV) (Reichling et al., 1994). This finding points to the importance of defect states in the band gap especially for (nonlinear) ablation of wide-band gap materials.

2. The etch curves reflect the absorption behavior (one- or multi-photon absorption, saturation, etc.), but they give no clear hints regarding the mechanism of energy transformation (thermal or direct). A more thorough understanding would ask for

- time-resolved surface temperature measurements in the course of the laser pulse; an interesting approach is presented by picosecond RHEED (Chapter 4, Section 4.5)
- theoretical models for the product distributions,
- systematic measurements of product distributions.

6.2. LASER ABLATION

At present investigations of this kind are performed by various research groups.

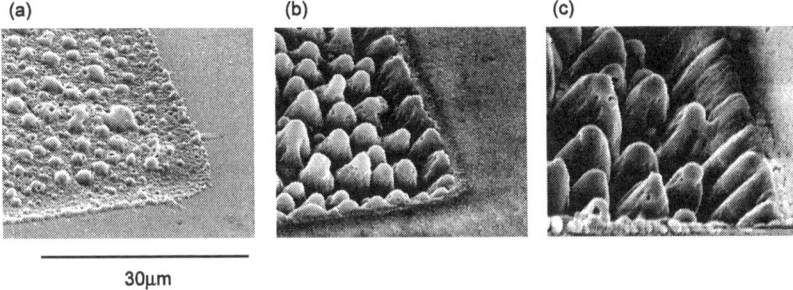

Figure 143 Scanning electron microscopy (SEM) micrographs of conical defects on a mica surface following irradiation with (a) 10, (b) 100 and (c) 1000 pulses of 193 nm radiation (1 J/cm²) (K.Rubahn and J.Ihlemann, private communication, 1998). Within the laser-irradiated spot, cones are growing opposite to the direction of the laser radiation. Similar structures are found following irradiation of polymers.

Ablation with multiple pulses

Up to now ablation following a single pulse has been discussed. Most technical applications, however, implement multiple *subsequent* laser pulses. In the case of energy accumulation the threshold fluence F_L^N for N pulses is related to the threshold fluence of a single pulse, F_L^S, by $F_L^N = N \cdot (F_L^S)^{y-1}$, where y is of the order of -0.05 for 400 nm irradiation of thin metal films with femtosecond pulses (Güdde et al., 1998). Thus the damage threshold gradually decreases with the number of shots.

For polymer ablation saturation of the ablation rate is often found, for example, for PI following irradiation with 500 pulses at 248 nm(Krajnovich and Vazquez, 1993). Electron microscopy reveals that conical defects are generated at the surface, which grow opposite to the direction of laser irradiation since the absorptivity of the material is highest in the direction normal to the surface. An example for a laser-irradiated mica surface is shown in Fig. 143. The cone formation is induced by laser beam inhomogeneities and diffraction at surface roughnesses (see below, laser-induced periodical structures, Section 6.4), which might also be due to intrinsic stress of the sample. Increasing fluences lead to a decrease in number density and an increase in diameter of (shallower) cones.

In the case of polymers conical defects often grow around carbon-enriched spots and have a higher ablation threshold (i.e., lower etch rate) compared with the surroundings, resulting in their growth at the expense of the surroundings. The phenomenon is somewhat similar to 'radiation hardening', which was observed very early in the history of UV laser treatment of polymers[10] (Andrew et al., 1983), occurs for most strong absorbers following irradiation with a large number of UV pulses, and results in an attenuation of the etch rate with increasing number of pulses. For weak absorbers (*e.g.*, PMMA) the opposite effect is observed, namely an 'incubation': the etch rate increases with increasing number of pulses.

It is noted that the generation of conical defects is one of the major problems in the course of laser material deposition methods such as PLD (pulsed laser deposition) (Venkatesan, 1994). On the other hand, cones can be useful in accomplishing more clean ablation patterns since they tend to enhance the power densities at the bottom of the cavities by reflective focusing, thus leading to nonthermal material ejection and chemical bond-breaking. In addition, the microstructuring of the surface accompanied by cone-formation is of some interest by itself. For example, the performance of high-transmittance, antireflection films can be improved by the formation of gradient-index (conical) films (Horwitz, 1980).

6.2.2 Insulators

Ceramics such as *aluminum-oxide* (Al_2O_3) are difficult to treat with chemical or mechanical methods due to their great hardness and their chemical, electrical and thermal resistivity. By the use of laser irradiation similar results as in the case of polymers can be obtained if irradiances an order of magnitude higher are used (Ihlemann et al., 1995). There are several motivations for the treatment of insulator surfaces, *e.g.*, microstructuring to obtain graded transmission dielectric masks, which in a subsequent step can be used to ablate three-dimensional structures in other polymers or dielectrics (Rubahn and Ihlemann, 1998). As an example, in Fig. 144 a Fresnel lens with micrometer dimensions is shown which has been generated via single-pulse laser ablation using a laser-structured dielectric mirror as a mask. Three-dimensional structures can be obtained by spatially varying the reflection coefficient in the mirror mask. Roughening of the surface, which results in a better growth behavior of thin metal layers (Polanski and Rubahn, 1996) and improved catalytic and adhesive properties, is another application.

---

[10] In this case PET was irradiated by light from a XeCl laser. The effect was tentatively attributed to preferential etching of amorphous regions in the PET. Nowadays the effect is thought to be more 'universal'.

6.2. LASER ABLATION

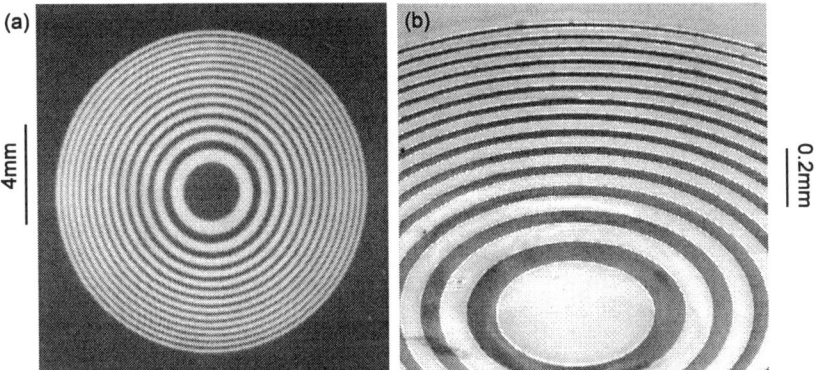

Figure 144 (a) Optical microscope view of a micro Fresnel lens, written in a dielectric mirror (highly reflecting for 248 nm). The lens structure has been obtained by scanning 193 nm laser ablation. (b) SEM picture of a micro Fresnel lens in a gold thin film on a dielectric substrate, written by using a single shot of 248 nm light and applying the mirror of (a). (K.Rubahn and J.Ihlemann, private communication, 1998.)

Projection photolithography
Recent work in projection photolithography aims to use shorter wavelengths (for example, 157 nm from an F_2 excimer laser) in order to obtain spatial resolution below 100 nm (Bloomstein et al., 1997). Due to strong absorption of 157 nm in air, the beam lines have to be purged with dry nitrogen.[11] The minimum obtainable feature size $d_{\min}$ scales with wavelength λ and is found from the Rayleigh criterion:

$$d_{\min} = k_1 \frac{\lambda}{NA}, \qquad (6.27)$$

where $NA = n \cdot \sin(\alpha)$ (2α is the full angle of the focused beam) is the numerical aperture of the projection objective (usually around 0.5) and k_1 is a coherence factor ranging between 0.55 and 0.8, depending on the exposure wavelength.[12] The values plotted in Fig. 145 may be fitted by a linear regression of the form $k_1 = 0.44 + 8\times10^{-4} \cdot \lambda$ [nm]. The resulting possible

---

[11] Laser damage to the surfaces of the optical elements is an obvious problem, too; cf. Chapter 5, Section 5.1.
[12] In the case of an Airy distribution k_1=0.61. This assumes point sources of light, which, however, is an inadequate approximation for photolithography.

lithographic resolution is also plotted in Fig. 145 for values of $NA=0.5$ to $NA=0.6$.

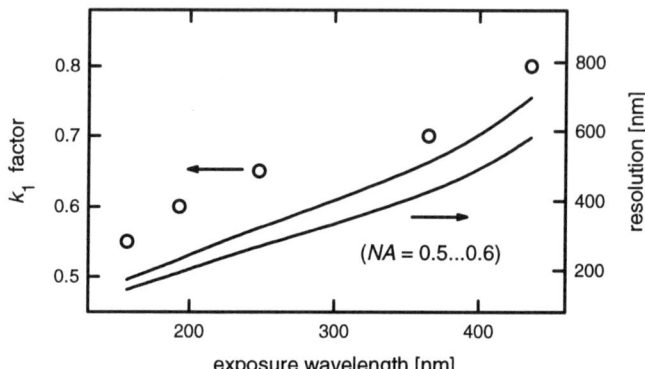

Figure 145 Values of the factor k_1 (○ (Elliott, 1995)) and range of possible lithographic resolution for NA between 0.5 and 0.6 according to Eq. (6.27).

Besides decreasing the wavelength, one might also increase the numerical aperture in order to obtain higher resolution (lower limiting curve in Fig. 145). However, with increasing NA the focus depth $f_{\min}$ and thus the depth of the structured region decreases as

$$f_{\min} = k_2 \frac{\lambda}{NA^2}, \qquad (6.28)$$

with k_2 another geometrical factor[13] with similar value as k_1. Thus the most useful parameter to be optimized is k_1, which can be reduced by applying, for example, phase-shift masks (Levenson et al., 1982; Ronse et al., 1994a). Suppose a grating to be imaged by coherent light. Due to diffraction of light from the light regions into regions that are nominally dark the image is degraded. This can be avoided by applying a phase shift layer (index of refraction n) on every second clear region with a thickness

$$d = \frac{\lambda}{2(n-1)}, \qquad (6.29)$$

which phase shifts the light by π and thus leads to destructive interference in the dark regions.

---

[13] In the Rayleigh limit one finds $k_2 = k_1/2$ (Peckerar et al., 1997), but this value, again, usually is too small in the case of photolithography.

6.2. LASER ABLATION

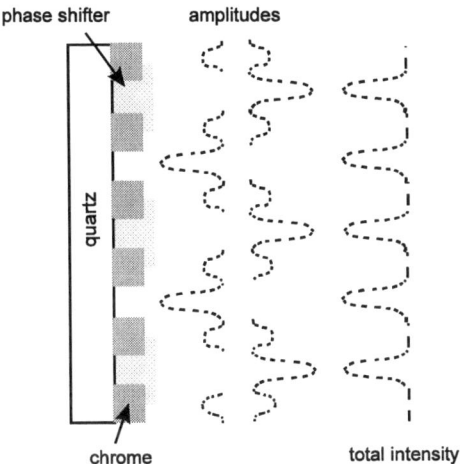

Figure 146 Principle of a phase shifting mask. The amplitudes of electromagnetic field strength from neighbouring transmitting areas are phase shifted by π, resulting in a decrease of diffraction intensity into the dark areas and thus an enhanced contrast ratio.

On the other hand, in real systems the resolution is also limited by wavefront aberrations between interfering partial beams as a result of different paths for zeroth and higher order diffracted beams. This problem can be overcome and resolution as well as focus depth can be improved by blocking the central part of the initial beam and letting zeroth and first- order beams interfere at the image plane under the same angle of incidence ('annular illumination' (Elliott, 1993)). In combination with 'soft', attenuated phase shift masks off axis illumination leads to even higher resolution, characterized by values of k_1 as low as 0.25 (Ronse et al., 1994b). An instructive introduction to that kind of problems in microlithography and microfabrication can be found in (Rai-Choudhury, 1997).

Surface structuring

A material with 'extreme' properties such as high hardness, enthalpy of evaporation and thermal conductivity is diamond. Even that material might be selectively structurized using UV laser light. The high thermal conductivity results in the generation of a graphite layer outside the ablation hole if one uses nanosecond pulses. However, if one employs femtosecond pulses, then the material outside the ablation hole is not influenced at all, and multiphoton absorption results in ablation even by use of pulses with frequencies below the indirect band gap of 5.4 eV (for example with 248 nm pulses) (Preuss and Stuke, 1995). By the same token, well-defined ablation of quartz is also possible with femtosecond pulses in the visible spectral range (790 nm) (Varel et al., 1997), where self-focusing phenomena even allow a three-dimensional microstructuring (Ashkenasi et al., 1998).

Depending on pulse energy and pulse duration surface structures of different qualities are generated. This is demonstrated in Fig. 147, which shows ablation holes generated by excimer laser irradiation of quartz using wavelengths of 193 nm and 248 nm and nanosecond ((a), (b)) as well as femtosecond (c) pulses. Apparently the cleanest ablation can be achieved by use of short wavelength light, whereas even ultrashort pulses result in the case of long wavelength radiation in a deformation of the edge of the ablation hole and in the generation of ripples inside the ablation hole.

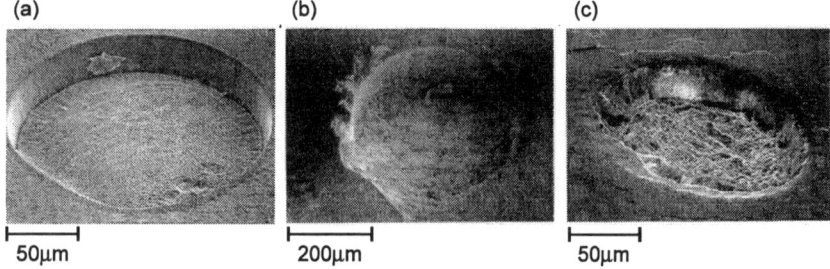

Figure 147 SEM micrographs of laser-induced holes in quartz following irradiation with (a) 193 nm, 22 ns, 50 pulses at 9.8 J/cm^2; (b) 248 nm, 22 ns, 125 pulses at 19 J/cm^2 and (c) 248 nm, 500 fs, 70 pulses at 8.2 J/cm^2. Reprinted with permission from (Ihlemann et al., 1992). Copyright 1992 Springer-Verlag.

Since the direct band gap energy of quartz (fused silica) is about 7.8 eV, light absorption is possible only via multiphoton excitation or via generation of color centers. Upon nanosecond irradiation analogous to the front side ablation shown in Fig. 147a, also rear side ablation off 1 or 2 nm thick quartz plates has been observed . This means that bulk absorption is small and the absorption is concentrated at defects on the surface ('laser damage'). Since the reflected light undergoes a phase shift at the front side whereas this does not occur for the diffracted light at the rear side, the probability for constructive interference is higher at the rear side and thus also the enhancement of intensity. From Fresnel's formulas one obtains for the ratio of laser intensities at the rear side, I_R, and at the front side, I_V : $I_R/I_V = 4n^2/(n+1)^2$ with the index of refraction n (Crisp et al., 1972). For $n = 1.5$ this results in an enhancement of intensity of 44% at the rear side compared with the front side.

6.2. LASER ABLATION

The structures seen in Fig.147 indicate that different mechanisms are responsible for the ablation. At 193 nm (Fig.147a) two-photon absorption dominates, leading to clean ablation holes and high probability for rear side ablation (negligible bulk absorption). On the other hand, in this case the ablation rates are small. At 248 nm an initial temporal regime with small ablation rate and clean holes exists, which transforms with increasing number of pulses in an explosive sputter phenomenon (Fig.147b). A positive feedback takes place: the light absorption induces surface irregularities, which lead to a higher absorption rate, resulting in an additional roughening of the surface etc. Since the plasma, which has been generated during the ablation, can no longer freely expand, additional melting and solidification processes occur at the walls.

In the case of femtosecond irradiation (Fig. 147c) the irradiance is so high that color centers are formed, which allow bulk absorption processes to occur. As a consequence, no rearside ablation is observed any longer. In this case too increasing numbers of pulses lead to explosive material removal, following an initial incubation phase. It is noted that recent time-resolved measurements of laser damage of dielectrics by visible laser pulses (526 nm and 1053 nm) have suggested that plasma formation is responsible for damage for pulse lengths below 10 ps, whereas melting and boiling processes are responsible for damage for pulse lengths above 100 ps (Stuart et al., 1995).

Surface roughening

Figure 148 shows, as an example of a laser-roughened surface, electron microscopy pictures of a mica surface which has been irradiated by 248 nm light with a fluence of a few hundred mJ/cm^2. On the left-hand side the surface is shown following irradiation in vacuum; on the right-hand side following irradiation in air. Obviously the irradiation in air leads to melting of the surface, combined with surface reactions, whereas in vacuum the sheet silicate mica is mechanically disrupted ('spallation').

6.2.3 Metals

Metal films are easily ablated upon irradiation with laser light as shown in Fig. 144b. In contrast to polymeric materials the thermal diffusion length L is, in the case of nanosecond pulses, significantly larger (≈ 1 μm) as compared with the optical penetration depth α^{-1} (≈ 10 nm (Matthias et al., 1994)). In that case thermal conduction results in melting of the borders of the ablation hole and in a loss of absorbed laser energy into the bulk of the material. The threshold irradiance increases linearly with thickness of the metal film as long as this thickness is smaller than the thermal diffusion length but larger than the optical penetration depth. If the thickness exceeds the thermal diffusion length, then uniform heating no longer takes place and the threshold becomes independent of film thickness.

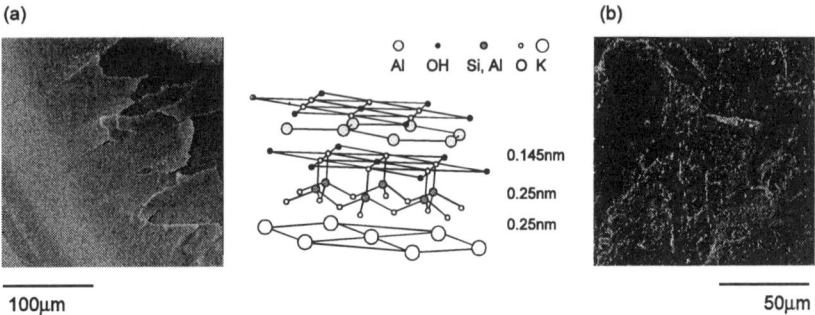

Figure 148 Scanning electron microscopy of excimer-laser-irradiated mica in vacuum (a) and in air (b). In between the SEM pictures the structure of the sheet silicate mica is plotted, where the numbers denote distances between planes of potassium atoms, silicon oxide tetraheders and hydroxylic groups.

Hence thermal ablation occurs if the energy density is higher than a critical energy density, which is determined by the ratio of threshold fluence and thermal diffusion length. Realistic values for the ablation thresholds might be estimated by reflectivity, thermal diffusion length and enthalpy of evaporation of the investigated materials in a stationary approach (Matthias et al., 1994) or by use of a thermal diffusion model (Siegel et al., 1997).

If one uses femtosecond pulses, then the thermal conductivity plays only a minor role and the ablation rate can be described just like the ablation rate for polymers with the simple Eq. (6.13) (Preuss and Stuke, 1995). The threshold irradiance for ablation with 500 fs pulses is two orders of magnitude smaller (20 mJ/cm^2 for Ni films) as compared with the threshold for ablation with 14 ns pulses. Also, since the thermal diffusion length depends on the square of the pulse length, the threshold fluence is independent of film thickness for thicknesses between 0.1 and 1 μm in the case of Ni films (Preuss et al., 1995). If the thickness of the film is sufficiently small (a few tens of nanometers), then the linear dependence of the damage threshold on thickness that has been found in the case of nanosecond pulses, is recovered (Güdde et al., 1998). However, femto- and nanosecond pulses differ in that in the former case energy transport is performed by diffusion of hot electrons (cf. Chapter 4, Section 4.1.3), whereas lattice heat diffusion is the dominant transport mechanism for nanosecond pulses. This is reflected in different critical thicknesses for Ni (50 nm) and Au (500 nm) thin films, above which the threshold becomes independent of thickness (Güdde et al., 1998). The two metals have similar thermal diffusion lengths, but the electron–phonon coupling constant for Ni is about 10^{18} W/m^{3}K, whereas it is only 10^{16} W/m^{3}K

6.2. LASER ABLATION

for Au, thus favoring lattice heating in the former case.

It is finally noted that the deposition of debris at the edge of the ablation hole is reduced too for ablation in air and one is able to machine even submicron structures (Simon and Ihlemann, 1996).

Generation of metal clusters

Laser-induced ablation of bulk metals is used to generate clusters of metals with high melting points within a molecular beam (Dietz et al., 1981). One mounts a rod of the material to be investigated directly before the exit of a pulsed nozzle (Fig. 149). A rare gas is then expanded through the nozzle under high pressure into a vacuum chamber (Chapter 1, Section 1.2.4). Upon opening of the nozzle the ablation laser is fired such that the rare gas pulse can transport the ablated material. In the course of the following expansion collisions between the expanding particles result in a cooling of internal degrees of freedom and a condensation process; i.e., large particles or clusters are formed.

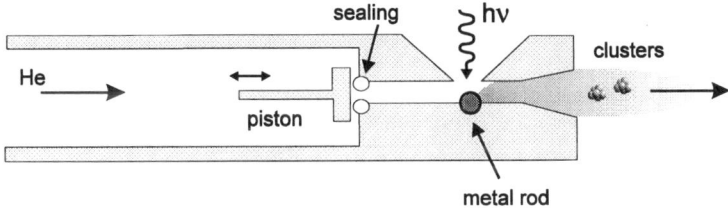

Figure 149 Pulsed nozzle source for the generation of metal clusters via laser ablation.

Generation of colloids

If the ablation of a metal is performed inside a fluid (water or organic solvents), then metallic colloids[14] might be generated with properties which are very different from those of the bulk metals or even those of thin films.

Within the colloids or colloid aggregates the electrons are collectively

---

[14] Colloidal solutions contain microparticles with diameters between nano- and micrometers, *e.g.*, soaps, ink, etc. They serve as models for molecular solutions on an enhanced, 'mesoscopic' length scale.

excited ('surface plasmon excitation', Chapter 1, Section 1.2.2).[15] If one adsorbs molecules on the colloids and performs spectroscopy, then the electromagnetic field enhancement that accompanies the plasmon excitation results in a Raman activity that is enhanced by up to a factor of 10^5 ('surface-enhanced Raman scattering', SERS, Chapter 3, Section 3.3). If one deposits the colloids onto glass substrates, then the electromagnetic field enhancement can be used to exploit the SERS effect without problems with laser-induced aggregation processes as are observed often on colloids that are immersed in solutions (Chumanov et al., 1995). Besides enhanced nonlinear optical properties, colloids (and also supported metal and semiconductor clusters, cf. Chapter 4, Section 4.1.3) are involved also as optical arrays of sub-wavelength dimensions in solar cell or photochromism applications.

The advantage of using laser ablation to generate these colloids as compared with, for example, chemical reduction of metallic salts (for example $AgNO_3$ for the generation of silver colloids) is the possibility to manipulate the size distribution and concentration of the colloids by variation of the irradiation parameters. A further advantage is that the surface of the generated colloids is wetted only from the solvent and not from additional chemicals. Recently, stable Ag, Au, Pt, Pd and Cu colloids have been generated via irradiation of an organic solution with a Nd:YAG laser (55 mJ). The colloids showed the expected SERS activity (Neddersen et al., 1993). In the course of the ablation a characteristic sound was heard — similar to the case of UV polymer ablation — which possibly resulted from mechanical breakage of the material and/or the supersonic velocity of the desorbing particles.

6.3 Laser Cleaning

In addition to structuring surfaces (Section 6.2), excimer laser radiation with moderate irradiances (10^8 W/cm^2) might also be used for the polishing and decoating of surfaces (Bergmann et al., 1995). For example, small dirt particles might efficiently be removed from fragile surfaces. Not more than five KrF excimer pulses at 248 nm with 460 mJ/cm^2 were necessary to clean the oil- and grease-covered surface of a copper single crystal, as deduced by inspection with a light microscope (Lu et al., 1994). Since the laser irradiation was performed in air, the adsorbed carbon/oxygen layer had to be removed by sputtering with high energetic ions before a microscopic analysis of the chemical surface structure via Auger spectroscopy could be performed. The

---

[15] Gold colloids of 10 nm diameter, for example, absorb at maximum near an excitation wavelength of 500 nm, silver colloids near 400 nm. Solutions with gold colloids thus appear yellow-reddish (only long wavelengths are reflected), those with silver colloids grey (all visible wavelengths are reflected). The large scattering cross section for visible light accompanied by the resonant excitation of surface plasmons results in very bright colors of the dyes or glasses in which they are diluted. This effect was used in medieval times for the fabrication of 'brilliant' windows for churches.

6.3. LASER CLEANING

analysis demonstrated that the laser treatment was able to re-establish the chemical characteristics of a clean copper surface.

Apparently the cleaning mechanism in the case of laser irradiation of metals is similar to ion sputtering. Due to the high absorptivity of the 248 nm radiation in metals the effective penetration depth is small and the method is very surface sensitive. If one applies low fluences (60 mJ/cm^2) and long irradiation times one gets better results as compared with short irradiation with high fluences (450 mJ/cm^2) since high fluences result in a strong thermal heating and this favors oxidation processes. The exact microscopic cleaning mechanism (a combination of photodecomposition, thermal laser ablation and spallation) is not known to date. Comparison with CO_2 laser irradiation, however, proves that UV light is a necessary prerequisite.

More clear conditions are given in the case of excimer laser ablation of micrometer-sized inorganic particles (*e.g.*, dust), which are bound by van der Waals or electrostatic forces on surfaces. These small particles are difficult to remove from the surface by conventional cleaning techniques such as ultrasonic cleaning, wiping, scrubbing or etching without affecting the underlying surface. Here, application of a thin liquid film (water or water/ethanol mixtures, applied via a pulsed nozzle) before laser exposing the sample proves to be very useful since the liquid will evaporate explosively. The corresponding relative pressure of 200 bar removes both the liquid and the dirt particles from the surface. As a result, a dry, clean surface is generated, which has been heated to less than 250°C (Tam et al., 1992). Cleaning occurs if the acceleration provided by the laser-induced heating exceeds the acceleration provided by the adhesion force F_{vdW}, which is dominated in the case of a dry surface by the van der Waals force. This force between a sphere of radius R and a plane surface at distance D might be written as (Israelachvili, 1992):

$$F_{vdW} = \frac{AR}{6D^2}, \qquad (6.30)$$

where A is the Hamaker constant. In the case of metals A is given essentially by the plasmon frequency (Eq. (5.8)): $A \approx 0.133 \cdot \hbar\omega_p$; hence in the case of silver one finds A=1.17 eV. For nonmetals, due to the smaller polarizability (i.e., the smaller value of the van der Waals coeffcient C_6, which enters the Hamaker constant directly), the constant has smaller values. For example, A=0.94 eV for water, A=0.4 eV for fused silica and A=0.24 eV for PTFE (Israelachvili, 1992). For a mixed system (such as aluminum particles deposited on quartz) one might use an average value of $A_{Al,Q} \approx \sqrt{A_{Al} \cdot A_Q}$=0.91 eV. Assuming an atomic separation of D=0.4 nm and a particle diameter of 1 μm, the attractive adhesion force is 7.6×10^{-8} N, giving an acceleration to a micron-sized particle of the order of 10^7 m/s^2.

The acceleration provided by the laser irradiation is approximately given by the laser-induced thermal expansion Δx of the particle (or the substrate surface), divided by the laser pulse length $\tau_p \approx 10^{-8}$ s. The thermal expansion is

$$\Delta x = \gamma L \Delta T, \qquad (6.31)$$

with ΔT the temperature rise in the particle (Eq. (5.24) since the absorption depth for metals is small and we are talking about a surface heat source), linear thermal expansion coefficient γ and thermal diffusion length $L = 2\sqrt{\kappa \tau_p}$ (cf. Chapter 5, Section 5.1) with κ the thermal diffusivity.

For example, one finds for irradiation of 1 μm diameter aluminum particles with 100 mJ/cm^2 KrF excimer laser light a temperature rise of about 40 K at the surface side due to absorption of light in the metal (Lu et al., 1997). Together with the thermal and mechanical constants of aluminum ($\gamma = 25 \times 10^{-6}$(°C)$^{-1}$, $\kappa = 10^8$ m^2/s (Weast, 1982)) this results in an acceleration of several 10^{12} m/s^2, which is five orders of magnitude larger compared with the adhesive acceleration. The strong acceleration is one of the main constituents of laser cleaning. Note also that the efficiency of cleaning can be increased if the laser absorption occurs mainly in the substrate or the substrate surface as compared to absorption in the dirt particle or the fluid that surrounds the particle.[16]

In general, excimer laser cleaning of surfaces is applied to metallic and non-metallic surfaces with equal satisfaction. In the industrial praxis it is used for the cleaning of difficult-to-approach or heat-sensitive samples. Examples include restoration of historical architecture, paintings or fabrics (Kautek and König, 1997), but also — in basic research — preparation of crystallographically ordered insulator surfaces. Commercially attractive restoration by use of Nd:YAG lasers of very dirty objects made of marble, sandstone, terracotta, plaster, metal, wood, paper and canvas has been reported. In all these cases the dominating mechanism is the generation of a surface shock wave, which results in a burst-off of the dirt particles.

Figure 150 shows an example of a *microscopic* surface treatment, namely LEED pictures of a mica surface before and after irradiation with a few 248 nm excimer laser pulses. An insulator such as the sheet crystal mica is, due to its flatness (roughness below 2 Å over 1 μm^2) and the simplicity of its preparation (cleaving along given cleavage planes such as the plane containing potassium atoms, cf. Fig. 148), an ideal substrate for epitaxial growth of thin metal films. Problems arise from adsorbates on the mica surface, which migrate out of the bulk onto the surface or adsorb from background gas. Due to charging effects insulators are sputtered only with difficulty, and thermal cleaning might lead

---

[16] An additional advantage of absorption in the substrate is that surface acoustic waves (SAWs) are generated if the laser pulse is sufficiently short and strongly focused (Gusev and Karabutov, 1993). These Rayleigh type waves extend over a large area and also lead to efficient cleaning by applying normal and tangential acceleration to the adsorbed dirt particles (Kolomenskii et al., 1998) (cf. Chapter 4, Section 4.2.2). Of course, the limiting factor here is the destruction of the substrate via laser damage.

6.3. LASER CLEANING

to the generation of crystal water or calcination processes, which destroy the ordered surface. Irradiation with UV light (248 nm) proves to be a valuable alternative since mica is opaque for this wavelength. An absorbed fluence of only a few tens of mJ/cm^2 results, within 20 ns, in an increase in surface temperature by several hundred degrees Kelvin. This high heating rate leads to an efficient desorption of adsorbates without destruction of the bulk structure of the crystal. At the same time such a laser-irradiated insulator surface forms an ideal basis for the growth of thin metal layers (Polanski and Rubahn, 1996).

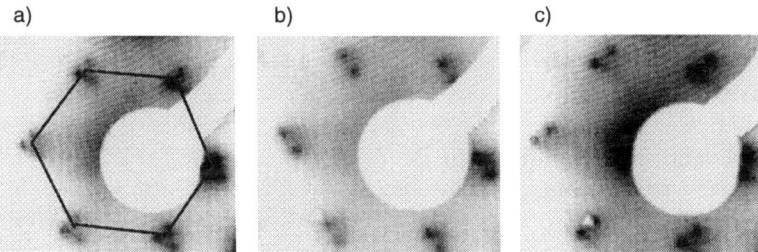

Figure 150 Low energy electron diffraction (LEED) from a mica surface (a) without laser irradiation, (b) and (c) following irradiation with two and ten laser pulses (248 nm) of 90 mJ/cm^2, respectively. Reprinted from *Surf. Sci.*, (Gerlach et al., 1996b), Copyright 1996, with permission from Elsevier Science.

The electron diffraction picture Fig. 150a shows a hexagonal structure (gray lines in Fig. 150a) with the lattice constant of mica (5.12 Å). The bright spot is the screening of the electron gun. Each of the six visible LEED spots is split into three smaller spots. This effect results from the generation of surface dipoles at the mica surface during the cleavage process. The dipoles are mutually rotated by 120° due to the bulk symmetry of mica. The effective dipole fields at the surface deflect the reflected electron beam from the specular direction, each domain into another angle. A symmetric picture is only obtained if the domains are smaller than the diameter of the electron beam of 1 mm. If one irradiates the surface with two excimer laser pulses of 90 mJ/cm^2 (Fig. 150b), then the hexagonal structure is conserved, meaning that the surface is not destroyed microscopically, but the splitting is reduced to two spots. A further eight laser pulses (Fig. 150c) result in a rotation of the spots. This kind of orientation and splitting remains even after further irradiation.

An explanation of this effect is that laser irradiation increases the size of the dipole domains via selective desorption, meaning that only two domains fit into the given diameter of the electron beam. Since the laser light is partially polarized, potassium ions are desorbed, which are oriented in a preferred direction with respect to the electric field vector of the laser. The remaining dipole domains, which are oriented perpendicular to them, contribute eventually to the splitting of the electron beam. Thus the excimer laser treatment allows a microscopic surface preparation.

6.4 Laser-Induced Periodic Structures

It is well known since 1965 that irradiation of a semiconductor surface with intense, *linearly* polarized light pulses from a ruby laser results in a periodic pattern on the surface, which is oriented perpendicular to the electrical field vector of the incoming light (Birnbaum, 1965). Experiments on metals, dielectrics, thin films and liquids have shown since then that this phenomenon is 'universal' (van Driel et al., 1982) and can occur on every surface. A comprehensive account of the literature up to 1986 is given in (Siegman and Fauchet, 1986). In that article the similarity of the phenomenon with the physical basis of the well-known Wood's anomalies (Wood, 1902) in light scattering from diffraction gratings is pointed out. It is suggested that the ripple formation effect be termed the 'stimulated Wood's anomaly', thus explaining its universality. Possible applications of the effect in laser annealing, photoetching, pulsed laser deposition, damage studies, optical grating fabrication and coupling to photodetectors are also discussed in (Siegman and Fauchet, 1986).

Figure 151 shows that a sinusoidal modulation of the surface structure can be generated over a spatially widely extended region that way. The distance between maxima corresponds, for irradiation under normal incidence, to the wavelength of the exciting laser. The orientation of the modulation is dominated by the orientation of the electric field vector of the irradiating light and in the case of very high laser intensities it depends only on the crystallographic structure of the surface.

This means that it is possible to choose the orientation of the spatial modulation with respect to the crystal axes — a side effect that might be relevant for possible applications. The modulation of the grating (defined by the diffraction efficiency I_1/I_0) increases with increasing number of applied laser pulses exponentially in the case of CO_2 laser irradiation on quartz (Keilmann and Bai, 1982). It becomes constant for $I_1/I_0 \approx 1\%$ and decreases afterwards. As a function of laser intensity one observes — depending on the initial roughness of the irradiated surface — transient, wavy structures for low intensities, which vanish if the laser is shut off. If the laser intensity is high enough to heat up the surface to the melting point, then permanent structures are generated. With a further increase in laser power the structures start to smear out until the start of nonlinear mechanisms results in quadratic surface

6.4. LASER-INDUCED PERIODIC STRUCTURES

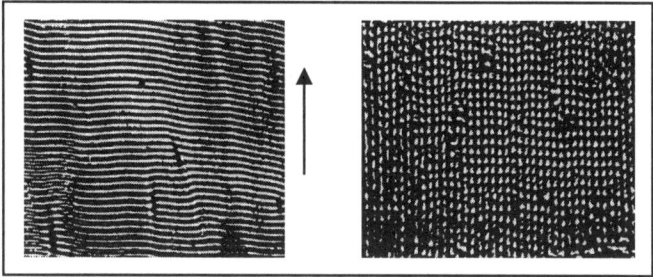

Figure 151 SEM micrographs of periodical surface structures on solid surfaces, induced by linear (left-hand side) and nonlinear light matter interactions (right-hand side). Left-hand side: written with a CO_2 laser and small intensity in suprasil. The arrow denotes the direction of the electric field, its length a scale of 100 μm. Right-hand side: written with a Nd:YAG laser at high intensity in germanium. In both cases the laser hits the surface under normal incidence. Left-hand side: Reprinted with permission from (Keilmann and Bai, 1982). Copyright 1982 Springer-Verlag. Right-hand side: Reprinted with permission from (Fauchet and Siegman, 1983). Copyright 1983 Springer-Verlag.

structures (Fig. 151, right-hand side). Such kind of structures are useful for the generation of electromagnetic field enhancement effects (*e.g.*, for surface-enhanced Raman scattering; see Chapter 3, Section 3.3).

In the case of strongly absorbing surfaces the effect might be described qualitatively in the following way: (a) Interference of the incoming light wave with the wave that is reflected at surface roughness, leads to the generation of a spatial grating on the surface. (b) At the position of the interference maxima the surface is strongly heated, resulting in melting and evaporation processes. A permanent grating is molten into the surface from which the incoming light wave is scattered since the corrugated surface acts as a periodic dielectric waveguide for the incident radiation (Hiraoka and Sendova, 1994). This kind of positive feedback results in a strong (approximately exponential in irradiation time) increase in the depth of the grating, until the lateral evolution of heat limits the lattice modulation by the occurrence of melting processes close to the interference minima. Hence the quality of the surface grating depends very much on the thermal material constants. On polymer surfaces gratings with amplitudes of greater than 100 nm and periods of 280 nm have been generated via irradiation with the fourth harmonic of a Nd:YAG laser[17] ($\lambda=266$ nm) (Hiraoka and Sendova, 1994).

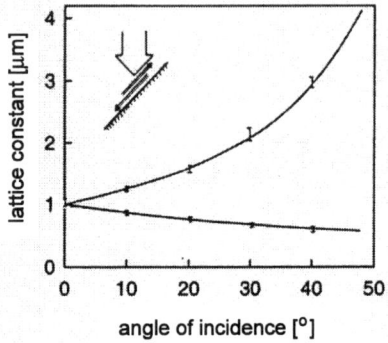

Figure 152 Periodicity of lattice structures, generated in silicon with p-polarized Nd:YAG laser light. The solid lines correspond to Eq. (6.32). Reprinted with permission from (Guosheng et al., 1982). Copyright 1982 American Physical Society.

This description of the effect does not explain the initial appearance of the surface grating and also does not account for the nonlinear interaction, which eventually results in a quadratic grating. Although the effect is 'universal', the initial generation of the grating depends on the irradiated surface. In cases where the surface supports excitation of plasmon polaritons or acoustic waves the most probable mechanism for ripple formation is coupling of the incoming light wave to those acoustic surface waves or to capillary waves. The corresponding interaction process is discussed in detail in (Akhmanov et al., 1990). For surfaces that do not support elementary excitations of the discussed kind the formation mechanism is based on the intrinsic lattice structures that are given by the roughness of a surface. Note that the importance of surface roughness for the process can be easily exploited by applying macroscopic scratches on the surfaces, which initiate the grating formation process.

A surface with statistically distributed roughness can be described as the superposition of a large number of surface gratings with different spatial

---

[17] A Nd:YAG laser is advantageous as compared with an excimer laser due to the better defined beam polarization.

6.4. LASER-INDUCED PERIODIC STRUCTURES

periodicities Λ_i.[18] As a result of the surface lattices the incoming light wave is diffracted into surface waves (Guosheng et al., 1982). If the laser irradiates the surface under an angle Θ, 'upward' and 'downward' running surface waves are generated (Fig. 152). The 'upward' running waves ('—') possess a larger periodicity Λ compared with the downward running waves:

$$\Lambda = \frac{\lambda}{1 \pm \sin\Theta}. \tag{6.32}$$

Interference of the incoming light wave with these surface waves results, via positive feedback, in an enhancement of that Fourier component of the rough surface that is in-phase with respect to the incoming light wave. Hence the absorbed laser power is modulated. If one irradiates at a given angle with respect to the surface, then one obtains lattice patterns with two different periodicities. If one irradiates along the surface normal, then a single grating is generated. The measured values (Fig. 152) show that this behavior is indeed observed for Nd:YAG laser irradiation of silicon.

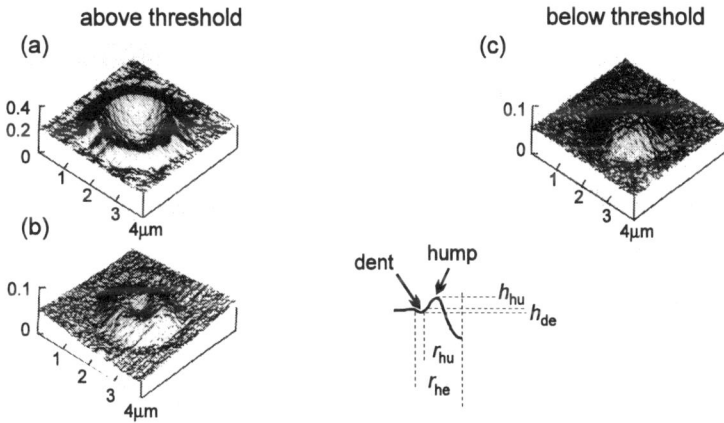

Figure 153 Topological changes in single shot UV laser (λ=302 nm, pulse length τ_l) irradiated PI surfaces, determined by force microscopy. The three figures correspond to different values of laser fluence F_0 with respect to the threshold fluence for ablation, $F_{th} = 37$ mJ/cm^2 at λ=248nm (Braun et al., 1989). (a) $F_0 \approx 1.25 \cdot F_{th}$, τ_l=800 ns; (b) $F_0 \approx 1.01 \cdot F_{th}$, τ_l=2.1 μs; (c) $F_0 \approx 0.96 \cdot F_{th}$, τ_l=5 μs. Reprinted with permission from (Himmelbauer et al., 1996). Copyright 1996 Springer-Verlag.

With increasing laser power the regime of linear interaction between

---

[18] This kind of roughness is not necessarily a height corrugation. It might also be a modulation of the dielectric function (which determines the optical response of the material on the incoming light wave) via fluctuations of temperature or electron density.

the incoming light wave and substrate is left and higher order nonlinear interactions between gratings having different wave vectors on the surface occur since the surface profile is changed by the initial interaction with the laser. This means that the spatial frequency components of higher order of the surface profile become important with increasing laser intensity. The summation of the corresponding wave vectors results in modulations that are rotated by 90° with respect to the initial modulation and have periodicities $\Lambda_\perp = \lambda/\cos\Theta$. The superposition of the initial with the subsequently generated gratings eventually results in a quadratic lattice pattern[19] (Fig. 151, right-hand side) (Fauchet and Siegman, 1983).

Recently, a more detailed picture of the microscopic mechanism of laser-induced morphology changes has evolved on the basis of force microscopy measurements on PI surfaces (Fig. 153) (Himmelbauer et al., 1996). Upon irradiation with a pulsed UV Ar$^+$ laser (λ=302 nm, τ_l=140 ns - 5 μs) just below the ablation threshold the formation of a hump on the surface was observed, surrounded by a dent. A further increase in the laser fluence led to the formation of a central hole, from which ablation of small carbon fragments (Küper et al., 1993) took place. Close to the threshold fluence, maximum values of both hump and dent of h_{hu}=18±10 nm and h_{de}=7±4 nm with radii r_{hu}=1.2±0.2 μm and r_{de}=1.7±0.3 μm were found. Note that the Gaussian radius of the laser beam was w=2.1 μm so that the fluence at r_{hu} was about 70% of the threshold fluence.

The volume increase that coincides with hump formation can be explained by the amorphization of crystalline domains, scission of polymer strands and possibly subsurface gas formation. This finding is in agreement with results from previous time-resolved light-scattering measurements on such gratings in polymers (Niino et al., 1990), which were interpreted in terms of a microstructure formation via the generation of a fluidized, well-localized subsurface layer. The dent might result from stress release, i.e., plastic deformations. Once a hump has been formed, the generation of an interference pattern either from two overlapping beams or from a single beam ('ripple' formation, at least in the case of multiple shot experiments) becomes possible, while the generation of humps can cause row doubling (Himmelbauer et al., 1997).

Due to the higher degree of spatial coherence of femtosecond laser light, enhanced ripple formation is expected (and observed) for interaction of ultrashort laser pulses with surfaces (Preuss et al., 1993). It is noted that the formation of transient ripples rather than a modulation of the dielectric function through elasto-optic coupling is assumed to be responsible for impulsive stimulated thermal scattering, a technique that is used to determine the energy of acoustic phonons on surfaces or in thin films (cf. Chapter 4,

---

[19] Other laser-assisted methods of formation of sub-micrometer periodic structures are denoted in Chapter 1, Section 1.2.3.

6.5 Laser–LIGA

The 'cold' UV laser ablation of polymers, described in Section 6.2, implies that one is able to UV laser-write structures precisely (with a resolution of micrometers or even better by use of holographic techniques) into polymers without destroying the surroundings.

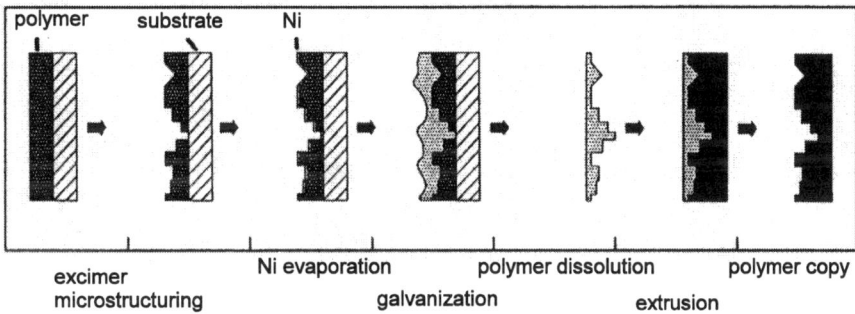

Figure 154 Laser–LIGA. Reprinted from *Appl.Surf. Sci.*, (Arnold et al., 1995), Copyright 1995, with permission from Elsevier Science.

The incubation of the polymer surface with a weak UV pulse or even with cw UV light results in unsaturated fragments of the initial polymer, the absorption bands of which are shifted to longer wavelengths, i.e., having higher absorption coefficients in the UV (Küper and Stuke, 1989). If one dopes the polymer with a few volume percent of an absorbing dye, then one can write, with a few laser pulses in a variety of polymers, structures with a quality which is nearly independent of the initial absorption characteristics of the polymer (Preuss et al., 1993).

By the use of fluences that change as a function of time, different focusing conditions and different angles of irradiation even *three-dimensional* structures can be generated. This is not possible with conventional techniques by the use of a mask and homogeneous irradation. On the other hand, mass fabrication of structures that have been directly written with an excimer laser is nearly impossible due to the extended manufacturing time. A way out of this dilemma is the initial generation of the structures of interest with the laser and their reproduction by a combination of lithography, galvanoforming and plastic molding ('LIGA', in German: 'Lithographie, Galvanoformung, Abformung') (Ehrfeld and Münchmeyer, 1991). The resulting 'laser-LIGA'-method is shown

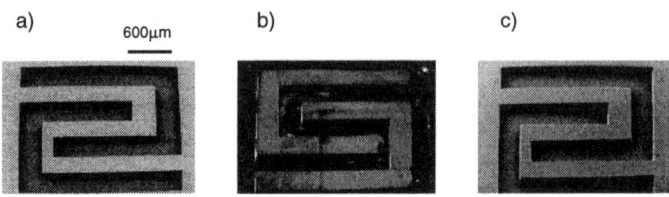

Figure 155 SEM micrographs as examples of Laser–LIGA: (a) Structure, generated in 200 μm PMMA by an excimer laser. (b) Nickel replica. (c) Injection-molded PMMA structure, using the replica. Total width of the structures is 3 mm. Reprinted from *Appl.Surf. Sci.*, (Arnold et al., 1995), Copyright 1995, with permission from Elsevier Science.

schematically in Fig. 154 and its applicability is demonstrated by an example in Fig. 155.

After generation of the master structure via excimer laser ablation the surface is coated by a thin metallic layer (50–100 nm thick) and is then electroplated with nickel (electric elimination of nickel from a wetting solution). The polymer layer is removed via shock-freezing at liquid nitrogen temperature and via chemical removal in trichloromethane ($CHCl_3$) and the rear side is polished down to a maximum thickness of 500 μm. Then the nickel structure serves as an inverse master sample, which can be used for polymer copies by melting in the polymer.

The implication of a UV laser (or equivalently an ultraviolet lithography system) to this method allows one to generate a polymer original quickly and reliably. In contrast to X-ray lithography, which allows one to obtain higher resolution and larger structure depths (aspect ratios of 100, compared to 10 in the case of laser–LIGA), here an improved three-dimensional structurization (different height structures and skewed walls) becomes possible, which is also significantly less time consuming.

6.6 Laser CVD

Laser chemical vapor deposition (LCVD) is a photochemical method that takes advantage of the fact that irradiation with laser light results in a local increase in temperature which facilitates the growth of thin films at significantly lower substrate temperatures as compared with methods employing no lasers. The resulting low process temperature avoids, in lithographic processes, the destruction of initially generated structures via

6.6. LASER CVD

thermal diffusion. This is especially important for the generation of structures in the micrometer to nanometer range. Extensive reviews of methods and processes can be found in (Bäuerle, 1986; Boyd, 1987; Letokhov, 1988; Herman, 1989; Bäuerle, 1996b; Bäuerle, 1996a).

Figure 156 exemplifies two possible ways of thin film or nanostructure generation via laser CVD. In the case of thermochemical processes (Fig. 156a, e.g., pyrolysis, oxidation or etching) the laser serves to increase locally the surface temperature. The reactants are applied to the surface by a gas jet and adsorb either before surface heating or induced by the surface heating. Within the laser focus the temperature-activated chemical reactions or the thermal fragmentation of the reactants take place. The resulting products desorb or diffuse out of the reaction zone. In the case of photochemical CVD (Fig. 156b) the reactants are excited in the gas phase by the laser light. Some of them relax, while the remaining ones are used as reactants on the surface, again some of them recombining and some reactively adsorbing. A competing process to the reaction is desorption of the reactants. After finishing the reaction, the byproducts desorb and diffuse out of the reaction zone just as in the case of a thermochemically induced reaction.

If the rate of breaking the bonds in the gas phase or the reaction kinetics on the surface is the rate-limiting step, then a linear increase in product yield ($\propto t \cdot \exp(-E_a/k_B T)$, t meaning the reaction time and E_a the activation energy) is to be expected. If the diffusion of the reactants along the surface is rate limiting, then one expects a parabolic increase ($\propto \sqrt{t} \cdot \exp(-E_a/k_B T)$. If diffusion can be neglected (e.g., for low irradiances where the laser does not significantly decrease the number of possible reactants), then the reaction yields from the gas phase, Y_{gas}, and on the surface, Y_{surf}, might be estimated as (Boyd, 1987)

$$Y_{gas} \propto n_r \sigma P w^{-1}, \tag{6.33}$$

and

$$Y_{surf} \propto \Theta \sigma P w^{-2}. \tag{6.34}$$

Thus the reaction rates are proportional to the density of the reactants, n_r, multiplied by the photochemical cross section σ or the product between surface coverage Θ and cross section, respectively. On the surface the reaction rate increases linearly with laser intensity (power P proportional to the beam waist-area w^2). In the gas phase the reaction rate is only proportional to w^{-1} due to geometrical reasons: the reaction products are formed in three-dimensional space above the surface and not only inside the irradiated two-dimensional area.

An examplary set-up for cw laser-induced microstructuring is shown in Fig. 157. Using this method, it is possible to generate three-dimensional structures in aluminum oxide, which can be metallized in a second step via

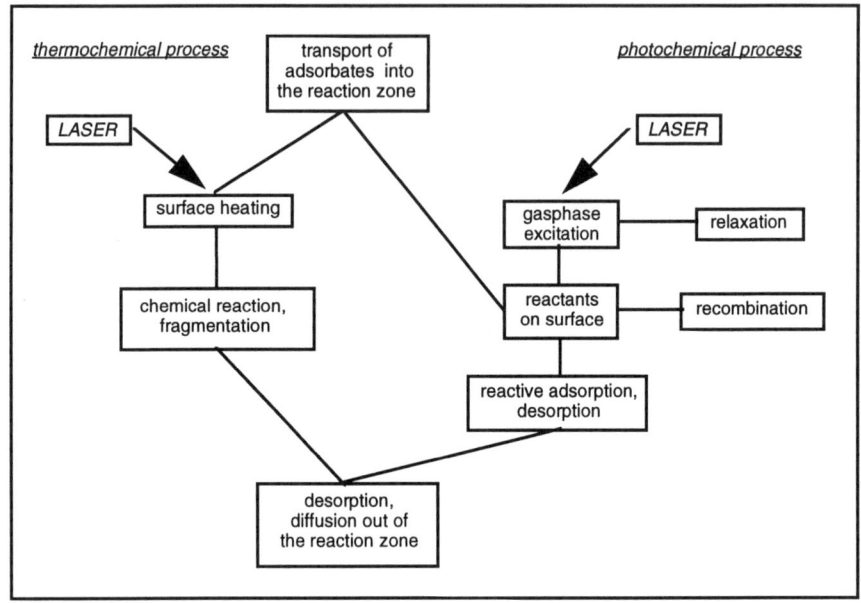

Figure 156 Possible flux plot of thermochemical and photochemical laser-induced surface reactions that might occur in the course of thin film fabrication.

laser-induced elimination of aluminum (Lehmann and Stuke, 1994). For this purpose within a reaction chamber a mixture of 65% trimethylamine alane (TMAA, $AlH_3N(CH_3)_3$) and 35% oxygen is inserted. An Ar^+ laser beam is divided into two beams with 0.3 mW laser power each, which are focused onto a common point at the substrate by the use of a microscope objective. Within the focus area with a diameter of 3 μm aluminum oxide is eliminated from the gas phase. If one moves the substrate in three dimensions, then an Al_2O_3 rod is grown in the laser focus. The power of both laser beams for the discussed experiment was so small that the activation threshold for the elimination-reaction could be overcome only by the application of both beams simultaneously. In that way a directed growth was induced which allowed the generation of complex three-dimensional structures as sketched on the left-hand side of Fig. 157. If one uses only a single laser beam, then the structure to be generated has to be written two-dimensionally into a three-dimensional support (*e.g.*, a polymer), which is removed by a chemical reaction in a subsequent process step (Lehmann and Stuke, 1991). An example of this latter method is shown in Fig. 158a.

6.7. PULSED LASER DEPOSITION

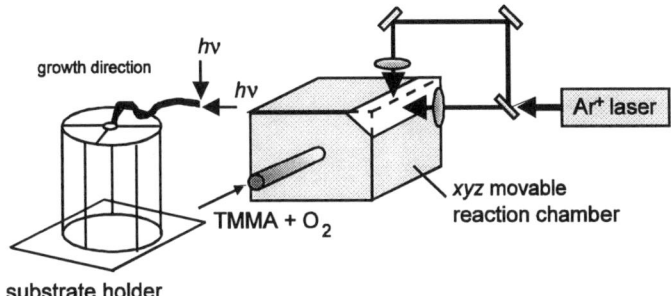

Figure 157 Three-dimensional laser CVD, induced by a cw Ar^+ laser in a reaction chamber. On the left-hand side a possible structure is shown, which grows in the focus region of the two laser beams.

Whereas free-standing structures with diameters of less than 10 μm can be manufactured from nonconducting materials by the use of a single laser beam (Fig. 158b), this becomes difficult with metal structures. A serious problem in the latter case is that the thermal conductance of aluminum depends only slightly on the sample temperature, whereas in the case of an insulator such as aluminum-oxide it decreases strongly with increasing sample temperature due to an increase in competing phonon–phonon scattering processes. Consequently, the necessary temperature increase in the case of fabrication of Al_2O_3 remains localized in the laser focus, whereas it becomes spatially smeared-out in the case of metals. As noted above, it is, however, quite possible to eliminate inside the laser focus a metallic coating on the oxide structure and hence to combine conducting and nonconducting parts in a microstructure.

Recently it has been shown that it is possible to generate in that way even mechanical elements such as tweezers or linear motors with characteristic dimensions in the 100 μm range (Lehmann and Stuke, 1995). Laser-induced thermal expansion of specified elements of these structures results in micromechanical motion, including forces in the nano- to micro-Newton regime.

6.7 Pulsed Laser Deposition

Thin film structures on surfaces, which cannot be produced via conventional sputter techniques (electron or ion beam sputtering), might be generated via *laser sputtering* or pulsed laser deposition (PLD). As shown in Fig. 159, the

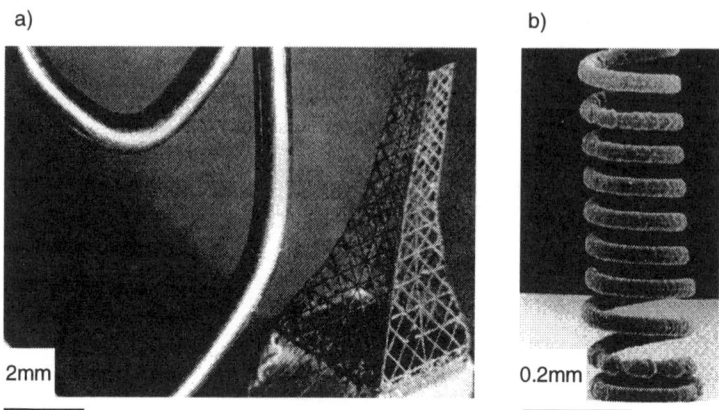

Figure 158 Structures that have been generated by three-dimensional laser CVD. (a) Grid structure made of aluminum, initially adsorbed on polycarbonate, which later was dissolved. (b) Microspring made of boron. (a) Reprinted with permission from (Lehmann and Stuke, 1991). Copyright 1991 Springer-Verlag. (b) Reprinted with permission from (Johansson et al., 1992). Copyright 1992 American Institute of Physics.

method is conceptually very simple. It is nowadays a well-established tool for the growth of dielectric, metallic, semiconductive and superconductive thin films. Prominent features of PLD include 10^5 times higher deposition rates as compared with conventional techniques, particle energies of the order of 100 eV due to the high laser power densities and consequently defect and stress formation in the generated films as well as a high implantation rate. Hence metastable phases and highly supersaturated solid solutions can be prepared (Krebs, 1997).

As an example, thin films made of one-dimensional conductors, e.g., Nb_5Te_4, might be generated that way. A rotating $NbTe_2$ target is irradiated by a KrF excimer laser (5 J/cm^2). The ablated Nb and Te atoms are deposited on a silicon substrate (T=680 K) and form a thin (300 nm thick) Nb_5Te_4 film, the roughness of which depends strongly on laser irradiance and surface temperature (Grangeon et al., 1995). The speed of the ablated atoms can be varied over a wide range (4 to 85 km/s, corresponding to kinetic energies of 9.375 eV to 37.5 eV for the tellur atoms at 3 J/cm^2 to 11 J/cm^2 fluence). It has been observed by measurement of the velocity distributions of the ablated atoms that optimum growth conditions of the film are provided if the atoms

6.7. PULSED LASER DEPOSITION

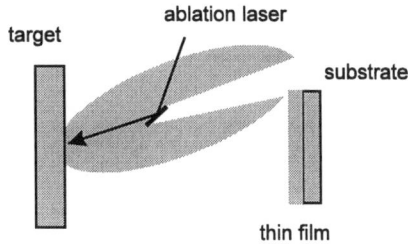

Figure 159 Set-up for pulsed laser deposition of thin films.

of the different species arrive at nearly the same time. Hence by changing the laser fluence a good control of the layer growth process becomes possible.

As noted above (Section 6.2), cone formation at the surface of the target, from which the material is evaporated, is a major problem in PLD, since it leads to an enhanced droplet formation, the deposition rate decreases as a function of time, and the deposition angle is no longer controllable. The cones grow in the direction of the laser beam and the laser beam hits the target under an angle of a few tens of degrees with respect to normal incidence. Thus the peak intensity of the desorbed particles misses the substrate (Fig. 159). This problem can be partially overcome by use of a rotating target and by rastering the laser beam. Another possibility is to avoid cone formation by use of ultrashort laser pulses and/or high fluences well above the ablation threshold. This is demonstrated in Fig. 160.

Theoretical models of laser sputtering for thin film deposition purposes are of course intimately related to the more general ablation models discussed in Section 6.2. However, as shown above, for PLD the regime of high fluences is of special interest. Here, a variety of models have emerged, including normal vaporization from the outer surface and down to a depth related to the absorption length as well as subsurface superheating[20] and explosive boiling or 'phase explosion'. A critical discussion of those models is provided in (Miotello and Kelly, 1995), denoting that the thermal part of the process is dominated by explosive boiling. In contrast to normal vaporization, which occurs at all fluences, for explosive boiling to occur the pulse length has to be sufficiently short and the fluence sufficiently high so that the surface temperature approaches the thermodynamic critical temperature. Here, the

---

[20] I.e., rapid heating so that the system reaches temperatures above the boiling temperature in the liquid phase.

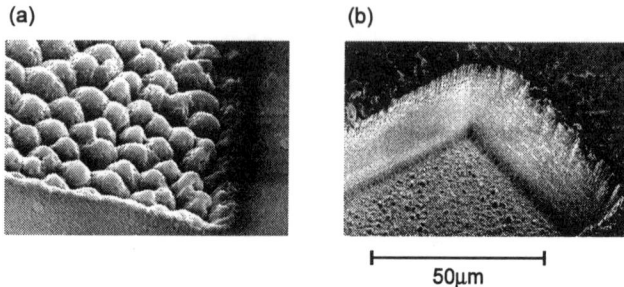

Figure 160 SEM pictures of mica samples, irradiated in ambient air with 248 nm excimer laser radiation. (a) 1000 shots at 0.3 J/cm^2. (b) 500 shots at 9.67 J/cm^2. In the latter case cone formation is avoided. (K.Rubahn and J.Ihlemann, private communication, 1998)

homogeneous nucleation rate rises strongly, and the substrate undergoes a rapid liquid/vapor (and equilibrium droplet) phase transition. The explosive boiling is accompanied by a strong pressure rise above the liquid and thus the emission of fairly large (up to 1 μm diameter) droplets. This explains the high probability for droplet formation during PLD.

7
Laser Medicine

7.1 Medical Laser Surface Treatment

Less than two years after the development of laser light sources the laser was applied to heat treatment of organic tissues (photocoagulation of retinal tissue, corresponding to a denaturation of the albumen (Zaret et al., 1961)) and shortly thereafter to drilling of holes into teeth in order to remove caries (Goldman et al., 1964). Important boundary conditions for the optimum application of laser radiation are the matching of the power and energy density of the light to the thermal properties of the tissue as well as the use of a wavelength that is absorbed solely at those points where it should. Use of continuous, low power light allows one to remove, over wide areas, metastases of tumors via laser-induced photoreactions. For this purpose porphyrin derivatives are enriched in the tumors, which absorb the laser photons and transfer the energy photochemically to molecular oxygen (Dougherty et al., 1979). In that way singlet oxygen atoms are generated that destroy the tumor tissue.[1] Similar to the process of laser ablation of tissue via UV-light this photochemical application does not generate much heat.

Short, intense laser pulses of nano- to picosecond length are used to generate, via nonlinear absorption processes, optical breakdown plasmas in tissue fluids, for example within the eye (Vogel et al., 1990). The increase in temperature of more than 10^4 K within a few nano- to picoseconds results in a strong increase in pressure of about 50 kbar, a supersonic expansion of the plasma and the generation of a shock wave, which leaves a cavitation bubble behind it. The collapsing cavitation bubble induces a secondary pressure rise within about a millisecond, which is used to open selectively cloudy lens membranes in the eye, for example resulting from a glaucoma operation.

Other important laser applications in medicine are of a spectroscopic nature.

---

[1] The disadvantage of porphyrins is their small absorption depth, which makes them usable mainly for surface tumors. Nowadays one tries to apply special, antibody-coupled dyes, which mark *all* tumor cells selectively at the cell-biological level and thus makes them treatable by laser radiation.

By the use of short pulse absorption spectroscopy short-living intermediates of biological processes might be observed. Tumors are easily identified by marking them with fluorescing dyes and subsequent fluorescence spectroscopy. They might be destroyed with light from the same laser source that identifies them.

Given the multitude of possible applications of lasers in medical diagnosis and therapy and its user friendliness (especially the accessibility of nearly all parts of the human body via endoscopic surgery), it is obvious that lasers have been applied nowadays to nearly all medical branches. Examples include cardiology (Horvath et al., 1995), dentistry (Featherstone et al., 1998), dermatology (Maser et al., 1983), gastroenterology (Sander et al., 1989), gynecology (Baggish, 1986), neurosurgery (Edwards et al., 1983; Loesel et al., 1998), ophthalmology (Wollensak, 1988; Ren et al., 1995), orthopedics (Horoszowski et al., 1987), otorhinolaryngology (Duncavage et al., 1985) and urology (Hofstetter, 1986). In Table 3 the medical branches are listed together with the most commonly used lasers and the main physical mechanism that is taken advantage of. The abbreviations of the lasers mean: Er:YAG, Ho:YAG - erbium and holmium doped yttrium aluminum garnet lasers; Nd:YLF - neodynium doped yttrium lithium fluoride laser; Er:YSGG - erbium doped yttrium scandium gallium garnet laser.

7.2 Lasers in Ophthalmology

In what follows some new findings in the field of *ophthalmology* are discussed with respect to the surface applications of lasers, with special emphasis on corrections of the cornea. More extended discussions of medical laser applications can be find, besides the above references, in (Hillenkamp and Sacchi, 1980), (Berlien and Müller, 1993), (Eichler and Seiler, 1993), (Niemz, 1996) or in special articles of the periodicals *Lasers in Surgery and Medicine* and *Lasers in Medical Science*.

Figure 161 is a schematic plot of the human eye. The numbers denote the regions where laser light is applied for medical purposes. The refractive properties of the eye are determined by the cornea, lens and the distance between both (image plane). About 80% of the refractive power D of the eye, or 49 diopters (=49 m^{-1}), are generated by the curved cornea (n_c=1.376), the final percentage by the lens ($n \approx 1.4$), which can be bent to a variable extent. The refractive power D is given by the radius of curvature of the anterior surface of the cornea, r=7.7 mm, and the index of refraction of the cornea, n_c: $D = (n - n_1)/r$.

7.2. LASERS IN OPHTHALMOLOGY

Table 3 List of medical branches where lasers are implemented. Exemplary references can be found in: a: (Horvath et al., 1995); b: (Goldman et al., 1964); c: (Niemz, 1995); d: (van Gemert et al., 1995); e: (Sander et al., 1989); f: (Bastert and Wallwiener, 1992); g: (Fischer et al., 1994); h: (Zweng, 1971); i: (Christ et al., 1995); j: (Ith et al., 1994); k: (Duncavage et al., 1985); l: (Dretler, 1988). The abbreviations mean: 'coag.'='coagulation'; 'vap.'='vaporization'.

Branch	Laser type	λ [nm]	Reference	Mechanism
Cardiology	CO_2	10600	a	excision
Dentistry	Er:YAG	2940	b	thermomechanical
Dentistry	Nd:YLF	1053	c	ablation
Dermatology	various	various	d	photolysis, coag., vap.
Gastro--enterology	Nd:YAG	1064	e	coagulation
Gynecology	CO_2	10600	f	coag., vap., excision
Neurosurgery	Nd:YLF	1053	g	plasmaind. ablation
Ophthalmology	various	various	h	various
Orthopedics	Ho:YAG	2120	i	thermal ablation
Orthopedics	Er:YSGG	2780	j	thermal ablation
Otorhino--laryngology	CO_2	10600	k	thermal ablation
Urology	Nd:YAG	1064	l	coag., disruption

Changes of curvature of the cornea[2] strongly influence the effective focal distance of the eye. On the other hand, weaknesses in the strength of vision, which are caused by a change of the distance between lens and image plane,[3] might be corrected at least partially by changing the curvature of the cornea. Here in recent years an important practical field of application for the excimer laser ablation has emerged (excimer photorefractive keratectomy, PRK) (Pettit et al., 1995).

Figure 162 shows simple set-ups for PRK.

Since one of the main goals is to ablate only the surface of the cornea and not to affect the surrounding tissue, lasers are used that ablate surface-sensitive and 'cold'. According to the experiments with polymers, discussed in Chapter 6, Section 6.2, and due to the large absorption coefficient of the cornea in the UV (Eichler and Seiler, 1991), excimer lasers are expected to be useful tools (Srinivasan, 1986). Especially the ArF laser (λ=193 nm) with

---

[2] For example astigmatism, i.e., nonspherical curvature, which results in optical aberrations such as focus lines instead of focus points.
[3] For example short-sightedness (myopia) or long-sightedness (hyperopia), corresponding to a strongly or weakly curved cornea.

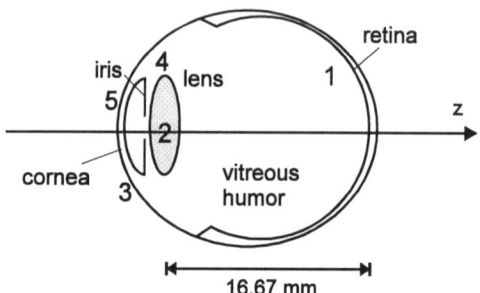

Figure 161 Sketch of the human eye and possible positions where laser radiation can be applied to correct disease-induced modifications. 1: Photocoagulation of the retina (Ar^+ laser). 2: Ablation of lens material (excimer laser). 3: Change of curvature of the cornea via ablation (excimer laser). 4: Generation of shock waves in order to tear-up cloudy lens membranes (short pulse Nd:YAG laser). 5: Perforation of the iris in order to decrease the pressure inside the eye (short pulse Nd:YAG laser).

a photon energy of 6.4 eV delivers enough energy to directly disrupt the protein bonds (E_b=3.5 eV). Compared to the KrF laser (λ=248 nm) it also has been shown to be less mutagenic since DNA UV-absorbs mainly in the wavelength range 240 nm to 260 nm. A huge advantage of the application of UV ablation is the nearly linear dependence between etch depth and number of irradiating pulses (Chapter 6, Section 6.2.1), which allows one to remove the material very precisely. Using an ArF laser, cold material removal is possible in collagen (the main ingredient of the cornea), whereas CO_2 and Nd:YAG lasers severely damage the tissue (Fig. 163). Since the fluences which are necessary to overcome the ablation thresholds (about 40 mJ/cm^2) are obtained relatively easily, even without focusing, the successful application of the 'cold' ablation only relies on the use of the correct wavelength. Hence in recent years the use of frequency quintupled Nd:YAG lasers (λ = 1064 nm $\rightarrow$ λ=213 nm) instead of the high power excimer lasers has been discussed. These lasers are easier to handle and possess smaller spatial dimensions and higher repetition rates (Ren et al., 1994).

By use of an adjustable aperture, which covers either the inner or the outer part of the cornea, the curvature of the cornea and thus the focal point of the retina can be changed. The change of the refractive power ΔD (in diopters) via laser irradiation can be calculated from the change in the radius of curvature of the cornea, Δr, via

$$\Delta D = -D \cdot \frac{\Delta r}{r}. \tag{7.1}$$

7.2. LASERS IN OPHTHALMOLOGY

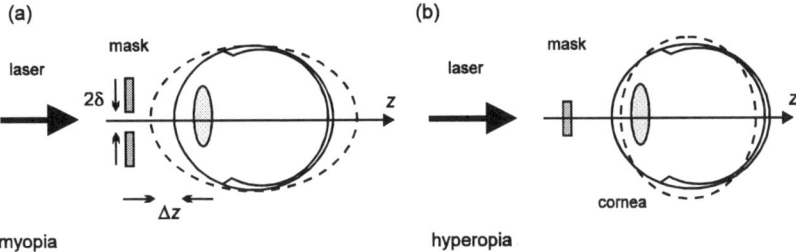

Figure 162 Set-up for laser correction of short- (a) and long-sightedness (b). The solid curves denote the normal eyes. The well-defined irradiation of the cornea is induced by a variable mask, the dimensions of which are adapted to the area of the eye to be irradiated.

A correction of the refractive power of the eye by -10 diopters makes a change in the radius of curvature by 1.58 mm necessary. This corresponds to an ablated cornea thickness of $\Delta z = 87$ μm, which follows from geometrical considerations (Niemz, 1996):

$$\Delta z = \frac{r + \Delta r}{\delta} \cdot \sin(\sin^{-1}(\frac{\delta}{r}) - \sin^{-1}(\frac{\delta}{r + \Delta r})) - \Delta r, \qquad (7.2)$$

where $\delta \approx 2.5$ mm is a typical ablated radius on the cornea (cf. Fig. 162). Note that the center thickness of the cornea is 550 μm. In addition to using the method shown in Fig. 162a this effect can also be achieved in a more flexible way that better fits the properties of the individual eye by the use of a mask that is erodible via excimer laser irradiation (Ren et al., 1995).

Important for a well-defined ablation of tissue are the homogeneity of the beam profile and the possibility to vary the beam diameter in a continuous manner. A set-up that is used very often for this purpose incorporates two axicons (Ren and Birngruber, 1990). By varying the distance between them, the diameter of the burn spot can be varied over a wide range. For fluences between 50 and 250 mJ/cm^2 ablation yields between 0.1 and 0.5 μm per laser pulse are obtained.

As in every application of lasers in medicine one faces also in the case of PRK detrimental by-reactions. These are mainly secondary reactions in the course of laser irradiation, e.g., the generation of UV fluorescence light in the tissue, which has longer wavelengths than the initial radiation and thus

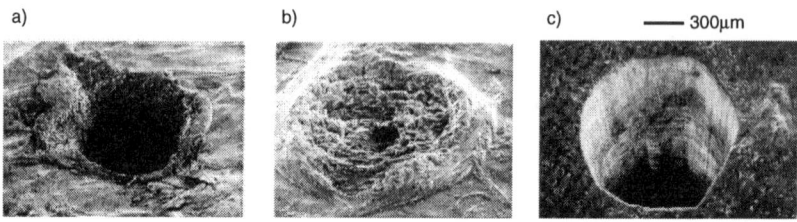

Figure 163 Laser-induced holes in collagen. (a) CO_2 laser (500 W, 1 Hz); (b) Nd:YAG laser (1.2 J, 1 Hz); (c) ArF laser (1.4 J/cm^2, 20 Hz). Reprinted with permission from (Dickmann, 1989).

a higher penetration depth into the surrounding tissue. As a result of this irradiation the lens might become cloudy or the DNA of the cornea becomes destroyed. However, extensive investigations have shown that the probability for such secondary effects or mutageneous and cancerogene influences can be minimized by use of appropriate wavelengths and fluences.[4] Long-time studies have shown that the laser-corrected value of refraction of myopic patients stayed within 10% of the optimum values over more than two years following PRK (Ediger and Durkin, 1997).

7.3 Basic Mechanisms

In principle a combination of photochemical, photothermal and photomechanical mechanisms is responsible for the ablation following absorption of laser light of wavelength λ_L. Binding energies of organic materials in the irradiated tissue are between 6 and 8 eV, corresponding to wavelengths of λ_b=206 nm to 155 nm. Since in a simple statistical approximation the quantum yield for molecular dissociation, Φ, decreases exponentially with increasing wavelength, $\Phi \propto \exp - (\lambda_L/\lambda_b)$, direct photochemical mechanisms are of importance for the ablation process only for irradiation with light in the far UV (VUV, $\lambda_L \leq$ 200 nm). The use of those wavelengths, however, suffers from heavy intricacies related to VUV beam shaping and optimization techniques (Bloomstein et al., 1997).

---

[4] Recently, treating mild to moderate myopia (−1.5 to −7.0 D) including astigmatism (up to 1.5 D) by excimer lasers has obtained governmental approval in the United States (April 1997). For the treatment of hyperopia, no lasers have been approved so far (November 1997).

7.3. BASIC MECHANISMS

In Chapter 6, Section 6.2 we discussed in some detail photochemical and photothermal models for ablation. In the case of irradiation of organic tissue the *photomechanical* components gain in importance due to the high water content of the tissue. Following absorption of laser light in the biomolecules (proteins) or in the water of the tissue, a fast, radiationless relaxation of the highly excited molecular states takes place (typical time constants 100 ps to 1 ns). If this superheating of the irradiated tissue volume occurs within a time interval that is small compared with the characteristic times of thermal diffusion, then bubbles of superheated vapor are generated in the tissue. The strong pressure rise in the superheated volume results in the generation of a shock wave, which expands explosively. The threshold temperature for this process of explosive pressure equalization is about 300°C (Oraevsky et al., 1991) in the case of human tissue. The shock wave that travels through the tissue finally leads to the ejaculation of superheated material. The collisions that occur during this process result in thermo-chemoluminescence, i.e., a self-contained glowing of the ablated tissue.

The expansion behavior depends on the duration τ_L of the irradiating laser pulse. The characteristic time constant is the hydrodynamic expansion time $\tau_\mathrm{hyd} = l_\mathrm{eff}/c_\mathrm{S}$, which is given by the time sound needs to travel through the irradiated part of the tissue, l_eff; here, c_S is the speed of sound in human tissue. In the gas-dynamic case ($\tau_\mathrm{L} \gg \tau_\mathrm{hyd}$) already during the excitation pulse a thermal pressure wave travels through the irradiated tissue, which is followed by a negative pressure wave, i.e., a pull within the tissue. The negative pressure wave compensates for the increase in pressure that is generated by the laser which is further irradiating the tissue, and thus it is not before the end of the laser pulse that a positive pressure rise occurs. Hence the necessary amount of pull in order to ablate the tissue cannot be generated in that way.

In the case of irradiation with short pulses ($\tau_\mathrm{L} \ll \tau_\mathrm{hyd}$, typical $\tau_\mathrm{L} \leq 10$ ns) the fast temperature rise results in a fast increase in pressure. The pressure wave that travels through the tissue is followed by a regime of under-pressure, which is no longer compensated by the laser, resulting in strong pulling forces.[5] For excitation in the UV this kind of photo-thermomechanical spallation is an important ablation mechanism, whereas in the visible or infrared spectral range the thermal explosion of inhomogeneously overheated tissue becomes often more important.

The maximum pull in the direction of laser irradiation, σ_p, can be estimated quantitatively if one assumes a laser pulse that starts and ends instantaneously (Dingus and Scammon, 1991)

$$\sigma_\mathrm{p} = \frac{1 - \exp(-\tau_\mathrm{L})}{2\tau_\mathrm{L}} \frac{1}{\kappa_\mathrm{T}} \beta \frac{\alpha P_\mathrm{L}}{\rho C_V}. \tag{7.3}$$

---

[5] Similar to the release of a contracted spring.

The pull is induced by the maximum laser irradiance that is absorbed per unit mass at the surface, $\alpha P_L/\rho$ (absorption coefficient α, irradiance P_L, density of the tissue ρ), which results in a maximum increase in temperature (division by the heat capacity C_V). There then follows a thermal expansion (multiplication with the expansion coefficient β) and a resulting pull (division by the isothermal compressibility of the tissue, κ_T). The exponential prefactor takes into account that the pull is attenuated by the interactions with the pressure generating laser pulse, as discussed above. The factor 2 results from the pull having the possibility to act in two directions.

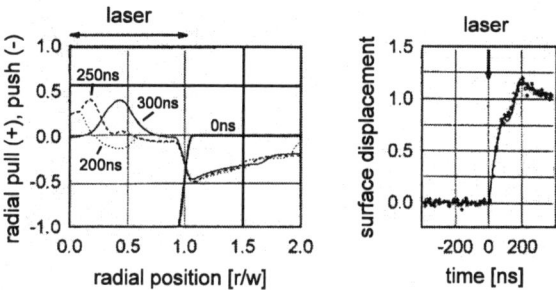

Figure 164 Left-hand side: Calculated pull and push forces acting on acrylic glass and following laser irradiation for different delay times. Right-hand side: Thermoelastic expansion of a glass filter as a function of time t following irradiation with a laser pulse of duration 10 ns, normalized to the value at $t = \infty$. The solid line results from a three-dimensional calculation. Reprinted with permission from (Albagli et al., 1994). Copyright 1994 Optical Society of America (OSA).

A spallation of the tissue occurs if $\sigma_p \geq \sigma_s$, where σ_s means the strength of the tissue ($\sigma_s \approx 20$ bar for the retina[6]) with respect to pull. The total ablation depth for that case is expected to be of the order of the laser absorption depth $1/\alpha$. However, in the experiment one finds a significantly smaller ablation depth, and hence ablation occurs mainly at the surface. This observation is not in agreement with the discussed one-dimensional model since from this model the magnitude of the pull is expected to increase with increasing depth in the tissue and approaches zero at the surface, since the laser-induced pressure wave is reflected at the surface.

---

[6] This value is reached following KrF laser irradiation with a threshold of 150 mJ/cm², an absorption coefficient of 330 cm^{-1} and a maximum increase in temperature of only 12 K.

7.3. BASIC MECHANISMS

Better agreement between theory and experimental findings is obtained if one takes into account the three-dimensionality of the process. In that case, however, similar to the calculation of heat evolution in laser-irradiated materials, solutions of the three-dimensional thermoelastic wave equation become possible only numerically. Fig. 164 shows, as a result of such calculations, the radial distribution of pull (to the top) and push forces (to the bottom) for different times following irradiation of an acrylic glass surface with a nanosecond laser pulse (Albagli et al., 1994). The position $r=0$ denotes the center of the laser beam, which is assumed to have the Gaussian radius w. Initially a pressure zone is formed up to the border of the laser beam corresponding to the laser-induced temperature distribution (thick line in the figure), which is via reflection at the surface transformed into a zone with pull forces. Outside of the laser-irradiated region ($r/w \geq 1$) push forces are dominating just as in the one-dimensional case since the compensating influence of the laser is missing.

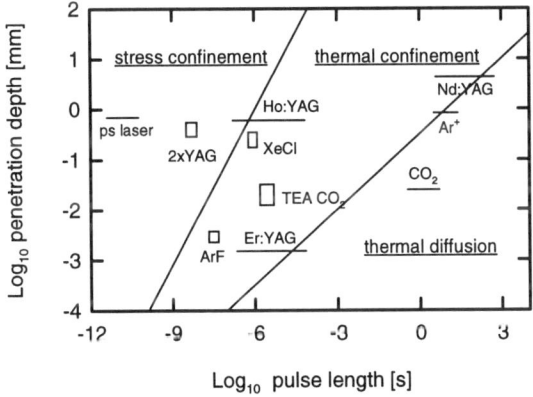

Figure 165 Spatial localization ('confinement') of thermal and push or pull energy for several lasers that are applied in medicine. For the abbreviations see Table 3. **2×YAG** means frequency-doubled Nd:YAG laser and **ps laser** picosecond laser. **TEA CO$_2$** is a high pressure CO$_2$ laser. For details of the lasers compare also Chapter 1, Section 1.1. Reprinted with permission from (Jacques, 1993). Copyright 1993 Optical Society of America (OSA).

On the right-hand side of Fig. 164 a comparison between calculated and measured values of the expansion of a laser-irradiated glass filter is shown. The good agreement between theory and experiment means that the three-dimensional spallation model accurately describes the photo-thermomechanical component of laser ablation. In the case of the experiment the laser-irradiated surface served as one arm of a Michelson interferometer. From the temporal change in the interference pattern, the direction and amplitude of the motion of the surface layer could be deduced. The first maximum in Fig. 164 at the right-hand side results from radial surface waves, which pass the point of interferometric observation; the temporal position is given by the ratio between laser beam radius and longitudinal speed of sound, c_s^l. The second maximum appears at a later point of time due to the smaller transversal speed of sound, c_s^t. The laser-induced expansion vanishes due to thermal diffusion processes after several hundreds of milliseconds.

Finally it is noted that a general trend deduced from a large number of experiments and calculations is that short-pulse (sub-nanosecond) lasers minimize deterioriation of the tissue that surrounds the laser burn spot as compared with lasers with longer pulse duration. At the same time the short-pulse laser allows a better defined localization or confinement of the point where the photomechanical pull is exerted. The deviation from total localization $L=1$ depends exponentially on the ratio between pulse length τ_L and absorption depth $1/\alpha$ in the case of a medium that does not scatter laser light (Jacques, 1993):

$$L = \frac{1 - \exp(-\tau_L c_s \alpha)}{\tau_L c_s \alpha}. \tag{7.4}$$

In Fig. 165 the penetration depths into organic tissue are plotted for commercial lasers that are used in laser medicine. In principle, three different regimes can be identified: (i) the regime without spatial localization of the laser energy, where thermal diffusion dominates; (ii) the regime of thermal localization, where the laser pulse length is short compared with the time constant for thermal diffusion; and (iii) the regime of stress confinement, where the laser-induced stress is released much faster compared with the time it takes to propagate out of the irradiated tissue. As can be seen from the figure, excimer lasers (XeCl and ArF) are *not* an optimal light source to avoid stress diffusion due to the small penetration depth of the UV light which counteracts the advantages provided by the short light pulses. A better choice are Nd:YAG and short-pulse (pico- or even femtosecond) lasers (Neev, 1998).

Bibliography

Abe, H., Charle, K.-P., Tesche, B., and Schulze, W. (1982). Surface plasmon absorption of various colloidal metal particles. *Chem. Phys.*, 68:137 – 141.

Abelés, F., editor (1972). *Optical Properties of Solids*. North-Holland, London.

Aeschlimann, M., Hull, E., Cao, J., Schmuttenmaer, C., Jahn, L., Gao, Y., Elsayed-Ali, H., Mantell, D., and Scheinfein, M. (1995). A picosecond electron gun for surface analysis. *Rev.Sci.Instrum.*, 66:1000 – 1009.

Ahrens, H., Welling, H., and Scheel, H. (1973). Measurement of optical absorption in dielectric reflectors. *Appl.Phys.*, 1:69 – 71.

Akhmanov, S., Emel'yanov, V., and Koroteev, N. (1990). *Interaction of Strong Laser Radiation with Solids and Nonlinear Optical Diagnostics of Surfaces*. Teubner, Leipzig.

Akhmanov, S., Koroteev, N., and Shumay, I. (1989). *Nonlinear Optical Diagnostics of Laser-Excited Semiconductor Surfaces*. Harwood Academic, London.

Al-Shamery, K. and Freund, H.-J. (1996). Laser-stimulated desorption from surfaces. *Curr.Opinion Sol.Stat. & Mat.Sci.*, 1:622 – 629.

Albagli, D., Dark, M., Perelman, L., Rosenberg, C. V., Itzkan, I., and Feld, M. (1994). Photomechanical basis of laser ablation of biological tissue. *Opt. Lett.*, 19:1684 – 1686.

Alivisatos, A. (1996). Perspectives on the physical chemistry of semiconductor nanocrystals. *J. Phys. Chem.*, 100:13226 – 13239.

Alivisatos, A., Waldeck, D., and Harris, C. (1985). Nonclassical behavior of energy transfer from molecules to metal surfaces: Biacetyl($^3n\pi^*$)/Ag(111). *J. Chem. Phys.*, 82:541 – 547.

Allara, D. and Nuzzo, R. (1983). The application of reflection infrared and surface enhanced Raman spectroscopy to the characterization of chemisorbed organic disulfides on Au. *J.Electron.Spectrosc.Relat.Phenom.*, 30:11.

Allmen, M. (1995). *Laser-Beam Interactions with Materials*. Springer Series in Materials Science, Vol.2. Springer-Verlag, Berlin.

Ambrose, W., Goodwin, P., Enderlein, J., Semin, D., Martin, J., and Keller, R. (1997). Fluorescence photon antibunching from single molecules on a surface. *Chem.Phys.Lett.*, 269:365 – 370.

Ambrose, W., Goodwin, P., Martin, J., and Keller, R. (1994a). Alterations of single molecule fluorescence lifetimes in near-field optical microscopy. *Science*, 265:364 – 366.

Ambrose, W., Goodwin, P., Martin, J., and Keller, R. (1994b). Fluorescence detection of single molecules using pulsed near-field optical excitation and time correlated photon counting. *Proc. SPIE*, 2125:2 – 11.

Amos, R. and Barnes, W. (1997). Modification of the spontaneous emission rate of Eu^{3+} ions close to a thin metal mirror. *Phys.Rev.B*, 55:7249 – 7254.

Anderson, J., Rubloff, G., and Stiles, P. (1973). Optical reflectance studies of chemisorption on a clean metal surface. *Solid State Commun.*, 12:825 – 828.

Andersson, T. and Granqvist, C. (1977). Morphology and size distributions of islands in discontinuous films. *J. Appl. Phys.*, 48:1673 – 1679.

Andresen, P., Faubel, M., Haeusler, D., Kraft, G., Luelf, H.-W., and Skofronick, J. (1985). Characteristics of a piezoelectric pulsed nozzle beam. *Rev. Sci. Instrum.*, 56:2038 – 2042.

Andrew, J., Dyer, P., Forster, D., and Key, P. (1983). Direct etching of polymeric materials using a XeCl laser. *Appl.Phys.Lett.*, 43:717 – 719.

Andrews, D. and Hands, I. (1996). Second harmonic generation in partially ordered media and at interfaces: Analysis of dynamical and orientational factors. *Chem.Phys.*, 213:277 – 294.

Anisimov, S., Kapeliovich, B., and Perel'man, T. (1974). Emission of electrons from the surface of metals induced by ultrashort laser pulses. *Sov.Phys.JETP*, 66:776 – 781.

Antoniewicz, P. (1980). Model for electron- and photon-stimulated desorption. *Phys. Rev. B*, 21:3811 – 3815.

Apell, P. and Penn, D. (1983). Optical properties of small metal spheres: Surface effects. *Phys.Rev.Lett.*, 50:1316 – 1319.

Arias, J., Aravind, P., and Metiu, H. (1982). The fluorescence lifetime of a molecule emitting near a surface with small, random roughness. *Chem. Phys. Lett.*, 85:404 – 408.

Arnold, J., Dasbach, U., Ehrfeld, W., Hesch, K., and Löwe, H. (1995). Combination of excimer laser micromachining and replication processes suited for large scale production. *Appl. Surf. Sci.*, 86:251 – 258.

Ashby, M. and Easterling, K. (1984). The transformation hardening of steel surfaces by laser beams — I. Hypo-eutectoid steels. *Acta Metall.*, 32:1935 – 1948.

Ashkenasi, D., Rosenfeld, A., Varel, H., Wähmer, M., and Campbell, E. (1997). Laser processing of sapphire with picosecond and sub-picosecond pulses. *Appl.Surf.Sci.*, 120:65 – 80.

Ashkenasi, D., Varel, H., Rosenfeld, A., Henz, S., Herrmann, J., and Campbell, E. (1998). Application of self-focusing of ps laser pulses for three-dimensional microstructuring of transparent materials. *Appl.Phys.Lett.*, 72:1442 – 1444.

Aussenegg, F., Leitner, A., and Gold, H. (1995). Optical second-harmonic generation of metal-island films. *Appl. Phys. A*, 60:97 – 101.

Aussenegg, F., Leitner, A., Lippitsch, M., Reinisch, H., and Riegler, M. (1987). Novel aspects of fluorescence lifetime for molecules positioned close to metal surfaces. *Surf. Sci.*, 189/190:935 – 945.

Avouris, P., Schmeisser, D., and Demuth, J. (1983). Nonradiative relaxation of electronically excited N_2 on Al(111). *J. Chem. Phys.*, 79:488 – 492.

Azzam, R. and Bashara, N. (1992). *Ellipsometry and Polarized Light*. North Holland, Amsterdam.

Baggish, M. (1986). The state of the art of laser surgery in gynecology. *Lasers Surg. Med.*, 6:390 – 395.

Balzer, F., Bammel, K., and Rubahn, H.-G. (1993a). Laser investigations of Na atoms deposited via inert spacer layers close to metal surfaces. *J. Chem. Phys.*, 98:7625 – 7635.

Balzer, F., Bordo, V., and Rubahn, H.-G. (1997a). Frequency shifts and lifetime changes of Na atoms near rough metal surfaces. *Opt. Lett.*, 22:1262 – 1264.

Balzer, F., Bordo, V., and Rubahn, H.-G. (1998a). Surface-induced changes of the optical response of particles in nanoscaled layered systems: A combined experimental and theoretical study. *Proc. SPIE*, 3272:42 – 50.

Balzer, F., Gerlach, R., Manson, J., and Rubahn, H.-G. (1997b). Photodesorption of Na atoms from rough Na surfaces. *J. Chem. Phys.*, 106:7995 – 8012.

Balzer, F., Gerlach, R., Schober, M., and Rubahn, H.-G. (1998b). A HAS study of laser irradiation of alkali covered dielectrics. *Surf. Sci.*, 402–404:841 – 844.

Balzer, F., Hartmann, M., Renger, M., and Rubahn, H.-G. (1993b). Kinetics of photo-induced dissociation of Na clusters deposited on mica. *Z. Phys. D*, 28:321 – 329.

Balzer, F., Jett, S., and Rubahn, H.-G. (1998c). Alkali cluster films on insulating surfaces: Comparison between scanning force microscopy and extinction data. *Chem.Phys.Lett.*, 297:273 – 280.

Balzer, F. and Rubahn, H.-G. (1994). Real time observation of heat transfer along laser-irradiated dielectric surfaces. *Surf. Sci.*, 307 – 309:367 – 371.

Balzer, F. and Rubahn, H.-G. (1995a). Influence of surface roughness on frequency shift and third order nonlinear susceptibility of adsorbed particles. *Proc. SPIE*, 2547:40 – 51.

Balzer, F. and Rubahn, H.-G. (1995b). Size effects and the determination of absolute temperature increases in laser heating of dielectrics. *Chem. Phys. Lett.*, 233:75 – 80.

Balzer, F. and Rubahn, H.-G. (1995c). Third-order nonlinear optics of Na clusters bound to dielectric surfaces. *Chem. Phys. Lett.*, 238:77 – 81.

Balzer, F. and Rubahn, H.-G. (1998). Nonlinear optics of rough cluster films. *Proc. SPIE*, 3272:23 – 34.

Bammel, K., Ellis, J., and Rubahn, H.-G. (1993). Two-photon laser observation of diffusion of Na atoms through self-assembled monolayers on a Au surface. *Chem. Phys. Lett.*, 201:101 – 107.

Barber, P. and Hill, S. (1990). *Light Scattering by Particles: Computational Methods*, volume 2 of *Advanced Series in Applied Physics*. World Scientific, Singapore.

Barker, J. and Auerbach, D. (1984). Gas-surface interactions and dynamics; thermal energy atomic and molecular beam studies. *Surf.Sci.Rep.*, 4:1 – 99.

Barman, S., Horn, K., Häberle, P., Ishida, H., and Liebsch, A. (1998). Photoinduced plasmon excitations in alkali-metal overlayers. *Phys.Rev.B*, 57:6662 – 6665.

Barnes, W. (1998). Fluorescence near interfaces: The role of photonic mode density. *J. Mod.Opt.*, 45:661 – 700.

Bartels, L., Meyer, G., Rieder, K.-H., Velic, D., Knoesel, E., Hotzel, A., Wolf, M., and Ertl, G. (1998). Dynamics of electron-induced manipulation of individual CO molecules on Cu(111). *Phys.Rev.Lett.*, 80:2004 – 2007.

Bastert, G. and Wallwiener, D. (1992). *Lasers in Gynecology — Possibilities and Limitations.* Springer-Verlag, Berlin, New York.

Bauer, M., Pawlik, S., and Aeschlimann, M. (1997). Resonance lifetime and energy of an excited Cs state on Cu(111). *Phys.Rev.B*, 55:10040 – 10043.

Bäuerle, D. (1986). *Chemical Processing with Lasers.* Springer Series in Materials Science, Vol.1. Springer, Berlin.

Bäuerle, D. (1996a). Laser-chemical processing: Recent developments. *Appl.Surf.Sci.*, 106:1 – 10.

Bäuerle, D. (1996b). *Laser Processing and Chemistry.* Springer-Verlag, Berlin, Heidelberg.

Benedek, G. and Toennies, J. (1994). Helium atom scattering spectroscopy of surface phonons: Genesis and achievments. *Surf.Sci.*, 299–300:587 – 611.

Bennett, A. (1970). Influence of the electron charge distribution on surface-plasmon dispersion. *Phys.Rev.B*, 1:203–207.

Bennett, T., Krajnovich, D., and Grigoropoulos, C. (1996). Separating thermal, electronic and topographic effects in pulsed laser melting and sputtering of gold. *Phys.Rev.Lett.*, 76:1659 – 1662.

Bergmann, H., Schutte, K., Schubert, E., and Emmel, A. (1995). Laser-surface processing of metals and ceramics for industrial applications. *Appl.Surf.Sci.*, 86:259 – 265.

Bergmann, K. (1988). State selection by optical methods. In Scoles, G., editor, *Atomic and Molecular Beam Methods*, pages 293 – 344. Oxford University Press, Oxford.

Berkovic, G. (1995). Strategies for measuring third harmonic generation from ultrathin films. *Chem. Phys. Lett.*, 241:355 – 359.

Berlien, H. and Müller, G. (1988-1993). *Angewandte Lasermedizin. Lehr- und Handbuch für Klinik und Praxis.* Ecomed.

Berman, P., editor (1994). *Cavity Quantum Electrodynamics.* Academic Press, Boston.

Betz, G. and Varga, P., editors (1990). *Desorption Induced by Electronic Transitions – DIET IV*. Springer-Verlag, Berlin.

Betzig, E. and Chichester, R. (1993). Single molecules observed by near-field scanning optical microscopy. *Science*, 262:1422 – 1424.

Bian, R., Dunn, R., Xie, X., and Leung, P. (1995). Single molecule emission characteristics in near-field microscopy. *Phys. Rev. Lett.*, 75:4772 – 4775.

Bingler, H.-G., Brunner, H., Leitner, A., Aussenegg, F., and Wokaun, A. (1995). Interference enhanced surface raman scattering of adsorbates on a silver-spacer-islands multilayer system. *Mol.Phys.*, 85:587 – 606.

Binnig, G., Rohrer, H., Gerber, C., and Weibel, E. (1982). Surface studies by scanning tunneling microscopy. *Phys. Rev. Lett.*, 49:57 – 61.

Birge, R. (1983). One-photon and two-photon excitation spectroscopy. In Kliger, D., editor, *Ultrasensitive Laser Spectroscopy*, pages 109 – 174. Academic Press, New York.

Birnbaum, M. (1965). Semiconductor surface damage produced by Ruby lasers. *J. Appl. Phys.*, 36:3688 – 3689.

Blodgett, K. and Langmuir, I. (1937). Built-up films of Barium stearate and their optical properties. *Phys. Rev.*, 51:964 – 982.

Bloembergen, N., Chang, R., Jha, S., and Lee, C. (1968). Optical second-harmonic generation in reflection from media with inversion symmetry. *Phys. Rev.*, 174:813 – 822.

Bloomstein, T., Horn, M., Rothschild, M., Kunz, R., Palmacci, S., and Goodman, R. (1997). Lithography with 157nm lasers. *J.Vac.Sci.Technol. B*, 15:2112 – 2116.

Bohren, C. and Huffman, D. (1983). *Absorption and Scattering of Light by Small Particles*. Wiley, New York.

Boneß, J., Marowsky, G., Braun, J., Witte, G., and Rubahn, H.-G. (1998). Second-harmonic generation from sodium covered Si(111)7×7 surfaces. *Surf. Sci.*, 402 – 404:513 – 517.

Bordo, V., Henkel, C., Lindinger, A., and Rubahn, H.-G. (1997). Evanescent wave fluorescence spectra of Na atoms. *Opt. Commun.*, 137:249 – 253.

Born, M. and Wolf, E. (1975). *Principles of Optics*. Pergamon Press, Oxford.

Böttcher, A., Morgante, A., Grobecker, R., Greber, T., and Ertl, G. (1994). Singlet-to-triplet conversion of metastable He atoms at alkali-metal overlayers. *Phys.Rev.B*, 49:10607 – 10612.

Bourdon, E., Cho, C.-C., Das, P., Polanyi, J., Stanners, C., and Xu, G.-Q. (1991). Photochemistry of adsorbed molecules. IX. Ultraviolet photodissociation and photoreaction of HBr on LiF(001). *J.Chem.Phys.*, 95:1361 – 1377.

Bourdon, E., Cowin, J., Harrison, I., Polanyi, J., Segner, J., Stanners, C., and Young, P. (1984). UV photodissociation and photodesorption of adsorbed molecules. 1. CH_3Br on LIF(001). *J.Phys.Chem.*, 88:6100 – 6103.

Boyd, G., Rasing, T., Leite, J., and Shen, Y. (1984). Local-field enhancement on rough surfaces of metals, semimetals, and semiconductors with the use of optical second-harmonic generation. *Phys. Rev. B*, 30:519 – 526.

Boyd, G., Shen, Y., and Hänsch, T. (1986). Continuous-wave second-harmonic generation as a surface microprobe. *Opt. Lett.*, 11:97 – 99.

Boyd, I. (1987). *Laser Processing of Thin Films and Microstructures*, volume 3 of *Springer Series in Materials Science*. Springer-Verlag, Berlin.

Bradley, J. (1988). A simplified correlation between laser processing parameters and hardened depth in steels. *J.Phys.D: Appl.Phys.*, 21:834 – 837.

Brand, J. and George, S. (1986). Effects of laser pulse characteristics and thermal desorption parameters on laser induced thermal desorption. *Surf. Sci.*, 167:341 – 362.

Brannon, J., Lankard, J., Baise, A., Burns, F., and Kaufman, J. (1985). Excimer laser etching of polyimide. *J.Appl.Phys.*, 58:2036 – 2043.

Bratz, A. and Marowsky, G. (1990). Angular dependence of surface second harmonic generation. *Mol. Eng.*, 1:59 – 65.

Braun, R., Nowak, R., Hess, P., Oetzmann, H., and Schmidt, C. (1989). Photoablation of polyimide with IR and UV laser radiation. *Appl.Surf.Sci.*, 43:352 – 357.

Brechignac, C. and Connerade, J. (1994). Giant resonances in free atoms and in clusters. *J. Phys. B: At. Mol. Opt. Phys.*, 27:3795 – 3828.

Breech, F. and Cross, L. (1962). Optical micromission stimulated by a ruby laser. *Appl. Spectrosc.*, 16:59 – 59.

Brenig, W. and Menzel, D., editors (1985). *Desorption Induced by Electronic Transitions — DIET II*. Springer-Verlag, Berlin.

Brevet, P.-F. (1997). *Surface Second Harmonic Generation*. Presses Polytechniques, Lausanne.

Brorson, S., Fujimoto, J., and Ippen, E. (1987). Femtosecond electronic heat-transport dynamics in thin gold films. *Phys.Rev.Lett.*, 59:1962 – 1965.

Brunco, D., Thompson, M., Otis, C., and Goodwin, P. (1992). Temperature measurements of polyimide during KrF excimer laser ablation. *J. Appl. Phys.*, 72:4344 – 4350.

Brune, H. (1998). Microscopic view of epitaxial metal growth: Nucleation and aggregation. *Surf.Sci.Rep.*, 31:121 – 229.

Brune, H., Giovannini, M., Bromann, K., and Kern, K. (1998). Self-organized growth of nanostructure arrays on strain-relief patterns. *Nature*, 394:451 – 452.

Bubenzer, A. and Koidl, P. (1984). Exact expressions for calculating thin-film absorption coefficients from laser calorimetric data. *Appl. Opt.*, 23:2886 – 2891.

Buck, M., Eisert, F., Grunze, M., and Träger, F. (1995). Second-order nonlinear susceptibilities of surfaces: A systematic study of the wavelength and coverage dependence of thiol adsorption on polycrystalline gold. *Appl.Phys.A*, 60:1 – 12.

Budde, F., Heinz, T., Kalamarides, A., Loy, M., and Misewich, J. (1993). Vibrational distributions in desorption induced by femtosecond laser pulses: Coupling of adsorbate vibration to substrate electronic excitation. *Surf.Sci.*, 283:143 – 157.

Budde, F., Heinz, T., Loy, M., Misewich, J., Rougemont, F. D., and Zacharias, H. (1991). Femtosecond time-resolved measurement of desorption. *Phys.Rev.Lett.*, 66:3024 – 3027.

Burns, A., Stechel, E., and Jennison, D., editors (1993). *Desorption Induced by Electronic Transitions — DIET V*. Springer-Verlag, Berlin.

Butcher, P. and Cotter, D. (1990). *The Elements of Nonlinear Optics*. Cambridge University Press, Cambridge.

Büttgenbach, S. (1991). *Mikromechanik*. Teubner, Stuttgart.

Byers, J., Yee, H., Petralli-Mallow, T., and Hicks, J. (1994). Second harmonic generation-circular dichroism spectroscopy from chiral monolayers. *Phys.Rev.B*, 39:14643 – 14647.

Cain, S. (1993). A photothermal model for polymer ablation: Chemical modification. *J.Phys.Chem.*, 97:7572 – 7577.

Cain, S., Burns, F., Otis, C., and Braren, B. (1992). Photothermal description of polymer ablation: Absorption behavior and degradation time scales. *J. Appl. Phys.*, 72:5172 – 5178.

Camillone, N., Chidsey, C., Liu, G., and Scoles, G. (1993). Substrate dependence of the surface structure and chain packing of docosyl mercaptan self-assembled on the (111), (110), and (100) faces of single crystal gold. *J.Chem.Phys*, 98:4234–4245.

Camillone, N., Leung, T., and Scoles, G. (1997). A low energy helium atom diffraction study of decanethiol sel-assembled on au(111). *Surf.Sci.*, 373:333–349.

Campbell, E., Ulmer, G., and Hertel, I. (1992). Thermionic emission from the fullerenes. *Z.Phys.D*, 24:81 – 85.

Cao, G. X., Nabighian, E., and Zhu, X. D. (1997). Diffusion of hydrogen on Ni(111) over a wide range of temperatures: Exploring quantum diffusion on metals. *Phys. Rev. Lett.*, 79:3696 – 3699.

Casimir, H. and Polder, D. (1948). The influence of retardation on the London–van der Waals forces. *Phys. Rev.*, 73:360 – 372.

Cavanagh, R., Germer, T., and Stephenson, J. (1995). Ultrafast time-resolved infrared probing of energy transfer at surfaces. *Vib.Spectrosc.*, 9:77 – 83.

Cavanagh, R., Heilweil, E., and Stephenson, J. (1993). Time-resolved probes of surface dynamics. *Surf.Sci.*, 283:226 – 232.

Cavanagh, R., Heilweil, E., and Stephenson, J. (1994). Time-resolved measurement of energy transfer at surfaces. *Surf. Sci.*, 299–300:643 – 655.

Cavanagh, R. and King, D. (1981). Rotational- and spin-state distributions: NO thermally desorbed from Ru(001). *Phys. Rev. Lett.*, 47:1829 – 1832.

Chabal, Y. (1988). Surface infrared spectroscopy. *Surf. Sci. Rep.*, 8:211 – 357.

Chai, J.-W. and Reilly, J. (1984). Multiple peak formation in laser ionization mass spectrometry. *Opt.Comm.*, 49:51 – 54.

Chance, R., Prock, A., and Silbey, R. (1975). Frequency shifts of an electric-dipole transition near a partially reflecting surface. *Phys. Rev. A*, 12:1448 – 1452.

Chance, R., Prock, A., and Silbey, R. (1978). Molecular fluorescence and energy transfer near interfaces. *Adv. Chem. Phys.*, 37:1 – 65.

Chang, Y., Xu, L., and Tom, H. (1997). Observation of coherent surface optical phonon oscillations by time-resolved surface second-harmonic generation. *Phys.Rev.Lett.*, 78:4649 – 4652.

Chase, L. (1994). Pump-probe method for investigating laser ablation and optical damage threshold mechanisms in optical materials. *Nucl. Instr. Meth. Phys. B*, 91:597 – 600.

Chemla, D., Heritage, J., Liao, P., and Isaacs, E. (1983). Enhanced four-wave mixing from silver particles. *Phys. Rev. B*, 27:4553 – 4558.

Chen, C. (1993). *Introduction to Scanning Tunneling Microscopy*. Oxford University Press, Oxford.

Chen, C., de Castro, A., and Shen, Y. (1981). Surface-enhanced second-harmonic generation. *Phys. Rev. Lett.*, 46:145 – 148.

Chen, C., de Castro, A., Shen, Y., and DeMartini, F. (1979). Surface coherent anti-Stokes Raman spectroscopy. *Phys. Rev. Lett.*, 43:946 – 949.

Chen, C., Heinz, T., Ricard, D., and Shen, Y. (1983). Surface-enhanced second-harmonic generation and Raman scattering. *Phys. Rev. B*, 27:1965 – 1979.

Chen, J., Bower, J., Wang, C., and Lee, C. (1973). Optical second-harmonic generation from submonolayer Na-covered Ge surfaces. *Opt. Commun.*, 9:132 – 134.

Chen, M., Tsai, S., Chen, M., Ou, S., Li, W.-H., and Lee, K. (1995). Effect of silver-nanoparticle aggregation on surface-enhanced Raman scattering from benzoic acid. *Phys. Rev. B*, 51:4507 – 4515.

Chevrollier, M., Bloch, D., Rahmat, G., and Ducloy, M. (1991). Van der Waals-induced spectral distortions in selective-reflection spectroscopy of Cs vapor: The strong atom-surface interaction regime. *Opt. Lett.*, 16:1879 – 1881.

Chevrollier, M., Fichet, M., Oria, M., Rahmat, G., Bloch, D., and Ducloy, M. (1992). High resolution selective reflection spectroscopy as a probe of long-range surface interaction: Measurement of the surface van der Waals attraction on excited Cs atoms. *J. Phys. II France*, 2:631 – 657.

Chiarello, G., Cupolillo, A., Amoddeo, A., Caputi, L., Papagno, L., and Colavita, E. (1997). Collective excitations of two layers of K on Ni(111). *Phys. Rev. B*, 55:1376 – 1379.

Cho, G., Kütt, W., and Kurz, H. (1990). Subpicosecond time-resolved coherent-phonon oscillations in GaAs. *Phys.Rev.Lett.*, 65:764 – 766.

Christ, M., Barton, T., Hörmann, K., Foth, H.-J., and Stasche, N. (1995). A new approach to determining laser effects on bone. *Proc.SPIE*, 2327:394 – 401.

Chuang, T. (1983). Laser-induced gas-surface interactions. *Surf. Sci. Rep.*, 3:1 – 105.

Chumanov, G., Sokolov, K., Gregory, B., and Cotton, T. (1995). Colloidal metal films as a substrate for surface-enhanced spectroscopy. *J. Phys. Chem.*, 99:9466 – 9471.

Collier, C., Vossmeyer, T., and Heath, J. (1998). Nanocrystal superlattices. *Annu.Rev.Phys.Chem.*, 49:371 – 404.

Comsa, G. and David, R. (1985). Dynamical parameters of desorbing molecules. *Surf. Sci. Rep.*, 5:145 – 198.

Condon, E. and Shortley, G. (1957). *The Theory of Atomic Spectra*. Cambridge University Press, New York.

Conrad, H., Ertl, G., Küppers, J., Sesselmann, W., and Haberland, H. (1982). Electron spectroscopy of surfaces by impact of metastable He atoms: CO on Pd(110). *Surf.Sci.*, 121:161 – 180.

Cook, R. and Hill, R. (1982). An electromagnetic mirror for neutral atoms. *Opt.Commun.*, 43:258 – 260.

Cremer, P., Stauners, C., Niemantsverdriet, J., Shen, Y., and Somorjai, G. (1995). The conversion of di-d bonded ethylene to ethylidyne of Pt(111) monitored with sum frequency generation: Evidence for an ethylidene (or ethyl) intermediate. *Surf.Sci.*, 328:111 – 118.

Crisp, M., Boling, N., and Dube, G. (1972). Importance of Fresnel reflections in laser surface damage of transparent dielectrics. *Appl. Phys. Lett.*, 21:364 – 366.

Cruz, L., Fonseca, L., and Gómez, M. (1989). T-matrix approach for the calculation of local fields in the neighborhood of small clusters in the electrodynamic regime. *Phys. Rev. B*, 40:7491 – 7500.

Culver, J., Li, M., Jahn, L., Hochstrasser, R., and Yodh, A. (1993). Vibrational response of surface adsorbates to femtosecond substrate heating. *Chem.Phys.Lett.*, 214:431 – 437.

Dai, H.-L. and Ho, W., editors (1995). *Laser Spectroscopy and Photochemistry on Metal surfaces, Parts I and II*. World Scientific, Singapore.

Dainty, J., editor (1975). *Laser Speckle and Related Phenomena*. Springer-Verlag, Berlin.

Danielzik, B., Fabricius, N., Röwekamp, M., and Von der Linde, D. (1986). Velocity distribution of molecular fragments from polymethylmethacrylate irradiated with UV laser pulses. *Appl. Phys. Lett.*, 48:212 – 214.

Daum, W., Krause, H.-J., Reichel, U., and Ibach, H. (1993). Identification of strained silicon layers at Si-SiO$_2$ interfaces and clean Si surfaces by nonlinear optical spectroscopy. *Phys. Rev. Lett.*, 71:1234 – 1237.

Davis, G., Gower, M., Fotakis, C., Efthimiopoulos, T., and Argurakis, P. (1985). Spectroscopic studies of ArF laser photoablation of PMMA. *Appl. Phys. A*, 36:27 – 30.

Davison, S. and Steslicka, M. (1992). *Basic Theory of Surface States*. Clarendon Press, Oxford.

Dekorsy, T., Kütt, W., Pfeiffer, T., and Kurz, H. (1993). Coherent control of LO-phonon dynamics in opaque semiconductors by femtosecond laser pulses. *Europhys.Lett.*, 23:223 – 228.

Delamarche, E. and Michel, B. (1996). Structure and stability of self-assembled monolayers. *Thin Solid Films*, 273:54 – 60.

Delamarche, E., Michel, B., Biebuyck, H., and Gerber, C. (1996). Golden interfaces: The surface of self-assembled monolayers. *Adv.Mater.*, 8:719 – 729.

Delsart, C. and Keller, J.-C. (1978). The optical Autler-Townes effect in Doppler-broadened three-level systems. *Le Journal de Physique*, 39:350 – 360.

Demtröder, W. (1996). *Laserspectroscopy*. Springer-Verlag, Berlin.

Dewitz, J., Hübner, W., and Bennemann, K. (1996). Theory for nonlinear Mie-scattering from spherical metal clusters. *Z. Phys. D*, 37:75 – 84.

Dickmann, K. (1989). Laser drilling of collagen: An exclusive application for excimer lasers. *Lambda Highlights*, 18:4 – 4.

Diels, J.-C., Fontaine, J., McMichael, I., and Simoni, F. (1985). Control and measurement of ultrashort pulse shapes (in amplitude and phase) with femtosecond accuracy. *Appl.Opt.*, 24:1270 – 1282.

Diels, J.-C. and Rudolph, W. (1996). *Ultrafast Laser Pulse Phenomena*. Academic Press, San Diego.

Dietz, T., Duncan, M., Powers, D., and Smalley, R. (1981). Laser production of supersonic metal cluster beams. *J. Chem. Phys.*, 74:6511 – 6512.

Dijkkamp, D., Gozdz, A., Venkatesan, T., and Wu, X. (1987). Evidence for the thermal nature of laser-induced polymer ablation. *Phys. Rev. Lett.*, 58:2142 – 2145.

Ding, L., Li, J., Wang, E., and Dong, S. (1997). K^+ sensors based on supported alkanethiol/phospholipid bilayers. *Thin Solid Films*, 293:153 – 158.

Dingus, R. and Scammon, R. (1991). Grueneisen-stress-induced ablation of biological tissue. In S.L.Jacques, editor, *Laser-Tissue Interaction II*, volume 1427, pages 45 – 54. Proc. SPIE.

Dixon-Warren, S., Heyd, D., Jensen, E., and Polanyi, J. (1993). Photochemistry of adsorbed molecules. XII. Photoinduced ion-molecule reactions at a metal surface for $CH_3X/RCl/Ag(111)$ (X=Br,I). *J.Chem.Phys.*, 98:5954 – 5960.

Domen, K., Fujino, T., Hirose, C., and Kano, S. (1998). Direct observation of short-lived unstable surface species by tunable picosecond infrared pulses. *Appl.Surf.Sci.*, 121:484 – 487.

Dougherty, T., Lawrence, G., Kaufman, J., Boyle, D., Weishaupt, K., and Goldfarb, A. (1979). Photoradiation in the treatment of recurrent breast carcinoma. *J. Natl. Cancer Inst.*, 62:231 – 237.

Douketis, C., Haslett, T., Wang, Z., Moskovits, M., and Iannotta, S. (1995a). Rough silver films studied by surface enhanced Raman spectroscopy and low temperature scanning tunneling microscopy. *Progr.Surf.Sci.*, 50:187 – 195.

Douketis, C., Wang, Z., Haslett, T., and Moskovits, M. (1995b). Fractal character of cold-deposited silver films determined by low-temperature scanning tunneling microscopy. *Phys. Rev. B*, 51:11022 – 11031.

Dretler, S. (1988). Laser lithotripsy: A review of 20 years of research and clinical applications. *Lasers Surg.Med.*, 8:341 – 356.

Drexhage, K. (1974). Interaction of light with monomolecular dye layers. In Wolf, E., editor, *Progress in Optics XII*, volume XII of *Progress in Optics XII*, chapter IV, pages 163 – 232. North-Holland.

Drexhage, K., Kuhn, H., and Schäfer, F. (1968). Variation of the fluorescence decay time of a molecule in front of a mirror. *Ber. Bunsenges. Phys. Chem.*, 72:329 – 329.

Drexler, K. (1992). *Nanosystems: Molecular Machinery, Manufacturing, and Computation*. Wiley, New York.

Dubois, L. (1992). Synthesis, structure, and properties of model organic surfaces. *Annu.Rev.Phys.Chem.*, 43:437–463.

Ducloy, M. (1993). Influence of atom-surface collisional processes in FM selective reflection spectroscopy. *Opt. Commun.*, 99:336 – 339.

Ducloy, M. and Fichet, M. (1991). General theory of frequency modulated selective reflection. Influence of atom surface interactions. *J. Phys. II France*, 1:1429 – 1446.

Duggal, A., Rogers, J., and Nelson, K. (1992). Real-time optical characterization of surface acoustic modes on polyimide thin-film coatings. *J.Appl.Phys.*, 72:2823 – 2839.

Duke, C., editor (1994). *Surface Science — The First Thirty Years.* North Holland.

Duley, W. (1996). *UV Lasers: Effects and Applications in Materials Science.* Cambridge University Press, Cambridge.

Duley, W. (1998). *Laser Welding.* John Wiley & Sons, New York.

Duncavage, J., Ossoff, R., and Toohill, R. (1985). Carbon dioxide laser management of laryngeal stenosis. *Ann.Otol.Rhinol.Laryngol.*, 94:565 – 569.

Duveneck, G., Oroszlan, P., Abel, A., Klee, B., Steiner, V., Ehrat, M., Gygax, D., and Widmer, H. (1996). Biochemical affinity sensing systems based on luminescence generation in the evanescent field of optical waveguides. *Proc.SPIE*, 2631:14 – 28.

Dvorak, J., Borguet, E., and Dai, H.-L. (1997). Monitoring adsorption and desorption on a metal surface by optical non-resonant reflectivity changes. *Surf.Sci.*, 369:L122.

Dyer, P. and Srinivasan, R. (1989). Pyroelectric detection of ultraviolet laser ablation products from polymers. *J.Appl.Phys.*, 66:2608 – 2612.

Ediger, M. and Durkin, A. (1997). Lasers in healthcare: Focus on corneal refractive surgery. *Optics & Photonics News*, November:18 – 22.

Edwards, M., Boggan, J., and Fuller, T. (1983). The laser in neurological surgery. *J. Neurosurg.*, 59:555 – 566.

Ehrfeld, W. and Münchmeyer, D. (1991). Three-dimensional microfabrication using synchrotron radiation. *Nucl. Instrum. Methods A*, 303:523 – 531.

Eichler, H., Guenter, P., and Pohl, D. (1986). *Laser-Induced Dynamic Gratings.* Springer-Verlag, Berlin, Heidelberg.

Eichler, J. and Seiler, T. (1991). *Lasertechnik in der Medizin.* Springer-Verlag, Berlin, New York.

Eichler, J. and Seiler, T. (1993). *Lasertechnik in der Medizin.* Springer, Berlin.

Elliott, D. (1993). Pathway to the gigabit memory. *Laser and Optronics*, June/July.

Elliott, D. (1995). *Ultraviolet Laser Technology and Applications.* Academic Press, New York.

Elsayed-Ali, H. and Herman, J. (1990). Ultrahigh vacuum picosecond laser-driven electron diffraction system. *Rev. Sci. Instrum.*, 61:1636 – 1647.

Elsayed-Ali, H. and Juhasz, T. (1993). Femtosecond time-resolved thermomodulation of thin gold films with different crystal structures. *Phys.Rev.B*, 47:13599 – 13610.

Enderlein, J., Goodwin, P., Orden, A. V., Ambrose, W., Erdmann, R., and Keller, R. (1997). A maximum likelihood estimator to distinguish single molecules by their fluorescence decays. *Chem.Phys.Lett.*, 270:464 – 470.

Engelmann, G., Ziegler, J., and Kolb, D. (1998). Electrochemical fabrication of large arrays of metal nanoclusters. *Surf.Sci.*, 401:L420 – L424.

Erley, G. and Daum, W. (1998). Silicon interband transitions observed at Si(100)-SiO_2 interfaces. *Phys.Rev.B*, 58:R1734.

Ertl, G. and Küppers, J. (1985). *Low Energy Electrons and Surface Chemistry.* VCH, Weinheim.

Ertl, G. and Neumann, M. (1972). Laser-induzierte schnelle thermische Desorption von Festkörper-Oberflächen. *Z.Naturforsch.*, 27a:1607 – 1610.

Esslinger, T., Weidemüller, M., Hemmerich, A., and Hänsch, T. (1993). Surface-plasmon mirror for atoms. *Opt.Lett.*, 18:450 – 452.

Eva, E. and Mann, K. (1996). Calorimetric measurement of two-photon absorption and color center formation in UV-window materials. *Appl.Phys.A*, 62:143 – 150.

Evans, S. and Ulman, A. (1990). Surface potential studies of alkyl-thiol monolayers adsorbed on gold. *Chem. Phys. Lett.*, 170:462 – 466.

Exter, M. V. and Lagendijk, A. (1988). Ultrashort surface-plasmon and phonon dynamics. *Phys.Rev.Lett.*, 60:49 – 52.

Falcone, R., Gordon, S., Hamster, H., and Sullivan, A. (1994). Laser ablation: Mechanisms and applications II. In Geohegan, D. and Miller, J., editors, *AIP Conference Proceedings*, New York.

Fann, W., Storz, R., Tom, H., and Bokor, J. (1992). Electron thermalization in gold. *Phys. Rev. B*, 46:13592 – 13595.

Fauchet, P. and Siegman, A. (1983). Observations of higher-order laser-induced surface ripples on <111> germanium. *Appl. Phys. A*, 32:135 – 140.

Fauster, T. and Steinmann, W. (1995). Two-photon photoemission spectroscopy of image states. In Halevi, P., editor, *Photonic Probes of Surfaces*, pages 347 – 411. Elsevier, Amsterdam.

Featherstone, J., Rechmann, P., and Fried, D., editors (1998). *Lasers in Dentistry IV*, volume 3248. Proc.SPIE.

Feder, R., editor (1985). *Polarized Electrons in Surface Physics*. World Scientific, Singapore.

Feldstein, M. and Scherer, N. (1998). Femtosecond correlated optical reactivity and scanning tunneling microscopy studies on metal surfaces. *Proc. SPIE*, 3272:58 – 66.

Feldstein, M., Vöhringer, P., Wang, W., and Scherer, N. (1996). Femtosecond optical spectroscopy and scanning probe microscopy. *J.Phys.Chem.*, 100:4739 – 4748.

Fendler, J., editor (1998). *Nanoparticles and Nanostructured Films*. Wiley-VCH, Weinheim, New York, Chichester.

Fenter, P., Eberhardt, A., and Eisenberger, P. (1994). Self-assembly of n-alkyl thiols as disulfides on au(111). *Science*, 266:1216 – 1218.

Fenter, P., Eberhardt, A., Liang, K., and Eisenberger, P. (1997). Epitaxy and chainlength dependent strain in self-assembled monolayers. *J. Chem. Phys.*, 106:1600 – 1608.

Fenter, P., Schreiber, F., Berman, L., Scoles, G., Eisenberger, P., and Bedzyk, M. (1998). On the structure and evolution of the buried S/Au interface in self-assembled monolayers: X-ray standing wave results. *Surf.Sci.*, 412–413:213 – 235.

Fichet, M., Schuller, F., Bloch, D., and Ducloy, M. (1995). van der Waals interactions between excited-state atoms and dispersive dielectric surfaces. *Phys. Rev. A*, 51:1553 – 1564.

Fischer, J., Dams, J., Götz, M., Kerker, E., Loesel, F., Messer, C., Niemz, M., Suhm, N., and Bille, J. (1994). Plasma-mediated ablation of brain tissue with picosecond laser pulses. *Appl.Phys.B*, 58:493 – 499.

Fischer, R., Fauster, T., and Steinmann, W. (1993). Three-dimensional localization of electrons on Ag islands. *Phys.Rev.B*, 48:15496 – 15499.

Fischer, U. and Pohl, D. (1989). Observation of single-particle plasmons by near-field optical microscopy. *Phys.Rev.Lett.*, 62:458 – 461.

Fisher, R., editor (1983). *Optical Phase Conjugation*. Academic Press, New York.

Fleischmann, M., Hendra, P., and McQuillan, A. (1974). Raman spectra of pyridine adsorbed at a silver electrode. *Chem. Phys. Lett.*, 26:163 – 166.

Flörsheimer, M., Steinfort, A., and Günter, P. (1993). Lattice constants of Langmuir-Blodgett films measured by atomic force microscopy. *Surf. Sci. Lett.*, 297:L39 – L42.

Fogarassy, E. and Lazare, S., editors (1992). *Laser Ablation of Electronic Materials*. North-Holland, Amsterdam.

Ford, G. and Weber, W. (1984). Electromagnetic interactions of molecules with metal surfaces. *Phys. Rep.*, 113:195 – 287.

Förster, T. (1965). Delocalized excitation and excitation transfer. *Mod. Quantum Chem.*, 3:93 – 137.

Fowler, W. (1968). *Physics of Color Centers*. Academic Press, New York.

Francisco, T., Camillone, N., and Miller, R. (1996). Rotationally inelastic scattering of C_2H_2 from LiF(100): Translational energy dependence. *Phys.Rev.Lett.*, 77:1402 – 1405.

Franken, P., Hill, A., Peters, C., and Weinreich, G. (1961). Generation of optical harmonics. *Phys. Rev. Lett.*, 7:118 – 119.

Frenken, J. and van der Veen, J. (1985). Observation of surface melting. *Phys. Rev. Lett.*, 54:134 – 137.

Freunscht, P., Duyne, R. V., and Schneider, S. (1997). Surface-enhanced Raman spectroscopy of trans-stilbene adsorbed on platinum- or self-assembled monolayer-modified silver film over nanosphere surfaces. *Chem.Phys.Lett.*, 281:372 – 378.

Frohn, J., Reynolds, J., and Engel, T. (1994). Surface morphology changes upon laser heating of Pt(111). *Surf. Sci.*, 320:93 – 104.

Fuchs, H., Ohst, H., and Prass, W. (1991). Ultrathin organic films: Molecular architectures for advanced optical, electronic and bio-related systems. *Adv. Mater.*, 3:10 – 18.

Furzikov, N. (1990). Approximate theory of highly absorbing polymer ablation by nanosecond laser pulses. *Appl.Phys.Lett.*, 56:1638 – 1640.

Gadzuk, J. (1995). Surface femtochemistry by laser-excited hot electrons. In Dai, H.-L. and Sibener, S., editors, *Laser Techniques for Surface Science*. Proc.SPIE.

Garcia-Vidal, F. and Pendry, J. (1996). Collective theory for surface enhanced Raman scattering. *Phys. Rev. Lett.*, 77:1163 – 1166.

Ge, N.-H., Wong, C., Lingle, R., McNeill, J., Gaffney, K., and Harris, C. (1998). Femtosecond dynamics of electron localization at interfaces. *Science*, 279:202 – 205.

Gentry, W. (1988). Low-energy pulsed beam sources. In Scoles, G., editor, *Atomic and Molecular Beam Methods*, pages 54 – 82. Oxford University Press, Oxford.

Gentry, W. and Giese, C. (1978). Ten-microsecond pulsed molecular beam source and a fast ionization detector. *Rev. Sci. Instrum.*, 49:595 – 600.

George, S., DeSantolo, A., and Hall, R. (1985). Surface diffusion of Hydrogen on Ni(100) studied using laser-induced thermal desorption. *Surf. Sci.*, 159:L425 – L432.

George, T. and Arnoldus, H. (1990). Observation of atomic relaxation near an interface through detection of emitted fluorescence. *Comments At. Mol. Phys.*, 24:109 – 117.

Gerlach, R., Manson, J., and Rubahn, H.-G. (1996a). Near-field time-of-flight spectroscopy of sodium atoms desorbing from surface-bound clusters. *Opt.Lett.*, 21:1183 – 1185.

Gerlach, R., Polanski, G., and Rubahn, H.-G. (1996b). Modification of electric dipole domains on mica by excimer laser irradiation. *Surf. Sci.*, 352 - 354:485 – 489.

Gerlach, R., Polanski, G., and Rubahn, H.-G. (1997). Structural manipulation of ultrathin organic films on metal surfaces: The case of decane thiol/Au(111). *Appl. Phys. A*, 65:375 – 378.

Gogoll, S., Stenzel, E., Reichling, M., Johansen, H., and Matthias, E. (1996). Laser damage of $CaF_2(111)$ surfaces at 248 nm. *Appl.Surf.Sci.*, 96–98:332 – 340.

Golan, Y., Margulis, L., and Rubinstein, I. (1992). Vacuum deposited gold films. *Surf.Sci.*, 264:312 – 326.

Goldman, L., Hornby, P., Mayer, R., and Goldman, B. (1964). Impact of the laser on dental caries. *Nature*, 203:417 – 417.

Goodwin, P. and Otis, C. (1989). Collisional cooling and ablation product dynamics observed by resonant ionization spectroscopy of nascent carbon monoxide from 193nm laser ablation of polyimide. *Appl. Phys. Lett.*, 55:2286 – 2288.

Göppert-Mayer, M. (1931). Über Elementarakte mit zwei Quantensprüngen. *Ann. Phys.*, 9:273 – 294.

Gorbunov, A., Mertig, M., Kirsch, R., Eichler, H., Pompe, W., and Engelhardt, H. (1997). Nanopatterning by biological templating and laser direct writing in thin laser deposited films. *Appl.Surf.Sci.*, 109/110:621 – 625.

Gorris-Neveux, M., Monnot, P., Fichet, M., Ducloy, M., Barbe, R., and Keller, J. (1997). Doppler-free reflection spectroscopy of rubidium D_1 line in optically dense vapour. *Opt.Commun.*, 134:85 – 90.

Gortel, Z., Kreuzer, H., Piercy, P., and Teshima, R. (1983a). Resonant heating in photodesorption via laser-adsorbate coupling. *Phys.Rev.B*, 28:2119 – 2124.

Gortel, Z., Kreuzer, H., Piercy, P., and Teshima, R. (1983b). Theory of photodesorption of molecules by resonant laser-molecular vibrational coupling. *Phys.Rev.B*, 27:5066 – 5083.

Gostein, M., Parhiktheh, H., and Sitz, G. (1995). Survival probability of $H_2(v=1, J=1)$ scattered from Cu(110). *Phys.Rev.Lett.*, 75:342 – 345.

Gotschy, W., Vonmetz, K., Leitner, A., and Aussenegg, F. (1996a). Optical dichroism of lithographically designed silver nanoparticle films. *Opt. Lett.*, 21:1099 – 1101.

Gotschy, W., Vonmetz, K., Leitner, A., and Aussenegg, F. (1996b). Thin films by regular patterns of metal nanoparticles: Tailoring the optical properties by nanodesign. *Appl.Phys.B*, 63:381 – 384.

Govorkov, S., Schröder, T., Shumay, I., and Heist, P. (1992). Transient gratings and second-harmonic probing of the phase transformation of a GaAs surface under femtosecond laser irradiation. *Phys. Rev. B*, 46:6864 – 6868.

Grafström, S., Blasberg, T., and Suter, D. (1996). Reflection spectroscopy of spin-polarized atoms near a dielectric surface. *J. Opt. Soc. Am. B*, 13:3 – 10.

Grangeon, F., Sassoli, H., Mathey, Y., Autric, M., Pailharey, D., and Marine, W. (1995). Pulsed laser deposition of NbTe thin films. *Appl. Surf. Sci.*, 86:160 – 164.

Green, J., Silfvast, W., and Wood II, O. (1977). Evolution of a CO_2-laser-produced cadmium plasma. *J.Appl.Phys.*, 48:2753 – 2761.

Griffiths, P. and Haseth, J. (1986). *Fourier Transform Infrared Spectrometry*. Wiley, New York.

Groeneveld, R., Sprik, R., and Lagendijk, A. (1995). Femtosecond spectroscopy of electron-electron and electron-phonon energy relaxation in Ag and Au. *Phys.Rev.B*, 51:11433 – 11445.

Gruhlke, R., Holland, W., and Hall, D. (1986). Surface-plasmon cross coupling in molecular fluorescence near a corrugated thin metal film. *Phys.Rev.Lett.*, 56:2838 – 2841.

Grunze, M. (1993). Preparation and characterization of self-assembled organic films on solid surfaces. *Physica Scripta*, T49:711 – 717.

Güdde, J., Hohlfeld, J., Müller, J., and Matthias, E. (1998). Damage threshold dependence on electron-phonon coupling in Au and Ni films. *Appl.Surf.Sci.*, 127:40 – 45.

Guosheng, Z., Fauchet, P., and Siegman, A. (1982). Growth of spontaneous periodic surface structures on solids during laser illumination. *Phys. Rev. B*, 26:5366 – 5381.

Gusev, V. and Karabutov, A. (1993). *Laser Optoacoustics*. American Institute of Physics, New York.

Guthrie, W., Lin, T., Ceyer, S., and Somorjai, G. (1982). The angular and velocity distributions of no scattered from the Pt(111) crystal surface. *J. Chem. Phys.*, 76:6398 – 6407.

Guyot-Sionnest, P. (1991a). Coherent processes at surfaces: Free-induction decay and photon echo of the Si-H stretching vibration for H/Si(111). *Phys.Rev.Lett.*, 66:1489 – 1492.

Guyot-Sionnest, P. (1991b). Two-phonon bound state for the hydrogen vibration on the H/Si(111) surface. *Phys.Rev.Lett.*, 67:2323 – 2326.

Guyot-Sionnest, P., Tadjeddine, A., and Liebsch, A. (1990). Electronic distribution and nonlinear optical response at the metal-electrolyte interface. *Phys. Rev. Lett.*, 64:1678 – 1681.

Hache, F., Ricard, D., and Flytzanis, C. (1986). Optical nonlinearities of small metal particles: Surface-mediated resonance and quantum size effects. *J. Opt. Soc. Am. B*, 3:1647 – 1655.

Haglund,Jr, R. (1996). Microscopic and mesoscopic aspects of laser-induced desorption and ablation. *Appl.Surf.Sci.*, 96–98:1 – 13.

Haglund,Jr, R. and Itoh, N. (1994). Electronic processes in laser ablation of semiconductors and insulators. In Miller, J., editor, *Laser Ablation: Principles and Applications*. Springer-Verlag, Berlin.

Hähner, G., Toennies, J., and Wöll, C. (1990). Normal modes of CO adsorbed on metal surfaces. *Appl. Phys. A*, 51:208 – 215.

Hall, R. (1987). Pulsed-laser-induced desorption studies of kinetics of surface reactions. *J. Phys. Chem.*, 91:1007 – 1015.

Halperin, W. (1986). Quantum size effects in metal particles. *Rev.Mod.Phys.*, 58:533 – 606.

Hamermesh, A. (1962). *Group Theory*. Addison Wesley, New York.

Hanbury-Brown, R. and Twiss, R. (1957a). Interferometry of the intensity fluctuations in light. I. Basic theory: The correlation between photons in coherent beams of radiation. *Proc.Roy.Soc.A*, 242:300 – 324.

Hanbury-Brown, R. and Twiss, R. (1957b). Interferometry of the intensity fluctuations in light. II. An experimental test of the theory for partially coherent light. *Proc.Roy.Soc.A*, 243:291 – 319.

Harris, C., Alivisatos, A., and Waldeck, D. (1985). Nonradiative damping of molecular electronic excited states by metal surfaces. *Surf. Sci.*, 158:103 – 125.

Harris, R. and Wilkinson, J. (1995). Waveguide surface plasmon resonance sensors. *Sensors and Actuators B*, 29:261 – 267.

Harris, S., Holloway, S., and Darling, G. (1995). Hot electron mediated photodesorption: A time-dependent approach applied to NO/Pt(111). *J. Chem. Phys.*, 102:8235 – 8248.

Harrison, I., Polanyi, J., and Young, P. (1988a). Photochemistry of adsorbed molecules. III. Photodissociation and photodesorption of CH_3Br adsorbed on LiF(001). *J.Chem.Phys.*, 89:1475 – 1497.

Harrison, I., Polanyi, J., and Young, P. (1988b). Photochemistry of adsorbed molecules. IV. Photodissociation, photoreaction, photoejection and photodesorption of H_2S on LiF(001). *J.Chem.Phys.*, 89:1498 – 1523.

Hasselbrink, E. (1994). Mechanisms in photochemistry on metal surfaces. *Appl. Surf. Sci.*, 79/80:34 – 40.

Hattori, K., Okano, A., Nakai, Y., and Itoh, N. (1992). Laser-induced electronic processes on GaP(110) surfaces: Particle emission and ablation initiated by defects. *Phys.Rev. B*, 45:8424 – 8436.

Head-Gordon, M. and Tully, J. (1992). Vibrational relaxation on metal surfaces: Molecular-orbital theory and application to CO/Cu(100). *J. Chem. Phys.*, 96:3939 – 3949.

Hefter, U. and Bergmann, K. (1988). Spectroscopic detection methods. In Scoles, G., editor, *Atomic and Molecular Beam Methods*, pages 193 – 253. Oxford University Press, New York.

Heidberg, J., Stein, H., and Weiss, H. (1987). Vibrational predesorption of carbon monoxide from sodium chloride at 20 K induced by resonant infrared laser excitation. *Surf. Sci.*, 184:L431 – L438.

Heinz, B. and Morgner, H. (1997). MIES investigation of alkanethiol monolayers self-assembled on Au(111) and Ag(111) surfaces. *Surf. Sci.*, 372:100 – 116.

Heinz, T. (1991). Second-order nonlinear optical effects at surfaces and interfaces. In Ponath, H.-E. and Stegeman, G., editors, *Nonlinear Surface Electromagnetic Phenomena*, chapter 5, pages 353 – 416. Elsevier, Amsterdam.

Heinz, T., Chen, C., Ricard, D., and Shen, Y. (1982). Spectroscopy of molecular monolayers by resonant second-harmonic generation. *Phys. Rev. Lett.*, 48:478 – 481.

Heinz, T., Loy, M., and Thompson, W. (1985). Study of Si(111) surfaces by optical second-harmonic generation: Reconstruction and surface phase transformation. *Phys. Rev. Lett.*, 54:63 – 66.

Hellsing, B., Chakarov, D., Österlund, L., Zhdanov, V., and Kasemo, B. (1997). Photoinduced desorption of potassium atoms from a two dimensional overlayer on graphite. *J.Chem.Phys.*, 106:982 – 1002.

Helvajian, H. and Welle, R. (1989). Threshold level laser photoablation of crystalline silver: Ejected ion translational energy distributions. *J.Chem.Phys.*, 91:2616 – 2626.

Herman, I. (1989). Laser-assisted deposition of thin films from gas phase and surface-adsorbed molecules. *Chem. Rev.*, 89:1323 – 1357.

Herman, J. and Elsayed-Ali, H. (1992a). Superheating of Pb(111). *Phys.Rev.Lett.*, 69:1228 – 1231.

Herman, J. and Elsayed-Ali, H. (1992b). Time-resolved study of surface disordering of Pb(110). *Phys. Rev. Lett.*, 68:2952 – 2955.

Herman, J., Elsayed-Ali, H., and Murphy, E. (1993). Time-resolved structural study of Pb(100). *Phys.Rev.Lett.*, 71:400 – 403.

Herrmann, J. and Wilhelmi, B. (1987). *Laser for Ultrashort Light Pulses*. North-Holland, Amsterdam.

Hertel, T., Knoesel, E., Wolf, M., and Ertl, G. (1996). Ultrafast electron dynamics at Cu(111): Response of an electron gas to optical excitation. *Phys.Rev.Lett.*, 76:535 – 538.

Herzberg, G. (1950). *Spectra of Diatomic Molecules*. Van Nostrand, New York.

Higashi, G. (1989). The chemistry of alkyl-aluminum compounds during laser-assisted chemical vapor deposition. *Appl.Surf.Sci.*, 43:6 – 10.

Hill, W., Wehling, B., Fallourd, V., and Klockow, D. (1995). Application of surface-enhanced Raman scattering for chemical sensors. *Spectroscopy Europe*, 7(5):20 – 24.

Hillenkamp, F., Karas, M., Beavis, R. C., and Chait, B. T. (1991). Matrix-assisted laser desorption/ionization mass spectrometry of biopolymers. *Anal.Chem.*, 63:1193A – 1202A.

Hillenkamp, F. and Sacchi, R. P. A. C., editors (1980). *Lasers in Biology and Medicine*. Plenum, New York.

Himmelbauer, M., Arenholz, E., Bäuerle, D., and Schilcher, K. (1996). UV-laser-induced surface topology changes in polyimide. *Appl.Phys.A*, 63:337 – 339.

Himmelbauer, M., Arnold, N., Bityurin, N., Arenholz, E., and Bäuerle, D. (1997). UV-laser-induced periodic surface structures on polyimide. *Appl.Phys.A*, 64:451 – 455.

Hinds, E. (1991). Cavity quantum electrodynamics. *Adv. At., Mol. Opt. Phys.*, 28:237 – 294.

Hinds, E. (1994). Perturbative cavity quantum electrodynamics. *Adv. At., Mol. Opt. Phys., Suppl.2*, pages 1 – 56.

Hiraoka, H. and Sendova, M. (1994). Laser-induced sub-half-micrometer periodic structure on polymer surfaces. *Appl.Phys.Lett.*, 64:563 – 565.

Ho, W. (1996). Reactions at metal surfaces induced by femtosecond lasers, tunneling electrons, and heating. *J.Phys.Chem.*, 100:13050 – 13060.

Hodak, J., Martini, I., and Hartland, G. (1998). Spectroscopy and dynamics of nanometer-sized noble metal particles. *J.Phys.Chem.B*, 102:6958 – 6967.

Hoffmann, F. (1983). Infrared reflection-absorption spectroscopy of adsorbed molecules. *Surf.Sci.Rep.*, 3:107 – 192.

Hofstetter, A. (1986). Lasers in urology. *Lasers Surg. Med.*, 6:412 – 414.

Hoheisel, W., Jungmann, K., Vollmer, M., Weidenauer, R., and Träger, F. (1988). Desorption stimulated by laser-induced surface-plasmon excitation. *Phys. Rev. Lett.*, 60:1649 – 1652.

Hohlfeld, J., Conrad, U., and Matthias, E. (1996). Does femtosecond time-resolved second-harmonic generation probe electron temperatures at surfaces? *Appl.Phys.B*, 63:541 – 544.

Hohlfeld, J., Matthias, E., Knorren, R., and Bennemann, K. (1997a). Nonequilibrium magnetization dynamics of nickel. *Phys.Rev.Lett.*, 78:4861 – 4864.

Hohlfeld, J., Müller, J., Wellershof, S.-S., and Matthias, E. (1997b). Time-resolved thermoreflectivity of thin gold films and its dependence on film thickness. *Appl.Phys.B*, 64:387 – 390.

Holland, W. and Hall, D. (1984). Frequency shifts of an electric-dipole resonance near a conducting surface. *Phys. Rev. Lett.*, 52:1041 – 1044.

Holtslag, A. (1989). Calculations on temperature profiles in optical recording. *J. Appl. Phys.*, 66:1530 – 1543.

Horoszowski, H., Heim, M., and Farine, I. (1987). The carbon dioxide laser in orthopedic surgery. In Ben-Hur, E. and Rosenthal, I., editors, *Photomedicine*, volume III, chapter 4, pages 61 – 65. CRC Press, Bota-Racon.

Horvath, K., Smith, W., Laurence, R., Schoen, F., Appleyard, R., and Cohn, L. (1995). Recovery and viability of an acute myocardial infarct after transmyocardial laser revascularization. *J.Am.Coll.Cardiol.*, 25:258 – 263.

Horwitz, C. (1980). A new vacuum-etched high-transmittance (antireflection) film. *Appl.Phys.Lett.*, 36:727 – 730.

Hotzel, A., Knoesel, E., Wolf, M., and Ertl, G. (1998). Controlled modification of surface state lifetimes by physisorbed adsorbates. *Proc. SPIE*, 3272:228 – 237.

Hou, H., Gulding, S., Rettner, C., Wodtke, A., and Auerbach, D. (1997). The stereodynamics of a gas-surface reaction. *Science*, 277:80 – 82.

Hove, M. V., Weinberg, W., and Chan, C.-M. (1986). *Low-Energy Electron Diffraction*. Springer-Verlag, Berlin.

Hövel, H., Fritz, S., Hilger, A., Kreibig, U., and Vollmer, M. (1993). Width of cluster plasmon resonances: Bulk dielectric functions and chemical interface damping. *Phys. Rev. B*, 48:18178 – 18188.

Hua, X. and Gersten, J. (1986). Theory of second-harmonic generation by small metal spheres. *Phys. Rev. B*, 33:3756 – 3764.

Huang, C., Asaki, M., Backus, S., Murnane, M., Kapteyn, H., and Nathel, H. (1992). 17-fs pulses from a self-mode-locked Ti:sapphire laser. *Opt. Lett.*, 17:1289 – 1291.

Huang, W. and Lue, J. (1994). Quantum size effect on the optical properties of small metallic particles. *Phys.Rev.B*, 49:17279 – 17285.

Hulpke, E., editor (1992). *Helium Atom Scattering from Surfaces*. Springer-Verlag, Berlin.

Hulteen, J. and Duyne, R. V. (1995). Nanosphere lithography: A materials general fabrication process for periodic particle array surfaces. *J.Vac.Sci.Technol.A*, 13:1553 – 1558.

Hunt, J., Guyot-Sionnest, P., and Shen, Y. (1987). Observation of C-H stretch vibrations of monolayers of molecules by optical sum-frequency generation. *Chem. Phys. Lett.*, 133:189 – 192.

Ibach, H. and Mills, D. (1982). *Electron Energy Loss Spectroscopy and Surface Vibrations*. Academic Press, New York.

Ihlemann, J., Scholl, A., Schmidt, H., and Wolff-Rottke, B. (1995). Nanosecond and femtosecond excimer-laser ablation of oxide ceramics. *Appl. Phys. A*, 60:411 – 417.

Ihlemann, J., Wolff, B., and Simon, P. (1992). Nanosecond and femtosecond excimer laser ablation of fused silica. *Appl. Phys. A*, 54:363 – 368.

Illgner, C., Schaaf, P., Lieb, K.-P., Queitsch, R., and Barnikel, J. (1998). Material transport during excimer-laser nitriding of iron. *J.Appl.Phys.*, 83:2907 – 2914.

Inouye, H., Tanaka, K., Tanahashi, I., and Hirao, K. (1998). Ultrafast dynamics of nonequilibrium electrons in a gold nanoparticle system. *Phys.Rev.B*, 57:11334 – 11340.

Ishida, H. and Liebsch, A. (1992). Electronic excitations in thin alkali-metal layers adsorbed on metal surfaces. *Phys.Rev.B*, 45:6171 – 6187.

Israelachvili, J. (1992). *Intermolecular and Surface Forces*. Academic Press, London.

Ith, M., Pratisto, H., Altermatt, H., Frenz, M., and Weber, H. (1994). Dynamics of laser-induced channel formation in water and influence of pulse duration on the ablation of biotissue under water with pulsed erbium-laser radiation. *Appl.Phys. B*, 59:621 – 629.

Jackson, J. (1975). *Classical Electrodynamics*. McGraw-Hill, New York.

Jackson, R., Polanyi, J., and Sjövall, P. (1995). Photodissociation dynamics of $(NO)_2$ on LiF(001): Characterization of vibrationally excited NO fragments. *J.Chem.Phys.*, 102:6308 – 6326.

Jackson, W., Amer, N., Boccara, A., and Fournier, D. (1981). Photothermal deflection spectroscopy and detection. *Appl.Opt.*, 20:1333 – 1344.

Jacobs, D., Kolasinski, K., Madix, R., and Zare, R. (1987). Rotational alignment of NO desorbing from Pt(111). *J.Chem.Phys.*, 87:5038 – 5039.

Jacques, S. (1993). Role of tissue optics and pulse duration on tissue effects during high-power laser irradiation. *Appl. Opt.*, 32:2447 – 2454.

Jaluria, Y. and Torrance, K. (1986). *Computational Heat Transfer.* Springer Series in Computational Methods in Mechanics and Thermal Sciences. Springer-Verlag, Berlin.

Jelski, D., Leung, P., and George, T. (1988). Photochemistry at structured surfaces: A classical electromagnetic approach. *Int. Rev. Phys. Chem.*, 7:179 – 207.

Jensen, H., Reinisch, R., and Coutaz, J. (1997). Hydrodynamic study of surface plasmon enhanced non-local second-harmonic generation. *Appl.Phys.B*, 64:57 – 63.

Jensen, P., Larralde, H., Meunier, M., and Pimpinelli, A. (1998). Growth of three-dimensional structures by atomic deposition on surfaces containing defects: Simulations and theory. *Surf.Sci.*, 412–413:458 – 476.

Jeon, D., Hashizume, T., Sakurai, T., and Willis, R. (1992). Structural and electronic properties of ordered single and multiple layers of Na on the Si(111) surface. *Phys.Rev.Lett.*, 69:1419 – 1422.

Jersch, J., Demming, F., Hildenhagen, L., and Dickmann, K. (1998). Field enhancement of optical radiation in the nearfield of scanning probe microscope tips. *Appl.Phys.A*, 66:29 – 34.

Jersch, J. and Dickmann, K. (1996). Nanostructure fabrication using laser field enhancement in the near field of a scanning tunneling microscope tip. *Appl. Phys. Lett.*, 68:868 – 870.

Jiang, X.-P., Shapiro, M., and Brumer, P. (1996). Electronic absorption spectroscopy of diatomics on a dynamic surface: IBr on MgO(001). *J.Chem.Phys.*, 105:3479 – 3485.

Johansson, S., Schweitz, J., Westberg, H., and Boman, M. (1992). Microfabrication of three-dimensional boron structures by laser chemical processing. *J.Appl.Phys.*, 72:5956 – 5963.

Jongma, R., Berden, G., Rasing, T., Zacharias, H., and Meijer, G. (1997). Scattering of vibrationally and electronically excited CO molecules from a LiF (100) surface. *Chem.Phys.Lett.*, 273:147 – 152.

Jordan, C., Marowsky, G., and Rubahn, H.-G. (1995). Coverage dependent changes in the azimuthal anisotropy of second harmonic generated from Na/Si(111) 7×7 interfaces. *Opt. Commun.*, 120:98 – 102.

Jost, W. (1960). *Diffusion*. Academic Press, New York.

Jostell, U. (1979). Plasmons in monolayer Na, K and Rb films. *Surf.Sci.*, 82:333 – 348.

Jung, D. and Czanderna, A. (1994). Chemical and physical interactions at metal/self-assembled organic monolayer interfaces. *Crit. Rev. Solid State Mater. Sci.*, 19:1 – 54.

Käding, O., Skurk, H., Maznev, A., and Matthias, E. (1995). Transient thermal gratings at surfaces for thermal characterization of bulk materials and thin films. *Appl.Phys.A*, 61:253 – 261.

Kahl, M., Voges, E., and Hill, W. (1998). Optimization of SERS substrates by electron-beam lithography. *Spectr.Europe*, October:8 – 13.

Kaiser, N. (1996). Interference coatings for the ultraviolet spectral region. *Laser und Optoelektronik*, 28(2):52 – 60.

Kapteyn, H. and Murnane, M. (1994). Femtosecond lasers: The next generation. *Opt.Phot.News*, 5(3):20 – 28.

Karas, M., Bachmann, D., Bahr, U., and Hillenkamp, F. (1987). Matrix-assisted ultraviolet laser desorption of non-volatile compounds. *Int.J.Mass Spectr.Ion Proc.*, 78:53 – 68.

Karatev, V. I., Mamyrin, B. A., Shmikk, D. V., and Zagulin, V. A. (1973). The mass-reflectron, a new non-magnetic time-of-flight mass spectrometer with high resolution. *Sov.Phys.JETP*, 37:45 – 48.

Kautek, W. and König, E., editors (1997). *Lasers in the Conservation of Artworks*, Wien. Mayer-Verlag.

Keilmann, F. and Bai, Y. (1982). Periodic surface structures frozen into CO_2 laser-melted quartz. *Appl. Phys. A*, 29:9 – 18.

Kelly, P., Tang, Z.-R., Woolf, D., Williams, R., and McGilp, J. (1991). Optical second harmonic generation from Si(111)-As and Si(100)-As. *Surf. Sci.*, 251–252:87 – 91.

Kelly, R., Miotello, A., Braren, B., Gupta, A., and Casey, K. (1992). Primary and secondary mechanisms in laser-pulse sputtering. *Nucl. Instrum. Methods Phys. Res.*, B65:187–199.

King, D. (1975). Thermal desorption from metal surfaces: A review. *Surf.Sci.*, 47:384 – 402.

Kirilyuk, A., Petukhov, A., Rasing, T., Megy, R., and Beauvillain, P. (1997a). Second harmonic generation study of quantum well states in thin noble metal overlayer films. *Surf.Sci.*, 377:409 – 413.

Kirilyuk, V., Kirilyuk, A., and Rasing, T. (1997b). A combined nonlinear and linear magneto-optical microscopy. *Appl.Phys.Lett.*, 70:2306 – 2308.

Kirilyuk, V., Kirilyuk, A., and Rasing, T. (1997c). New mode of domain imaging: Second harmonic generation microscopy. *J.Appl.Phys.*, 81:5014.

Kirpekar, F., Nordhoff, E., Larsen, L. K., Kristiansen, K., Roepstorff, P., and Hillenkamp, F. (1998). DNA sequence analysis by MALDI mass spectrometry. *Nucleic Acids Res.*, 26:2554 – 2559.

Kittl, J., Reitano, R., Aziz, M., Brunco, D., and Thompson, M. (1993). Time-resolved temperature measurements during rapid solidification of Si-As alloys induced by pulsed-laser melting. *J.Appl.Phys.*, 73:3725 – 3733.

Klar, T., Perner, M., Grosse, S., Plessen, G. V., Spirkl, W., and Feldmann, J. (1998). Surface-plasmon resonances in single metallic nanoparticles. *Phys.Rev.Lett.*, 80:4249 – 4252.

Klein-Wiele, J.-H., Simon, P., and Rubahn, H.-G. (1998). Size-dependent plasmon lifetimes and electron-phonon coupling time constants for surface bound Na clusters. *Phys.Rev.Lett.*, 80:45 – 48.

Kneipp, K., Wang, Y., Kneipp, H., Perelman, L., Itzkan, I., Dasari, R., and Feld, M. (1997). Single molecule detection using surface-enhanced Raman scattering (SERS). *Phys.Rev.Lett.*, 78:1667 – 1670.

Knoesel, E., Hotzel, A., and Wolf, M. (1998). Temperature dependence of surface state lifetimes, dephasing rates and binding energies on Cu(111) studied with time-resolved photoemission. *J.Electron.Spectrosc.Relat.Phenom.*, 88-91:577 – 584.

Knoll, W. (1998). Interfaces and thin films as seen by bound electromagnetic waves. *Annu.Rev.Phys.Chem.*, 49:569 – 638.

Kolomenskii, A., Schuessler, H., Mikhalevich, V., and Maznev, A. (1998). Interaction of laser-generated surface acoustic pulses with fine particles: Surface cleaning and adhesion studies. *J.Appl.Phys.*, 84:2404 – 2410.

Koopmans, B., F., der Woude, and Sawatzky, G. (1992). Surface symmetry resolution of nonlinear optical techniques. *Phys. Rev. B*, 46:12780 – 12783.

Koopmans, B., Janner, A.-M., Jonkman, H., Sawatzky, G., and van der Woude, F. (1993). Strong bulk magnetic dipole induced second-harmonic generation from C_{60}. *Phys. Rev. Lett.*, 71:3569 – 3572.

Koopmans, B., Koerkamp, M., Rasing, T., and Berg, H. V. D. (1995). Observation of large Kerr angles in the nonlinear optical response from magnetic multilayers. *Phys.Rev.Lett.*, 74:3692 – 3695.

Koren, G. (1987). Observation of shock waves and cooling waves in the laser ablation of kapton films in air. *Appl.Phys.Lett.*, 51:569 – 571.

Koren, G. (1988). Plume temperature in the laser ablation of polyimide films measured by infrared emission spectroscopy. *Appl. Phys. B*, 46:147 – 149.

Kozlowski, M. and Thomas, I. (1994). Future trends in optical coatings for high-power laser applications. *Proc.SPIE*, 2262:54 – 59.

Krajnovich, D. and Vazquez, J. (1993). Formation of 'intrinsic' surface defects during 248nm photoablation of polyimide. *J.Appl.Phys.*, 73:3001 – 3008.

Krebs, H.-U. (1997). Characteristic properties of laser-deposited metallic systems. *Int.J.Non-Equilibr.Proc.*, 10:3 – 24.

Kreibig, U. (1997). Optics of nanosized metals. In Hummel, R. and Wißmann, P., editors, *Handbook of Optical Properties, Vol.II, Optics of Small Particles, Interfaces, and Surfaces*, pages 145 – 190. CRC Press, Boca Raton.

Kreibig, U., Gartz, M., and Hilger, A. (1997). Mie resonances: Sensors for physical and chemical cluster interface properties. *Ber.Bunsenges.Phys.Chem.*, 101:1 – 12.

Kreibig, U. and Vollmer, M. (1995). *Optical Properties of Metal Clusters*, volume 25 of *Springer Series in Materials Science*. Springer-Verlag, Berlin.

Krenn, J., Wolf, R., Leitner, A., and Aussenegg, F. (1997). Near-field optical imaging the surface plasmon fields of lithographically designed nanostructures. *Opt.Commun.*, 137:46 – 50.

Kress, W. and Wette, F. D., editors (1991). *Surface Phonons*. Springer-Verlag, Berlin.

Kretschmann, E. (1971). Die Bestimmung optischer Konstanten von Metallen durch Anregung von Oberflächenplasmaschwingungen. *Z. Phys.*, 241:313 – 324.

Kretschmann, E. and Raether, H. (1968). Radiative decay of non-radiative surface plasmons excited by light. *Z. Naturforschung*, 23a:2135 – 2136.

Kroo, N., Krieger, W., Lenkefi, Z., Szentirmay, Z., Thost, J., and Walther, H. (1995). A new optical method for investigation of thin metal films. *Surf.Sci.*, 331-333:1305 – 1309.

Kroo, N. and Szentirmay, Z. (1988). Decay time measurement of surface plasmons on silver gratings. *Hung.Acad.Sci.KFKI*, 1988-18/E:1 – 13.

Kroto, H., Fischer, J., and Cox, D., editors (1993). *The Fullerenes*, Oxford. Pergamon Press.

Kubiak, G., Berger, K. W., Haney, S., Tichenor, D., Stuler, R., Rockett, P., and Hunter, J. (1994). Laser ablation: Mechanisms and applications II. In Geohegan, D. and Miller, J., editors, *AIP Conference Proceedings*, New York.

Kuhn, H. (1970). Classical aspects of energy transfer in molecular systems. *J. Chem. Phys.*, 33:101 – 108.

Kuhn, H., Möbius, D., and Bücher, H. (1972). Spectroscopy of monolayer assemblies. In Weissberger, A. and Rossiter, B., editors, *Physical Methods of Chemistry, Vol.1*, volume 1b, Pt.3B, Ch.VII, pages 577 – 702. Wiley, New York.

Kuhnke, K., Becker, R., Epple, M., and Kern, K. (1997). C_{60} exciton quenching near metal surfaces. *Phys.Rev.Lett.*, 79:3246 – 3249.

Kumagai, H., Ezaki, M., Toyoda, K., and Obara, M. (1992). Fabrication of periodic submicron dot structures of n-InP by laser-induced surface electromagnetic wave etching. *Jap.J.Appl.Phys.*, 31:928 – 930.

Kümmerlein, J., Leitner, A., Brunner, H., Aussenegg, F., and Wokaun, A. (1993). Enhanced dye fluorescence over silver island films: Analysis of the distance dependence. *Mol. Phys.*, 80:1031 – 1046.

Küper, S., Brannon, J., and Brannon, K. (1993). Threshold behavior in polyimide photoablation: Single-shot rate measurements and surface-temperature modeling. *Appl. Phys. A*, 56:43 – 50.

Küper, S. and Stuke, M. (1989). Ablation of polytetrafluoroethylene (teflon) with femtosecond UV excimer laser pulses. *Appl. Phys. Lett.*, 54:4 – 6.

Lahee, A. and Toennies, P. (1993). Surface studies on the rebound. *Physics World*, pages 61 – 66.

Laitenberger, P., Claessens, C., Kuipers, L., Raymo, F., Palmer, R., and Stoddart, J. (1997). Building supramolecular nanostructures on surfaces: The influence of the substrate. *Chem.Phys.Lett.*, 279:209 – 214.

Lamb, W. and Retherford, R. (1947). Fine structure of the hydrogen atom by a microwave method. *Phys. Rev.*, 72:241 – 243.

Lambacher, A. and Fromherz, P. (1996). Fluorescence interference-contrast microscopy on oxidized silicon using a monomolecular dye layer. *Appl. Phys. A*, 63:207 – 216.

Lambert, J. (1977). *Vibrational and Rotational Relaxation in Gases.* Clarendon Press, New York.

Lamprecht, B., Leitner, A., and Aussenegg, F. (1997). Femtosecond decay-time measurement of electron plasma oscillation in nanolithographically designed silver particles. *Appl. Phys. B*, 64:269 – 272.

Lang, N. and Kohn, W. (1970). Theory of metal surfaces: Charge density and surface energy. *Phys. Rev. B*, 1:4555 – 4568.

Lang, R. and Fukui, M. (1994). Selvedge treatment of the aluminum and silver bilayer system. *J. Phys. Soc. Jap.*, 63:1848 – 1860.

Larsson, L., Heimann, P., Lindenberg, A., Schuck, P., Bucksbaum, P., Lee, R., Padmore, H., Wark, J., and Falcone, R. (1998). Ultrafast structural changes measured by time-resolved X-ray diffraction. *Appl.Phys.A*, 66:587 – 591.

Lee, I., Callcott, T., and Arakawa, E. (1993). Desorption studies of metal atoms using laser-induced surface-plasmon excitation. *Phys. Rev. B*, 47:6661 – 6666.

Lee, Y.-S., Anderson, M., and Downer, M. (1997). Fourth-harmonic generation at a crystalline GaAs(001) surface. *Opt.Lett.*, 22:973 – 975.

Lehmann, O. and Stuke, M. (1991). Generation of three-dimensional free-standing metal micro-objects by laser chemical processing. *Appl. Phys. A*, 53:343 – 345.

Lehmann, O. and Stuke, M. (1994). Three-dimensional laser direct writing of electrically conducting and isolating microstructures. *Mater. Lett.*, 21:131 – 136.

Lehmann, O. and Stuke, M. (1995). Laser-driven movement of three-dimensional microstructures generated by laser rapid prototyping. *Science*, 270:1644 – 1646.

Lennard-Jones, J. (1932). Processes of adsorption and diffusion on solid surfaces. *Trans. Faraday Soc.*, 28:333 – 359.

Letokhov, V. (1988). Laser-induced chemistry — basic nonlinear processes and applications. *Appl. Phys. B*, 46:237 – 251.

Leung, P. (1997). Emission frequency of single molecules at a metallic aperture: The applicability of the image theory. *Opt.Commun.*, 136:360 – 364.

Leung, P. and George, T. (1986). Photodissociation of molecules at structured metallic surfaces. *J. Chem. Phys.*, 85:4729 – 4733.

Leung, P. and George, T. (1989). Molecular spectroscopy at corrugated metal surfaces. *Spectroscopy*, 4(1):35 – 41.

Leung, P., George, T., and Lee, Y. (1987). Limit of the image theory for the classical decay rates of molecules at surfaces. *J. Chem. Phys.*, 86:7227 – 7229.

Leung, P. and Hider, M. (1993). Nonlocal electrodynamic modeling of frequency shifts for molecules at rough surfaces. *J. Chem. Phys.*, 98:5019 – 5022.

Leung, P., Kim, Y., and George, T. (1988). Roughness-induced resonance for molecular fluorescence near a corrugated metallic surface. *Phys. Rev. B*, 38:10032 – 10034.

Leung, P., Pollard-Knight, D., Malan, G., and Finlan, M. (1994). Modelling of particle-enhanced sensitivity of the surface-plasmon-resonance biosensor. *Sensors and Actuators B*, 22:175 – 180.

Levenson, M., Viswanathan, N., and Simpson, R. (1982). Improving resolution in photolithography with a phase-shifting mask. *IEEE Trans.Electr.Dev.*, ED-29:1828 – 1836.

Levinson, H., Plummer, E., and Feibelman, P. (1979). Effects on photoemission of the spatially varying photon field at a metal surface. *Phys. Rev. Lett.*, 43:952 – 955.

Levis, R. (1994). Laser desorption and ejection of biomolecules from the condensed phase into the gas phase. *Annu.Rev.Phys.Chem.*, 45:483 – 518.

Levlin, M., Laakso, A., Niemi, H.-M., and Hautojärvi, P. (1997). Evaporation of gold thin films on mica: Effect of evaporation parameters. *Appl. Surf. Sci.*, 115:31 – 38.

Liebsch, A. (1989). Second-harmonic generation from alkali-metal overlayers. *Phys. Rev. B*, 40:3421 – 3424.

Liebsch, A. (1991). Electronic excitations in adsorbed alkali-metal layers. *Phys. Rev. Lett.*, 67:2858 – 2861.

Liebsch, A. (1997). *Electronic Excitations at Metal Surfaces*. Plenum Press, New York.

Lingle, Jr, R., Ge, N.-H., Jordan, R., McNeill, J., and Harris, C. (1996). Femtosecond studies of electron tunneling at metal-dielectric interfaces. *Chem. Phys.*, 205:191 – 203.

Lingle, Jr, R., Padowitz, D. F., Jordan, R., McNeill, J., and Harris, C. (1994). Two-dimensional localization of electrons at interfaces. *Phys.Rev.Lett.*, 72:2243 – 2246.

Lison, F., Adams, H.-J., Haubrich, D., Kreis, M., Nowak, S., and Meschede, D. (1997). Nanoscale atomic lithography with a cesium atom beam. *Appl.Phys.B*, 65:419 – 421.

Loesel, F., Fischer, J., Götz, M., Horvath, C., Juhasz, T., Noack, F., Suhm, N., and Bille, J. (1998). Non-thermal ablation of neural tissue with femtosecond laser pulses. *Appl.Phys.B*, 66:121 – 128.

Long, D. (1977). *Raman Spectroscopy*. McGraw Hill, New York.

Lu, C. and Czanderna, A. (1984). *Application of Piezoelectric Quartz Microbalances*. Elsevier, New York.

Lu, Y., Loh, T., Teo, B., and Low, T. (1994). Effect of polarization on laser-induced surface-temperature rise. *Appl. Phys. A*, 58:423 – 429.

Lu, Y. F., Song, W. D., Ang, B. W., Hong, M. H., Chan, D. S. H., and Low, T. S. (1997). A theoretical model for laser removal of particles from solid surfaces. *Appl. Phys. A*, 65:9 – 13.

Lubman, D., editor (1990). *Lasers and Mass Spectrometry*, New York. Oxford University Press.

Lucchese, R. and Tully, J. (1984). Laser induced thermal desorption from surfaces. *J. Chem. Phys.*, 81:6313 – 6319.

Luce, T., Hübner, W., and Bennemann, K. (1997). Theory for the nonlinear optical response at noble-metal surfaces with nonequilibrium electrons. *Z.Phys.B*, 102:223 – 232.

Lüpke, G., Bottomley, D., and van Driel, H. (1994). Resonant second-harmonic generation on Cu(111) by a surface-state to image-potential-state transition. *Phys. Rev. B*, 49:17303 – 17306.

Ma, Z., Sun, W., Sou, I., and Wong, G. (1998). Atomic force microscopy studies of ZnSe self-organized dots fabricated on ZnS/GaP. *Appl.Phys.Lett.*, 73:1340 – 1342.

Machlab, H., McGahan, W., Woollam, J., and Cole, K. (1993). Thermal characterization of thin films by photothermally induced laser beam deflection. *Thin Solid Films*, 224:22 – 27.

Magonov, S. and Whangbo, M.-H. (1996). *Surface Analysis with STM and AFM*. VCH, Weinheim.

Maiman, T. (1960). Stimulated optical radiation in ruby. *Nature*, 187:493–494.

Mandel, L. and Wolf, E. (1995). *Optical Coherence and Quantum Optics*. Cambridge Universtiy Press, New York.

Mann, K. and Hopfmüller, A. (1994). Characterization and shaping of excimer laser radiation. In *Proceedings of 2nd Workshop on Laser Beam Characterization*, pages 347 – 358.

Manson, J. (1991). Inelastic scattering from surfaces. *Phys.Rev.B*, 43:6924 – 6937.

Manson, J. (1994). Multiphonon atom-surface scattering. *Comp.Phys.Commun.*, 80:145 – 167.

Manson, J., Renger, M., and Rubahn, H.-G. (1996). Subthermal kinetic energy distributions of neutral atoms photodesorbed from Na cluster surfaces. *Phys. Lett. A*, 224:121 – 126.

Marowsky, G., Chi, L., Möbius, D., Steinhoff, R., Shen, Y., Dorsch, D., and Rieger, B. (1988). Non-linear optical properties of hemicyanine monolayers and the protonation effect. *Chem.Phys.Lett.*, 147:420–424.

Marshall, A. G. and Schweikhardt, L. (1992). Fourier transform ion cyclotron resonance mass spectrometry: Technique developments. *Int.J.Mass Spectrom.Ion.Proc.*, 118/119:37 – 70.

Masel, R. (1996). *Principles of Adsorption and Reaction on Solid Surfaces*. Wiley, New York.

Maser, M., Apfelberg, D., and Lash, H. (1983). Clinical applications of the Argon and carbon dioxide lasers in dermatology and plastic surgery. *World J. Surg.*, 7:684 – 691.

Matthias, E. and Dreyfus, R. (1989). From laser-induced desorption to surface damage. In Hess, P., editor, *Photoacoustic, Photothermal and Photochemical Processes at Surfaces and in Thin Films*, volume 47 of *Topics in Current Physics*, chapter 4, pages 89 – 128. Springer-Verlag, Berlin.

Matthias, E., Reichling, M., Siegel, J., Käding, O., Petzoldt, S., Skurk, H., Bizenberger, P., and Neske, E. (1994). The influence of thermal diffusion on laser ablation of metal films. *Appl. Phys. A*, 58:129 – 136.

Matthias, E., Siegel, J., Petzoldt, S., Reichling, M., Skurk, H., Käding, O., and Neske, E. (1995). In-situ investigation of laser ablation of thin films. *Thin Solid Films*, 254:139 – 146.

Mazumder, J. (1983). Laser heat treatment: The state of the art. *J.Met.*, 35:18 – 26.

McGahan, W. and Cole, K. (1992). Solutions of heat conduction equation in multilayers for photothermal deflection experiments. *J. Appl. Phys.*, 72:1362 – 1373.

McGilp, J. (1987). Determining metal-semiconductor interface structure by optical second-harmonic-generation. *J. Vac. Sci. Technol. A*, 5:1442 – 1446.

McLeod, J. (1954). The axicon: A new type of optical element. *J. Opt. Soc. Am.*, 44:592 – 597.

Menzel, D. (1995). Thirty years of MGR: How it came about, and what came of it. *Nucl.Instrum.Meth.Phys.Res.B*, 101:1 – 10.

Menzel, D. and Gomer, R. (1964). Desorption from metal surfaces by low-energy electrons. *J. Chem. Phys.*, 41:3311 – 3328.

Metev, S. and Veiko, V. (1994). *Laser-Assisted Microtechnology*, volume 29 of *Springer Series in Materials Science*. Springer-Verlag, Berlin.

Metiu, H. (1984). Surface enhanced spectroscopy. *Progr. Surf. Sci.*, 17:153 – 320.

Mie, G. (1908). Beiträge zur Optik trüber Medien speziell kolloidaler Metallösungen. *Ann. Phys. (Leipzig)*, 25:377 – 445.

Mihailov, S. and Duley, W. (1991). Study of the ablation threshold of polyimide (kapton h) utilizing double-pulsed XeCl excimer laser radiation. *J. Appl. Phys.*, 69:4092 – 4102.

Miller, J., editor (1994). *Laser Ablation, Principles and Applications*. Springer-Verlag, Berlin.

Miller, J. and Geohegan, D., editors (1994). *Laser Ablation: Mechanisms and Applications II*, New York. AIP Press.

Miotello, A. and Kelly, R. (1995). Critical assessment of thermal models for laser sputtering at high fluences. *Appl.Phys.Lett.*, 67:3535 – 3537.

Misewich, J., Heinz, T., and Newns, D. (1992). Desorption induced by multiple electronic transitions. *Phys.Rev.Lett.*, 68:3737 – 3740.

Misewich, J., Kalamarides, A., Heinz, T., Höfer, U., and Loy, M. (1994). Vibrationally assisted electronic desorption: Femtosecond surface chemistry of O_2/Pd(111). *J.Chem.Phys.*, 100:736 – 739.

Mizrahi, V. and Sipe, J. (1988). Phenomenological treatment of surface second-harmonic generation. *J. Opt. Soc. Am. B*, 5:660 – 667.

Möbius, D. and Bücher, H. (1972). Spectroscopy of monolayer assemblies, part II. In Weissberger, A. and Rossiter, B., editors, *Physical Methods of Chemistry*. Wiley-Interscience, New York.

Moini, S., Puri, A., and Das, P. (1993). Photochemistry near a semiconductor surface. *J.Chem.Phys.*, 98:746 – 752.

Moison, J. and Bensoussan, M. (1982). Laser-induced order-disorder transition on the (100) InP surface. *J. Vac. Sci. Technol.*, 21:315 – 318.

Monreal, R. and Apell, S. (1990). Electromagnetic-field-enhanced desorption of atoms. *Phys. Rev. B*, 41:7852 – 7855.

Morawitz, H. (1978). Surface plasmon effects on molecular decay processes near metallic interfaces. In Arecchi, F., Bonifacio, R., and Scully, M., editors, *Coherence in Spectroscopy and Modern Physics*, pages 261 – 300. Plenum Press, New York.

Moresco, F., Rocca, M., Zielasek, V., Hildebrandt, T., and Henzler, M. (1996). Evidence for the presence of the multipole plasmon mode on Ag surfaces. *Phys. Rev. B*, 54:R14333 – R14336.

Morin, M., Jakob, P., Levinos, N., Chabal, Y., and Harris, A. (1992a). Vibrational energy transfer on hydrogen-terminated vicinal Si(111) surfaces: Interadsorbate energy flow. *J. Chem. Phys.*, 96:6203 – 6212.

Morin, M., Levinos, N., and Harris, A. (1992b). Vibrational energy transfer of CO/Cu(100): Nonadiabatic vibration/electron coupling. *J. Chem. Phys.*, 96:3950 – 3956.

Moskovits, M. (1985). Surface-enhanced spectroscopy. *Rev. Mod. Phys.*, 57:783 – 826.

Mourou, G. (1997). The ultrahigh-peak-power laser: Present and future. *Appl. Phys. B*, 65:205 – 211.

Mourou, G. and Williamson, S. (1982). Picosecond electron diffraction. *Appl. Phys. Lett.*, 41:44 – 45.

Müller, E. (1951). Das Feldionenmikroskop. *Z. Phys.*, 131:136 – 142.

Müller, T., Vaccaro, P., Balzer, F., and Rubahn, H.-G. (1997). Size dependent optical second harmonic generation from surface bound Na clusters: Comparison between experiment and theory. *Opt. Commun.*, 135:103 – 108.

Murphy, R., Yeganeh, M., Song, K., and Plummer, E. (1989). Second-harmonic generation from the surface of a simple metal, Al. *Phys. Rev. Lett.*, 63:318 – 321.

Neddersen, J., Chumanov, G., and Cotton, T. (1993). A new method for preparing SERS active colloids. *Appl. Spectrosc.*, 47:1959 – 1964.

Neev, J., editor (1998). *Applications of Ultrashort-Pulse Lasers in Medicine and Biology*, volume 3255. Proc. SPIE.

Neuschäfer, D., Preiswerk, H., Spahni, H., Konz, E., and Marowsky, G. (1994). Second-harmonic generation using planar waveguides with consideration of pump depletion and absorption. *J.Opt.Soc.Am.B*, 11:649–654.

Niemz, M. (1995). Cavity preparation with the Nd:YLF picosecond laser. *J.Dent.Res.*, 74:1194 – 1199.

Niemz, M. (1996). *Laser-Tissue Interactions*. Springer-Verlag, Berlin.

Niino, H., Shimoyama, M., and Yabe, A. (1990). XeCl excimer laser ablation of a polyethersulfone film: Dependence of periodic microstructures on a polarized beam. *Appl.Phys.Lett.*, 57:2368 – 2370.

Nissim, Y., Lietoila, A., Gold, R., and Gibbons, J. (1980). Temperature distributions produced in semiconductors by a scanning elliptical or circular cw laser beam. *J. Appl. Phys.*, 51:274 – 279.

Nitzan, A. and Brus, L. (1981a). Can photochemistry be enhanced on rough surfaces? *J. Chem. Phys.*, 74:5321 – 5322.

Nitzan, A. and Brus, L. (1981b). Theoretical model for enhanced photochemistry on rough surfaces. *J. Chem. Phys.*, 75:2205 – 2214.

Nowacki, W. (1986). *Thermoelasticity*. Pergamon Press, New York.

Nowak, S., Pfau, T., and Mlynek, J. (1996). Nanolithography with metastable helium. *Appl.Phys.B*, 63:203 – 205.

Ogawa, S., Nagano, H., Petek, H., and Heberle, A. (1997). Optical dephasing in Cu(111) measured by interferometric two-photon time-resolved photoemission. *Phys.Rev.Lett.*, 78:1339 – 1342.

Ogilvy, J. (1991). *Theory of Wave Scattering from Random Rough Surfaces*. Adam Hilger, Bristol.

Oraevsky, A., Esenaliev, R., and Letokhov, V. (1991). Pulsed laser ablation of biological tissue: Review of the mechanisms. In Miller, R. and Haglund, R., editors, *Laser Ablation: Mechanism and Applications*, pages 112 – 122. Springer-Verlag, New York.

Östling, D., Stampfli, P., and Bennemann, K. (1993). Theory of nonlinear optical properties of small metallic spheres. *Z. Phys. D*, 28:169 – 175.

Otto, A. (1968). Wechselwirkung elektromagnetischer Oberflächenwellen. *Z. Angew. Physik*, 27:207 – 209.

Owrutsky, J., Culver, J., Li, M., Kim, Y., Sarisky, M., Yeganeh, M., Yodh, A., and Hochstrasser, R. (1992). Femtosecond coherent transient infrared spectroscopy of CO on Cu(111). *J.Chem.Phys.*, 97:4421 – 4426.

Özişik, M. (1985). *Heat Transfer*. McGraw-Hill, Inc., New York.

Padowitz, D., Harris, C., Jordan, R., Lingle Jr, R., McNeill, J., and Merry, W. (1994). Two-photon photoemission and the dynamics of electrons at interfaces. *Proc. SPIE*, 2125:88 – 106.

Padowitz, D., Merry, W., Jordan, R., and Harris, C. (1992). Two-photon photoemission as a probe of electron interactions with atomically thin dielectric films on metal surfaces. *Phys. Rev. Lett.*, 69:3583 – 3586.

Pagnot, T., Barchiesi, D., Labeke, D. V., and Pieralli, C. (1997). Use of a scanning near-field optical microscope architecture to study fluorescence and energy transfer near a metal. *Opt.Lett.*, 22:120 – 122.

Palasantzas, G. (1997). Roughness effects on the electrostatic-image potential near a dielectric interface. *J.Appl.Phys.*, 82:351 – 355.

Pan, R.-P., Wei, H., and Shen, Y. (1989). Optical second-harmonic generation from magnetized surfaces. *Phys. Rev. B*, 39:1229 – 1234.

Papadogiannis, N., Moustaizis, S., Loukakos, P., and Kalpouzos, C. (1997). Temporal characterization of ultra short laser pulses based on multiple harmonic generation on a gold surface. *Appl.Phys.B*, 65:339 – 345.

Papageorgopoulos, C. and Kamaratos, M. (1992). The behaviour of Na on 1×1 and 7×7 structures of Si(111) and its effect on the oxidation of these structures. *J.Phys.: Condens.Matter*, 4:1935 – 1945.

Parks, J. and McDonald, S. (1989). Evolution of the collective-mode resonance in small adsorbed sodium clusters. *Phys. Rev. Lett.*, 62:2301 – 2304.

Pascal, R., Zarnitz, C., Bode, M., and Wiesendanger, R. (1997). Fabrication of atomic gratings based on self-organization of adsorbates with repulsive interaction. *Appl.Phys.A*, 65:81 – 83.

Peckerar, M., Perkins, F., Dobisz, E., and Glembocki, O. (1997). Issues in nanolithography for quantum effect device manufacture. In Rai-Choudhury, P., editor, *Handbook of Microlithography, Micromachining and Microfabrication*, pages 681 – 763. IEEE Materials and Devices Series, London, UK.

Perner, M., Bost, P., Lemmer, U., Plessen, G. V., Feldmann, J., Becker, U., Mennig, M., Schmitt, M., and Schmidt, H. (1997). Optically induced damping of the surface plasmon resonance in gold colloids. *Phys.Rev.Lett.*, 78:2192 – 2195.

Perry, M. and Mourou, G. (1994). Terawatt to petawatt subpicosecond lasers. *Science*, 264:917 – 924.

Persson, B. and Lang, N. (1982). Electron-hole-pair quenching of excited states near a metal. *Phys. Rev. B*, 26:5409 – 5415.

Petek, H., Heberle, A., Nessler, W., Nagano, H., Kubota, S., Matsunami, S., Moriya, N., and Ogawa, S. (1997). Optical phase control of coherent electron dynamics in metals. *Phys.Rev.Lett.*, 79:4649 – 4652.

Petek, H. and Ogawa, S. (1997). Femtosecond time-resolved two-photon photoemission studies of electron dynamics in metals. *Progr.Surf.Sci.*, 56:239 – 310.

Pettit, G., Ediger, M., and Weiblinger, R. (1995). Excimer laser ablation of the cornea. *Opt. Eng.*, 34:661 – 667.

Pettit, G. and Sauerbrey, R. (1993). Pulsed ultraviolet laser ablation. *Appl. Phys. A*, 56:51 – 63.

Petzoldt, S., Elg, A., Reichling, M., Reif, J., and Matthias, E. (1988). Surface laser damage thresholds determined by photoacoustic deflection. *Appl.Phys.Lett.*, 53:2005 – 2007.

Pines, D. and Nozieres, P. (1966). *The Theory of Quantum Liquids*. Benjamin, New York.

Pipino, A., Hudgens, J., and Huie, R. (1997). Evanescent wave cavity ring-down spectroscopy for probing surface processes. *Chem.Phys.Lett.*, 280:104 – 112.

Plaja, L., Roso, L., Rzazewski, K., and Lewenstein, M. (1998). Generation of attosecond pulse trains during reflection of a very intense laser on a solid surface. *J.Opt.Soc.Am.B*, 15:1904 – 1911.

Pockrand, I., Brillante, A., and Möbius, D. (1980). Nonradiative decay of molecular excitation at a metal interface. *Il Nuovo Cimento B*, 23:350.

Poelsema, B. and Comsa, G. (1989). *Scattering of Thermal Energy Atoms*. Springer-Verlag, Berlin.

Polanski, G. and Rubahn, H.-G. (1996). Excimer laser assisted growth of Au thin films on mica (001). *J. Vac. Sci. Technol. A*, 14:110 – 114.

Porter, M., Bright, T., Allara, D., and Chidsey, C. (1987). Spontaneous organized molecular assemblies. IV. Structural characterization of n-alkyl thiol monolayers on gold by optical ellipsometry, infrared spectroscopy, and electrochemistry. *J.Am.Chem.Soc.*, 109:3559 – 3568.

Postawa, Z., Maboudian, R., El-Maazawi, M., Ervin, M., Wood, M., and Winograd, N. (1992). Electronic and nuclear effects in ion-induced desorption from NaCl (100). *J.Chem.Phys.*, 96:3298 – 3305.

Preuss, S., Demchuk, A., and Stuke, M. (1995). Sub-picosecond UV laser ablation of metals. *Appl. Phys. A*, 61:33 – 37.

Preuss, S., Späth, M., Zhang, Y., and Stuke, M. (1993). Time resolved dynamics of subpicosecond laser ablation. *Appl. Phys. Lett.*, 62:3049 – 3051.

Preuss, S. and Stuke, M. (1995). Subpicosecond Ultraviolet laser ablation of diamond: Nonlinear properties at 248nm and time-resolved characterization of ablation dynamics. *Appl. Phys. Lett.*, 67:338 – 340.

Prybyla, J., Tom, H., and Aumiller, G. (1992). Femtosecond time-resolved surface reactions: Desorption of CO from Cu(111) in < 325 fsec. *Phys.Rev.Lett.*, 68:503 – 506.

Qian, J.-P. and Wang, G.-C. (1990). A simple ultrahigh vacuum surface magneto-optical Kerr effect setup for the study of surface magnetic anisotropy. *J. Vac. Sci. Technol. A*, 8:4117 – 4119.

Quiniou, B., Bulovic, V., Wu, Z., Wang, X., and Osgood Jr, R. (1994). Nonlinear photoemission of image states: A new high-resolution surface spectroscopy. *Proc. SPIE*, 2125:78 – 87.

Raether, H. (1984). Roughness on silver films. *Surf. Sci.*, 140:31 – 36.

Raether, M. (1988). *Surface Plasmons*. Springer Tracts in Modern Physics. Springer-Verlag, Berlin.

Rai-Choudhury, P., editor (1997). *Handbook of Microlithography, Micromachining and Microfabrication*. IEEE Materials and Devices Series, London, UK.

Raizer, Y. (1970). Subsonic propagation of a light spark and threshold conditions for maintenance of a plasma by radiation. *Sov.Phys.JETP*, 58:2127 – 2138.

Rampi, M., Schueller, O., and Whitesides, G. (1998). Alkanethiol self-assembled monolayers as the dielectric of capacitors with nanoscale thickness. *Appl.Phys.Lett.*, 72:1781 – 1783.

Rasigni, M. and Rasigni, G. (1973). Anomalies of the optical properties of thin lithium layers and their relation to similar anomalies observed with other alkali metals. *J. Opt. Soc. Am.*, 63:775 – 785.

Rasigni, M., Rasigni, G., Gasparini, J., and Fraisse, R. (1976). Structure and optical conductivity of thin lithium deposits prepared at 6 K. *J. Appl. Phys.*, 47:1757 – 1761.

Ready, J. (1997). *Industrial Applications of Lasers*. Academic Press, New York.

Redhead, P. (1964). Interaction of slow electrons with chemisorbed oxygen. *Can. J. Phys.*, 42:886 – 905.

Reichelt, K. (1988). Nucleation and growth of thin films. *Vacuum*, 38:1083 – 1099.

Reichling, M., Siegel, J., Matthias, E., Lauth, H., and Hacker, E. (1994). Photoacoustic studies of laser damage in oxide thin films. *Thin Solid Films*, 253:333 – 338.

Reif, J., Rau, C., and Matthias, E. (1993). Influence of magnetism on second harmonic generation. *Phys. Rev. Lett.*, 71:1931 – 1934.

Reif, J., Zink, J., Schneider, C., and Kirschner, J. (1991). Effects of surface magnetism on optical second harmonic generation. *Phys. Rev. Lett.*, 67:2878 – 2881.

Ren, Q. and Birngruber, R. (1990). Axicon: A new laser beam delivery system for corneal surgery. *IEEE J.Quantum Electron.*, 26:2305 – 2308.

Ren, Q., Keates, R., Hill, R., and Berns, M. (1995). Laser refractive surgery: A review and current status. *Opt. Eng.*, 34:642 – 660.

Ren, Q., Simon, G., Legeais, J., Parel, J., Culberson, W., Shen, J., Takesue, Y., and Savoldelli, M. (1994). Ultraviolet solid-state laser (213nm) photorefractive keratectomy: In vivo study. *Ophthalmology*, 101:883 – 889.

Renger, M. and Rubahn, H.-G. (1993). Comment on the substrate dependence of the activation energy for photodesorption of Na atoms from large Na clusters. *Faraday Disc. Chem. Soc.*, 94:197 – 199.

Rettner, C., Auerbach, D., Tully, J., and Kleyn, A. (1996). Chemical dynamics at the gas-surface interface. *J.Phys.Chem.*, 100:13021 – 13033.

Richardson, N. and Bradshaw, A. (1979). The frequencies and amplitudes of vibrations at a metal surface from model cluster calculations. *Surf. Sci.*, 88:255 – 268.

Roeder, H., Bromann, K., Brune, H., and Kern, K. (1997). Strain mediated two-dimensional growth kinetics in metal heteroepitaxy. *Surf.Sci.*, 376:13 – 31.

Romana, V., Zweig, A., Frenz, M., and Weber, H. (1989). Time-resolved thermal microscopy with fluorescent films. *Appl. Phys. B*, 49:527 – 533.

Ronse, K., de Beeck, M. O., and Hove, L. V. D. (1994a). Fundamental principles of phase shifting masks by fourier optics: Theory and experimental verification. *J.Vac.Sci.Tech.B*, 12:589 – 600.

Ronse, K., Pforr, R., Jonckheere, R., and Hove, L. V. D. (1994b). Attenuated phase shifting masks in combination with off-axis illumination: Towards quarter-micron DUV lithography for random logic applications. *Microel. Eng.*, 23:133 – 138.

Rosenzweig, Z., Farbman, I., and Asscher, M. (1993). Diffusion of ammonia on Re(001): A monolayer grating optical second harmonic diffraction study. *J. Chem. Phys.*, 98:8277 – 8283.

Rosker, M., Marcy, H., Chang, T., Khoury, J., Hansen, K., and Whetten, R. (1992). Time-resolved degenerate four-wave mixing in thin films of C_{60} and C_{70} using femtosecond optical pulses. *Chem. Phys. Lett.*, 196:427 – 432.

Rossetti, R. and Brus, L. (1980). Time resolved molecular electronic energy transfer into a silver surface. *J. Chem. Phys.*, 73:572 – 577.

Rothenhäusler, B., Rabe, J., Korpiun, P., and Knoll, W. (1984). On the decay of plasmon surface polaritons at smooth and rough Ag-air interfaces: A reflectance and photo-acoustic study. *Surf. Sci.*, 137:373 – 383.

Royer, P., Bijeon, J., Goudonnet, J., Inagaki, T., and Arakawa, E. (1989). Optical absorbance of silver oblate particles. *Surf. Sci.*, 217:384 – 402.

Rubahn, H.-G. (1997). Time constants for the decay of elementary optical excitations in surface bound Na clusters. *Appl.Surf.Sci.*, 109–110:575 – 578.

Rubahn, K. and Ihlemann, J. (1998). Graded transmission dielectric optical masks by laser ablation. *Appl. Surf. Sci.*, 127 – 129:881 – 884.

Rudolf, H. and Steinmann, W. (1977). Two photon photoelectric effect in the surface plasma resonance of aluminum. *Phys. Lett.A*, 61:471 – 472.

Safaeinili, A., McKie, A., and Addison, Jr, R. (1996). Noncontact surface-hardness measurement using laser-based ultrasound. *MRS Bulletin*, October:53 – 57.

Sahoo, N. and Apparao, K. (1992). Laser calorimeter for UV absorption measurement of dielectric thin films. *Appl.Opt.*, 31:6111 – 6116.

Sander, S., Poesl, H., Zuern, W., Spuhler, A., and Braida, M. (1989). The water jet-guided Nd:YAG laser in the treatment of gastroduodenal ulcer with a visible vessel — a randomized, controlled and prospective study. *Endoscopy*, 21:217 – 220.

Sandoghdar, V., Sukenik, C., Haroche, S., and Hinds, E. (1996). Spectroscopy of atoms confined to the single node of a standing wave in a parallel-plate cavity. *Phys. Rev. A*, 53:1919 – 1922.

Sandoghdar, V., Sukenik, C., Hinds, E., and Haroche, S. (1992). Direct measurement of the van der Waals interaction between an atom and its image in a micron-sized cavity. *Phys. Rev. Lett.*, 68:3432 – 3435.

Sappey, A. and Nogar, N. (1994). Diagnostic studies of laser ablation for chemical analysis. In Miller, J., editor, *Laser Ablation*, pages 157–184. Springer-Verlag, Berlin.

Schawlow, A. and Townes, C. (1958). Infrared and optical masers. *Phys.Rev.*, 112:1940 – 1949.

Schmeisser, H. (1974). Growth and mobility effects of gold clusters on rocksalt (100) surfaces studied with the method of quantitative image analysis. Part I: Cluster size distributions. *Thin Solid Films*, 22:83 – 97.

Schmeisser, H. and Harsdorff, M. (1970). Investigation of the nucleation of gold on ultra high vacuum cleaved NaCl single crystals. *Z. Naturforsch. A*, 25:1896 – 1905.

Schmidt, H., Ihlemann, J., Wolff-Rottke, B., Luther, K., and Troe, J. (1998). Ultraviolet laser ablation of polymers: Spot size, pulse duration, and plume attenuation effects explained. *J. Appl. Phys.*, 83:5458 – 5468.

Schmuttenmaer, C., Aeschlimann, M., Elsayed-Ali, H., Miller, R., Mantell, D., Cao, J., and Gao, Y. (1994). Time-resolved two-photon photoemission from Cu(100): Energy dependence of electron relaxation. *Phys.Rev.B*, 50:8957 – 8960.

Schoenlein, R., Fujimoto, J., Eesley, G., and Capehart, T. (1988). Femtosecond studies of image-potential dynamics in metals. *Phys.Rev.Lett.*, 61:2596 – 2599.

Scholl, A., Baumgarten, L., Jacquemin, R., and Eberhardt, W. (1997). Ultrafast spin dynamics of ferromagnetic thin films observed by f-sec spin resolved two photon photoemission. *Phys.Rev.Lett.*, 79:5146 – 5149.

Schönflies, A. (1891). *Kristallsysteme und Kristallstruktur*. Leipzig.

Schreck, E., Hiller, B., and Singh, G. (1993). Calibration of micron-size thermocouples for measurements of surface temperature. *Rev. Sci. Instrum.*, 64:218 – 220.

Schröder, T., Schinke, R., Krohne, R., and Buck, U. (1997). Vibrational dynamics of large clusters from helium atom scattering: Calculations of Ar_{55}. *J.Chem.Phys.*, 106:9067 – 9077.

Schwarzenbach, A., Weber, H., and Balmer, J. (1984). Laser damage test on Balzers thin film coatings. *Applied Optics*, 23:3764 – 3766.

Shah, J. (1996). *Ultrafast Spectroscopy of Semiconductors and Semiconductor Nanostructures*. Springer-Verlag, Berlin, New York.

Shank, C., Yen, R., and Hirlimann, C. (1983). Femtosecond-time-resolved structural dynamics of optically excited silicon. *Phys.Rev.Lett.*, 51:900 – 902.

Shapiro, M. and Brumer, P. (1997). Quantum control of chemical reactions. *J.Chem.Soc.Faraday Trans.*, 93:1263 – 1277.

Shea, M. and Compton, R. (1993). Surface-plasmon ejection of Ag^+ ions from laser irradiation of a roughened silver surface. *Phys.Rev.B*, 47:9967 – 9970.

Shen, Y. (1984). *The Principles of Nonlinear Optics*. Wiley, New York.

Shen, Y. (1986). Surface second harmonic generation. *Ann. Rev. Mater. Sci.*, 16:69 – 86.

Shinojima, H., Yumoto, J., and Uesugi, N. (1992). Size dependence of optical nonlinearity of CdSSe microcrystallites doped in glass. *Appl. Phys. Lett.*, 60:298 – 300.

Shockley, W. (1939). On the surface states associated with a periodic potential. *Phys. Rev.*, 56:317 – 323.

Siegel, J., Ettrich, K., Welsch, E., and Matthias, E. (1997). UV-laser ablation of ductile and brittle metal films. *Appl.Phys.A*, 64:213 – 218.

Siegman, A. and Fauchet, P. (1986). Stimulated Wood's anomalies on laser-illuminated surfaces. *IEEE J.Quantum Electron.*, QE-22:1384 – 1403.

Simon, H., Mitchell, D., and Watson, J. (1975). Second harmonic generation with surface plasmons in alkali metals. *Opt. Commun.*, 13:294 – 298.

Simon, P. and Ihlemann, J. (1996). Machining of submicron structures on metals and semiconductors by ultrashort UV-laser pulses. *Appl.Phys. A*, 63:505 – 508.

Simoneau, P., Boiteaux, S. L., Aranjo, C. B. D., Bloch, D., Leite, J. R., and Ducloy, M. (1986). Doppler-free evanescent wave spectroscopy. *Opt. Commun.*, 59:103 – 106.

Singer, R., Leitner, A., and Aussenegg, F. (1995). Structure analysis and models for optical constants of discontinuous metallic silver films. *J. Opt. Soc. Am. B*, 12:220 – 228.

Sipe, J. (1979). The ATR spectra of multipole surface plasmons. *Surf. Sci.*, 84:75 – 105.

Sipe, J. (1980). Bulk-selvedge coupling theory for the optical properties of surfaces. *Phys. Rev. B*, 22:1589 – 1599.

Sipe, J., Moss, D., and van Driel, H. (1987). Phenomenological theory of optical second- and third-harmonic generation from cubic centrosymmetric crystals. *Phys. Rev. B*, 35:1129 – 1141.

Smilowitz, L., Jia, Q., Yang, X., Li, D., McBranch, D., Buelow, S., and Robinson, J. (1997). Imaging nanometer-thick patterned self-assembled monolayers via second-harmonic generation microscopy. *J.Appl.Phys.*, 81:2051 – 2054.

Sochacki, J., Jaroszewicz, Z., Staronski, L., and Kolodziejczyk, A. (1993). Annular-aperture logarithmic axicon. *J.Opt.Soc.Am. A*, 10:1765 – 1769.

Sokolowski-Tinten, K., Bialkowski, J., and der Linde, D. V. (1995). Ultrafast laser-induced order-disorder transitions in semiconductors. *Phys.Rev.B*, 51:14186 – 14198.

Sommerfeld, A. (1909). Über die Ausbreitung der Wellen in der drahtlosen Telegraphie. *Ann. Phys. Leipz.*, 28:665 – 737.

Somorjai, G. (1994). *Introduction to Surface Chemistry and Catalysis*. Wiley, New York.

Song, K., Heskett, D., Dai, H.-L., Liebsch, A., and Plummer, E. (1988). Dynamical screening at a metal surface probed by second-harmonic generation. *Phys. Rev. Lett.*, 61:1380 – 1383.

Soukissian, P., Bakshi, M., Hurych, Z., and Gentle, T. (1989). Electronic properties of alkali metal/silicon interfaces: A new picture. *Surf.Sci.*, 221:L759 – L768.

Spierings, G., Koutsos, V., Wierenga, H., Prins, M., Abraham, D., and Rasing, T. (1993). Optical second harmonic generation study of interface magnetism. *Surf. Sci.*, 287–288:747 – 749.

Srinivasan, R. (1986). Ablation of polymers and biological tissue by ultraviolet lasers. *Science*, 234:559 –565.

Srinivasan, R. (1993a). Ablation of polyimide (Kapton) films by pulsed (ns) ultraviolet and infrared lasers. *Appl. Phys. A*, 56:417 – 423.

Srinivasan, R. (1993b). Ablation of polymethyl methacrylate films by pulsed (ns) ultraviolet and infrared lasers: A comparative study by ultrafast imaging. *J. Appl. Phys.*, 73:2743 – 2750.

Srinivasan, R., Braren, B., Dreyfus, R., Handel, L., and Seeger, D. (1986). Mechanism of the ultraviolett laser ablation of polymethylmethacrylate at 193 and 248nm: Laser-induced fluorescence analysis, chemical analysis and doping studies. *J.Opt.Soc.Am. B*, 3:785 – 791.

Srinivasan, R., Casey, K., Braren, B., and Yeh, M. (1990). The significance of a fluence threshold for ultraviolet laser ablation and etching of polymers. *J. Appl. Phys.*, 67:1604 – 1606.

Srinivasan, R. and Mayne-Banton, V. (1982). Self-developing photoetching of poly(ethylene terephthalate) films by far-ultraviolet excimer laser radiation. *Appl. Phys. Lett.*, 41:576 – 578.

Stampfli, P. and Bennemann, K. (1994). Time dependence of the laser-induced femtosecond lattice instability of Si and GaAs: Role of longitudinal optical distortions. *Phys.Rev.B*, 49:7299 – 7305.

Steen, W. (1991). *Laser Material Processing.* Springer-Verlag, London.

Stegemann, G., Fortenberry, R., Karaguleff, C., Moshrefzadeh, R., III, W. H., Wyck, N. V., and Sipe, J. (1983). Coherent anti-Stokes Raman scattering in thin-film dielectric waveguides. *Opt. Lett.*, 8:295 – 297.

Steiger, R. (1971). Studies of oriented monolayers on solid surfaces by ellipsometry. *Helv. Chim. Acta*, 54:2645 – 2658.

Steinmann, W. (1989). Spectroscopy of image-potential states by two-photon photoemission. *Appl. Phys. A*, 49:365 – 377.

Steinmueller-Nethl, D., Höpfel, R., Gornik, E., Leitner, A., and Aussenegg, F. (1992). Femtosecond relaxation of localized plasma excitations in Ag islands. *Phys.Rev.Lett.*, 68:389 – 392.

Stipe, B., Rezaei, M., and Ho, W. (1998). Inducing and viewing the rotational motion of a single molecule. *Science*, 279:1907 – 1909.

Stoneham, A. (1994). Radiation effects in insulators. *Nucl. Instrum. Methods Phys. Res. B*, 91:1 – 11.

Strong, L. and Whitesides, G. (1988). Structures of self-assembled monolayer films of organosulfur compounds adsorbed on gold single crystals: Electron diffraction studies. *Langmuir*, 4:546 – 558.

Stuart, B., Feit, M., Rubenchik, A., Shore, B., and Perry, M. (1995). Laser-induced damage in dielectrics with nanosecond to subpicosecond pulses. *Phys.Rev.Lett.*, 74:2248 – 2251.

Stulen, R. and Knotek, M., editors (1986). *Desorption Induced by Electronic Transitions — DIET III.* Springer-Verlag, Berlin.

Suarez, C., Bron, W., and Juhasz, T. (1995). Dynamics and transport of electronic carriers in thin gold films. *Phys.Rev.Lett.*, 75:4536 – 4539.

Sukenik, C., Boshier, M., Cho, D., Sandoghdar, V., and Hinds, E. (1993). Measurement of the Casimir-Polder force. *Phys. Rev. Lett.*, 70:560 – 563.

Sun, C.-K., Vallee, F., Acioli, L., Ippen, E., and Fujimoto, J. (1994). Femtosecond-tunable measurement of electron thermalization in gold. *Phys.Rev.B*, 50:15337 – 15348.

Swalen, J., Gordon, J., Philpott, M., Brillante, A., Pockrand, I., and Santo, R. (1980). Plasmon surface polariton dispersion by direct optical observation. *Am. J. Phys.*, 48:669 – 672.

Szymonski, M. and Postawa, Z., editors (1995). *Desorption Induced by Electronic Transitions — DIET VI.* Nucl.Instrum.Methods Phys.Res. 101.

Tajima, H., Haraguchi, M., and Fukui, M. (1995). Surface plasmon polariton radiation from silver films on fluoride films and surface roughness parameters of those silver films. *Surf. Sci.*, 323:282 – 287.

Tam, A., Leung, W., Zapka, W., and Ziemlich, W. (1992). Laser-cleaning techniques for removal of surface particulates. *J. Appl. Phys.*, 71:3515 – 3523.

Tamm, I. (1932). Über eine mögliche Art der Elektronenbindung an Kristalloberflächen. *Z. Phys.*, 76:849 – 850.

Tanimura, K. and Kanasaki, J. (1998). Laser-induced bond breaking and structural changes on Si(111)-7×7 surfaces. *Appl.Surf.Sci.*, 127–129:33 – 39.

Tillman, N., Ulman, A., and Penner, T. (1989). Formation of multilayers by self-assembly. *Langmuir*, 5:101 – 111.

Toennies, J. and Winkelmann, K. (1977). Theoretical studies of highly expanded free jets: Influence of qunatum effects and a realistic intermolecular potential. *J. Chem. Phys.*, 66:3965 – 3979.

Tolk, N., Traum, M., Tully, J., and Madey, T., editors (1983). *Desorption Induced by Electronic Transitions — DIET I.* Springer-Verlag, Berlin.

Tom, H., Aumiller, G., and Brito-Cruz, C. (1988). Time-resolved study of laser-induced disorder of Si surfaces. *Phys. Rev. Lett.*, 60:1438 – 1441.

Tom, H., Mate, C., Zhu, X., Crowell, J., Shen, Y., and Somorjai, G. (1986). Studies of alkali adsorption on Rh(111) using optical second-harmonic generation. *Surf. Sci.*, 172:466 – 476.

Tompkin, H. (1993). *A User's Guide to Ellipsometry.* Academic Press, London.

Tong, W. and Williams, R. (1994). Kinetics of surface growth: Phenomenology, scaling and mechanisms of smoothening and roughening. *Annu. Rev. Phys. Chem.*, 45:401 – 438.

Touloukian, Y. and Ho, C., editors (1970 – 1973). *Thermophysical Properties of Matter*, volume 1, 3 und 10. IFI/Plenum, New York.

Trautman, J., Macklin, J., Brus, L., and Betzig, E. (1994). Near-field spectroscopy of single molecules at room temperature. *Nature*, 369:40 – 42.

Tredgold, R. (1994). *Order in Thin Organic Films*. Cambridge University Press, Cambridge.

Tsang, T. (1995). Optical third-harmonic generation at interfaces. *Phys. Rev. A*, 52:4116 – 4125.

Tsuei, K.-D., Plummer, E., Liebsch, A., Pehlke, E., Kempa, K., and Bakshi, P. (1991). The normal modes at the surface of simple metals. *Surf. Sci.*, 247:302 – 326.

Tully, J. (1981). Dynamics of gas-surface interactions: Thermal desorption of Ar and Xe from platinum. *Surf.Sci.*, 111:461 – 478.

Ulman, A. (1991). *An Introduction to Ultrathin Organic Films*. Academic Press, New York.

Ulman, A. and Tillman, N. (1989). Self-assembling double layers on gold surfaces: The merging of two chemistries. *Langmuir*, 5:1420 – 1422.

Urbach, L., Percival, K., Hicks, J., Plummer, E., and Dai, H.-L. (1992). Resonant surface second-harmonic generation: Surface states on Ag(110). *Phys. Rev. B*, 45:3769 – 3772.

van Driel, H., Sipe, J., Hache, A., and Atanasov, R. (1997). Coherence control of photocurrents in semiconductors. *Phys.Status Solidi B*, 204:3 – 8.

van Driel, H., Sipe, J., and Young, J. (1982). Laser-induced periodic surface structure on solids: A universal phenomenon. *Phys. Rev. Lett.*, 49:1955 – 1958.

van Gemert, M., Welch, A., Pickering, J., Tan, O., and Gijsbers, G. (1995). Wavelengths for laser treatment of port wine stains and telangiectasia. *Lasers Surg.Med.*, 16:147 – 155.

Varel, H., Ashkenasi, D., Rosenfeld, A., Wähmer, M., and Campbell, E. (1997). Micromachining of quartz with ultrashort laser pulses. *Appl.Phys.A*, 65:367 – 373.

Venables, J. (1994). Atomic processes in crystal growth. *Surf. Sci.*, 299–300:798 – 817.

Venkatesan, T. (1994). Pulsed-laser deposition of high-temperature superconducting thin films. In Miller, J., editor, *Laser Ablation: Principles and Applications*, pages 85 – 106. Springer-Verlag, Berlin.

Ventzek, P., Gilgenbach, R., Heffelfinger, D., and Sell, J. (1991). Laser-beam deflection measurements and modeling of pulsed laser ablation rate and near-surface plume densities in vacuum. *J.Appl.Phys.*, 70:587 – 593.

Verbiest, T., van Elshocht, S., Kauranen, M., Hellemans, L., Snauwaert, J., Nuckolls, C., Katz, T., and Persoons, A. (1998). Strong enhancement of nonlinear optical properties through supramolecular chirality. *Science*, 282:913 – 915.

Viereck, J., Stietz, F., Stuke, M., Wenzel, T., and Träger, F. (1997a). The role of surface defects in laser-induced thermal desorption from metal surfaces. *Surface Science*, 383:L749 – L754.

Viereck, J., Stuke, M., and Träger, F. (1997b). Laser-induced desorption of Na-dimers. *Appl. Phys. A*, 64:149 – 153.

Viswanathan, R., Burgess Jr, D., Stair, P., and Weitz, E. (1982). Summary abstract: Laser flash desorption of CO from clean copper surfaces. *J. Vac. Sci. Technol.*, 20:605 – 606.

Vogel, A., Schweiger, P., Frieser, A., Asiyo, M., and Birngruber, R. (1990). Intraocular Nd:YAG laser surgery: Light-tissue interaction, damage range, and reduction of collateral effects. *IEEE J. Quantum Electron.*, 26:2240 – 2260.

von der Linde, D. and Rzazewski, K. (1996). High order optical harmonic generation from solid surfaces. *Appl.Phys.B*, 63:499 – 506.

von der Linde, D., Sokolowski-Tinten, K., and Bialkowski, J. (1997). Laser-solid interaction in the femtosecond time regime. *Appl.Surf.Sci.*, 109–110:1 – 10.

Waldeck, D., Alivisatos, A., and Harris, C. (1985). Nonradiative damping of molecular electronic excited states by metal surfaces. *Surf. Sci.*, 158:103 – 125.

Walls, J. and Smith, R. (1994). *Surface Science Techniques*. Pergamon Press, Oxford.

Wang, C. and Duminski, A. (1968). Second-harmonic generation of light at the boundary of alkali halides and glasses. *Phys. Rev. Lett.*, 20:668 – 671.

Wang, W., Feldstein, M., and Scherer, N. (1996). Observation of coherent multiple scattering of surface plasmon polaritons on Ag and Au surfaces. *Chem.Phys.Lett.*, 262:573 – 582.

Wang, Y. and Herron, N. (1991). Nanometer-sized semiconductor clusters: Materials synthesis, quantum size effects, and photophysical properties. *J. Phys. Chem.*, 95:525 – 532.

Weast, R., editor (1982). *CRC Handbook of Chemistry and Physics.* CRC Press, Boca Raton.

Weber, W. and Eagen, C. (1979). Energy transfer from an excited dye molecule to the surface plasmons of an adjacent metal. *Opt. Lett.*, 4:236 – 238.

Wei, Z., Verhoef, R., Asscher, M., Farbman, I., and Ben-Shaul, A. (1996). Effect of lateral repulsion on desorption and diffusion kinetics: SHG experiments and MC simulations. *Appl.Surf.Sci.*, 106:80 – 89.

Weis, R., Müller, B., and Fromherz, P. (1996). Neuron adhesion on a silicon chip probed by an array of field-effect transistors. *Phys.Rev.Lett.*, 76:327 – 330.

Welsch, E. and Ristau, D. (1995). Photothermal measurements on optical thin films. *Appl.Opt.*, 34:7239 – 7253.

Whitmore, P., Robota, H., and Harris, C. (1982a). Electronic energy transfer from pyrazine to a silver(111) surface between 10 and 400 Å. *J. Chem. Phys.*, 76:740 – 741.

Whitmore, P., Robota, H., and Harris, C. (1982b). Mechanisms for electronic energy transfer between molecules and metal surfaces: A comparison of silver and nickel. *J. Chem. Phys.*, 77:1560 – 1568.

Wierenga, H., Jong, W. D., Prins, M., Rasing, T., Vollmer, R., Kirilyuk, A., Schwabe, H., and Kirschner, J. (1995). Interface magnetism and possible quantum well oscillations in ultrathin Co/Cu films observed by magnetization induced second harmonic generation. *Phys.Rev.Lett.*, 74:1462 – 1465.

Wijekoon, W., Ho, Z., and Hetherington, W. (1987). Ethylene adsorption on ZnO: CARS spectroscopy with optical waveguides. *J. Chem. Phys.*, 86:4384 – 4390.

Williamson, J., Cao, J., Ihee, H., Frey, H., and Zewail, A. (1997). Clocking transient chemical changes by ultrafast electron diffraction. *Nature*, 386:159 – 161.

Woelker, M., Bein, B., Pelzl, J., and Walther, H. (1991). Contributions to the technique and interpretation of the photothermal beam deflection experiment. *J.Appl.Phys.*, 70:603 – 610.

Wokaun, A. (1985). Surface enhancement of optical fields. Mechanism and applications. *Mol. Phys.*, 56:1 – 33.

Wokaun, A., Bergman, J., Heritage, J., Glass, A., Liao, P., and Olson, D. (1981). Surface second-harmonic generation from metal island films and microlithographic structures. *Phys. Rev. B*, 24:849 – 856.

Wokaun, A., Lutz, H.-P., King, A., Wild, U., and Ernst, R. (1983). Energy transfer in surface enhanced luminescence. *J. Chem. Phys.*, 79:509 – 514.

Wolf, M. (1997). Femtosecond dynamics of electronic excitations at metal surfaces. *Surf.Sci.*, 377 – 379:343 – 349.

Wollensak, J., editor (1988). *Laser in der Ophthalmologie*. F. Enke Verlag, Stuttgart.

Wood, R. (1902). On a remarkable case of uneven distribution of light in a diffraction grating spectrum. *Proc.Phys.Soc.(London)*, 18:269 – 275.

Wood, R. (1909). Selective reflection of monochromatic light by mercury vapour. *Phil. Mag.*, 18:187 – 193.

Wood, R. (1986). *Laser Damage in Optical Materials*. Hilger, Boston.

Wright, O. and Kawashima, K. (1992). Coherent phonon detection from ultrafast surface vibrations. *Phys.Rev.Lett.*, 69:1668 – 1671.

Wylie, J. and Sipe, J. (1984). Quantum electrodynamics near an interface. *Phys. Rev. A*, 30:1185 – 1193.

Wylie, J. and Sipe, J. (1985). Quantum electrodynamics near an interface. II. *Phys. Rev. A*, 32:2030 – 2043.

Xiao, X.-D., Xie, Y., and Shen, Y. (1992). Surface diffusion probed by linear optical diffraction. *Surf. Sci.*, 271:295 – 298.

Xiao, X.-D., Xie, Y., and Shen, Y. (1993). Coverage dependence of anisotropic surface diffusion: CO/Ni(110). *Phys. Rev. B*, 48:17452 – 17462.

Xie, X. and Dunn, R. (1994). Probing single molecule dynamics. *Science*, 265:361 – 363.

Xie, X. and Trautman, J. (1998). Optical studies of single molecules at room temperature. *Annu.Rev.Phys.Chem.*, 49:441 – 480.

Xiong, T., Leung, P., and George, T. (1995). Modeling of decay rates for molecules at an island surface. *J.Chin.Chem.Soc.*, 42:249 – 254.

Yang, M. and Reilly, J. (1989). Comparison of two methods of UV laser-induced surface ionization. *Opt. Commun.*, 71:193 – 196.

Yang, M. and Reilly, J. (1990). Kinetic energy distributions of aniline molecules and cations following their UV laser-induced desorption from a metal surface. *J. Phys. Chem.*, 94:6299 – 6305.

Yannouleas, C. (1998). Microscopic description of the surface dipole plasmon in large Na_n clusters. *Phys.Rev.B*, 58:6748 – 6751.

Yannouleas, C., Vigezzi, E., and Broglia, R. (1993). Evolution of the optical properties of alkali-metal microclusters towards the bulk: The matrix random-phase approximation description. *Phys. Rev. B*, 47:9849 – 9861.

Yariv, A. (1985). *Optical Electronics*. Holt-Saunders, New York.

Yariv, A. (1989). *Quantum Electronics*. Wiley, New York.

Yariv, A. and Yeh, P. (1984). *Optical Waves in Crystals*. Wiley, New York.

Yates, Jr, J. (1998). *Experimental Innovations in Surface Science: A Guide to Practical Laboratory Methods and Instruments*. Springer-Verlag, New York.

Ye, P. and Shen, Y. (1983). Local-field effect on linear and nonlinear optical properties of adsorbed molecules. *Phys.Rev.B*, 28:4288 – 4294.

Ye, Q., Fang, J., and Sun, L. (1997). Surface-enhanced Raman scattering from functionalized self-assembled monolayers. 2. Distance dependence of enhanced Raman scattering from an azobenzene terminal group. *J.Phys.Chem.B*, 101:8221 – 8224.

Yee, H., Byers, J., and Hicks, J. (1994). A nonlinear optical study of chiral surfaces. *Proc.SPIE*, 2125:119 – 131.

Ying, Z., Wang, J., Andronica, G., Yao, J.-Q., and Plummer, E. (1993). Azimuthal and incident angle dependences in the second-harmonic generation from aluminum. *J. Vac. Sci. Technol. A*, 11:2255 – 2259.

Zangwill, A. (1988). *Physics at Surfaces*. Cambridge University Press, Cambridge.

Zaret, M., Breinin, G., Schmidt, H., Ripps, H., Siegel, I., and Solon, L. (1961). Ocular lesions produced by an optical maser (laser). *Science*, 134:1525 – 1525.

Zenobi, R., Hahn, J., and Zare, R. (1988). Surface temperature measurement of dielectric materials heated by pulsed laser radiation. *Chem. Phys. Lett.*, 150:361 – 365.

Zewail, A. (1994). *Femtochemistry: Ultrafast Dynamics of the Chemical Bond.* World Scientific, Singapore.

Zhang, J., Sugioka, K., Wada, S., Tashiro, H., Toyoda, K., and Midorikawa, K. (1998). Precise microfabrication of wide band gap semiconductors SiC and GaN by VUV-UV multiwavelength laser ablation. *Appl.Surf.Sci.*, 127-129:793 – 799.

Zhou, C., Deshpande, M., Reed, M., Jones II, L., and Tour, J. (1997). Nanoscale metal/self-assembled monolayer/metal heterostructures. *Appl. Phys. Lett.*, 71:611 – 613.

Zhu, T., Yu, H., Wang, J., Wang, Y., Cai, S., and Liu, Z. (1997). Two-dimensional surface enhanced Raman mapping of differently prepared gold substrates with an azobenzene self-assembled monolayer. *Chem.Phys.Lett.*, 265:334 – 340.

Zhu, X., Rasing, T., and Shen, Y. (1988). Surface diffusion of CO on Ni(111) studied by diffraction of optical second-harmonic generation off a monolayer grating. *Phys. Rev. Lett.*, 61:2883 – 2885.

Zhu, X., Rasing, T., and Shen, Y. (1989). Laser-induced thermal desorption of CO on Ni(111): Determination of pre-exponential factor and heat of desorption. *Chem.Phys.Lett.*, 155:459 – 462.

Zhu, X., Shen, Y., and Carr, R. (1985). Correlation between thermal desorption spectroscopy and optical second harmonic generation for monolayer surface coverages. *Surf. Sci.*, 163:114 – 120.

Zhu, X.-Y. (1994). Surface photochemistry. *Annu. Rev. Phys. Chem.*, 45:113 – 144.

Zimmermann, F. and Ho, W. (1994). Velocity distributions of photochemically desorbed molecules. *J. Chem. Phys.*, 100:7700 – 7706.

Zimmermann, F. and Ho, W. (1995). State resolved studies of photochemical dynamics at surfaces. *Surf. Sci. Rep.*, 22:127 – 247.

Zink, J., Reif, J., and Matthias, E. (1992). Water adsorption on (111) surfaces of BaF_2 and CaF_2. *Phys.Rev.Lett.*, 68:3595 – 3598.

Zinke-Allmang, M., Feldman, L., and Grabow, M. (1992). Clustering on surfaces. *Surf. Sci. Rep.*, 16:377 – 463.

Zweng, H. (1971). Lasers in opthalmology. In Wolbarsht, M., editor, *Laser Applications in Medicine and Biology*, pages 239 – 253. Plenum Press, New York.

Index

ablation dynamics, 234
ablation temperatures, 226
ablation threshold, 220, 240
ablation, photochemical, 230
ablation, transient, 232
absorber saturation, 238
absorption coefficient, 186
absorption depth, 194
absorption spectrum, 89
acceleration via ablation, 251
acoustic surface phonons, 174
acousto-optic modulator, 16
activation energy, desorption, 54, 73
activation energy, diffusion, 61
adhesion force, 251
adiabatic potential, 82
adsorbate diffusion, 60
adsorbate modes, 84
adsorbate orientation, IR, 87
adsorbate symmetries, 87, 133
adsorbate–adsorbate relaxation, 151
adsorbate–substrate relaxation, 151
adsorption rate, 29
AES, 50
angular dependence of reflected light, 116
anharmonicity, 83
annular illumination, 245
anomalous absorption, 190
anti-Stokes line, 115
Antoniewicz model, 66
Ar^+ laser, 14
atomic force microscope, 47
ATR, 27, 155
Auger intensity, 130
Autler-Townes effect, 92
autocorrelator, 152
axicon, 210

ballistic electrons, 157
bandwidth limited pulse, 154
bandwidth, 12

beam waist, 3, 11
Beers law, 85, 237
Born–Oppenheimer approximation, 90
Bose–Einstein distribution, 7
Brewster angle, 204
bridge position, 84
bulk susceptibility, 122
buried interface, 136, 141

C_{60}, 157
carrier density, 216
carrier diffusion, 218
carrier generation rate, 216
carrier relaxation rate, 216
CARS, 142
Casimir–Polder potential, 103
cavitation bubble, 267
cavity quantum electrodynamics, 104
cavity ring-down spectroscopy, 97
charge transfer mediated dissociation, 179
chemisorption, 24
chirality, 134
chirp, 154
chromophore, 222
classical dipole lifetime, 104
classical frequency shift, 102
classical size effect, 166
cloud attenuation, 238
CLSM, 114
cluster size distribution, 163
cluster–cluster interactions, 169
CO_2 laser, 13
coherence length, 1
coherence, spatial, 2
coherence, temporal, 1
coherent control, 160, 175
coherent LO phonons, 175
cold melting, 175

colloid generation, 249
colloidal suspensions, 161
color center, 188
confocal parameter, 3
conical defects, 241
conical structures, 265
contact angle, 31
convection, 191
cooling rate, 207
cornea ablation, 268
corrugation depth, 116
CORSTM, 182
Coulomb well, 98
critical absorbed energy density, 220
critical angle, 94
crossing angle, 143
crystal microbalance, 223
CVD reaction rate, 261

damage threshold, 211
dangling bonds, 21
de Broglie wavelength, 39
Debye temperature, 192
Debye–Waller exponent, 201
Debye–Waller factor, 40
dephasing time constant, 171
dephasing, 86
depth of absorption, 86
desorption barrier, 58
desorption rate, 54
DFWM, 142, 158
dielectric masks, 242
DIET, 67
diffuse scattering, 116
diffusion rate, 61
diffusion, anisotropic, 75
DIMET, 181
dipole domains, 254
dipole surface potential, 106
dipole–dipole coupling, 102, 173
discontinuous metal films, 161
dissociative adsorption, 24
divergence, 2
Doppler broadening, 91
Drude damping, 166
Drude equation, 110
dynamic Stark splitting, 92

EELS, 87

effective absorption coefficient, 220
Einstein coefficient, 9
electromagnetic field induced desorption, 67
electron density distribution, 124
electron–hole pair excitation, 106
electron localization, 159
electron microscopy, 46
electron plasma, 215
electron spill-out, 123
electron–electron collisions, 169
electron–phonon coupling, 157, 215, 248
electrostatic image potential, 108
Eley–Rideal mechanism, 180
ellipsometry, 96
emissivity, 191
energy density, 5
energy shift, 112
erodible mask, 271
etch curves, 223
etch depth, 222
evanescent wave detector, 68
evanescent wave, 93, 104, 145
evaporation front, 197
excimer laser cleaning, 252
excimer laser, 17
excimer photorefractive keratectomy, 269
excimer laser nitriding, 208
exciton dynamics, 161
explosive boiling, 265
extinction spectrum, 89
extinction, 185
extrinsic size effect, 166

Fabry–Perot resonator, 11
far field, 3
femtosecond excitation, 151
femtosecond laser-induced desorption, 180
Fermi energy, 23
Fermi wavevector, 123
field enhancement, 178
field ion microscope, 46
field strength distribution, 11
fluence, 5
focus depth, 244
four-wave mixing, 142
Fouriers law, 192

fourth-harmonic generation, 148
fractal dimension, 117
fragmentation rate, 221
Franck–Condon factor, 89
Frank–Van der Merwe growth, 30
Fresnel factors, 125, 185
Fresnel lens, 242
Friedel oscillations, 123
frustrated rotation, 84
frustrated translation, 84
FTIR, 87
FTMS, 62
fullerenes, 157

Gaussian beam, 193
Gaussian distribution, 11
Green's function, 102
growth modes, 30

Hamaker constant, 251
HAS, 41
heat conduction, 192
heat radiation, 191
height–height correlation, 116
helium atomic beam, 41
higher harmonics, 211
holographic grating, 143
homogeneous broadening, 87
hopping frequency, diffusion, 61
hot electron diffusion, 248
hot electron dynamics, 159
hot electron induced desorption, 63
hot electron mediated photodissociation, 180
human eye, 268
hump formation, 258
hydrodynamic dielectric function, 111
hydrodynamic expansion time, 273
hyperpolarizability, 137
Hönl–London factor, 89

image dipole, 102
image potential spectroscopy, 158
image potential state, 23, 98
incubation, 223, 234, 242
inhomogeneous broadening, 87
injection seeding, 16
intensity autocorrelation, 153
interband transitions, 138

interferometric second order autocorrelation, 153
inversion symmetry, 122
inversion, 10
ionic radius, 163
IRAS, 87
irradiance, 5
ISTS, 174

jellium model, 139

k_1, 243
k_2, 244
Knudsen layer, 212
Knudsen number, 44

Lamb shift, 103
Lambert's scatterer, 117
Langmuir, 30
Langmuir–Blodgett method, 34
Langmuir–Hinschelwood mechanism, 180
laser beam homogenization, 210
laser beam intensity, 4
laser conditioning, 211
laser damage, 188, 210, 246
laser efficiency, 13
laser LIGA, 259
laser liquid-phase epitaxy, 209
laser medicine, 268
laser restauration, 252
laser sputtering, 265
laser-assisted microstructuring, 215
laser-induced bond breaking, 6
laser-induced fluorescence, 88
laser-induced grating, 73
laser-induced periodical structures, 254
laser-induced phase transition, 175
laser-induced pressure, 208
laser-induced temperature increase, 194
laser-induced thermal expansion, 251
lattice gas formula, 42
LCVD, 203, 260
LEED, 40
light microscopy, 46
Lindemann criterion, 216
linear absorption coefficient, 189
linear reflectivity change, 130
linewidth, 86
LITD, 53

local approximation, 107
local field enhancement, 128, 145
local field factor, 127
localized plasmon, 155
longitudinal modes, 10
Lorentz force, 135
Lorentz profile, 86

macroscopic ablation threshold, 239
magnetic domain control, 203
MALDI, 62
masterplot transformation depth, 206
material ejection, 235
material hardening, 203
materials diffusion, 74, 191
Maxwell–Boltzmann velocity distribution, 70, 229
melt hardening, 209
melt quenching, 209
melt trough, 206
melting temperature, 216
melting zone, 197
metal cluster generation, 249
metallic nanoparticles, 161
method of finite differences, 197
MGR model, 65
mica, 248
microscopic ablation threshold, 239
microstructuring, 246
Mie theory, 162
Miller's indices, 21
mirage effect, 240
mode cannibalism, 14
mode coupling, 18
mode density, 10
mode number, 11
mode radius, 11
mode selection, 11
mode spacing, 10
molecular architecture, 34
molecular hyperpolarizability, 137
monochromasy, 1
MSHG, 134
multiple pulse ablation, 241
multipole surface plasmons, 111
mylar, 219

nanocapacitor, 33
nanoantenna, 161
nanocrystallites, 161
nanometer scaled structures, 33
natural lifetime, 9
natural linewidth, 86
Nd:YAG laser, 16
near field, 3
NEXAFS, 50
nonlinear absorption coefficient, 189
nonlinear optical film, 38
nonlinear polarizability, 124
nonlinear surface susceptibilities, 125
nonlocal interaction, 106
nonlocal optical response, 110
nonlocalized plasmon, 155
normal mode description, 84
numerical aperture, 243

ophthalmology, 268
optical breakdown, 267
optical pumping, 83
order of desorption, 56
organic films, 34
Ostwald ripening, 32

Pauli principle, 24
periodic structures, 255
PET, 231
phase conjugation, 143
phase explosion, 265
phase matching, 94
phase matching, DFWM, 142
phase matching, SHG, 123
phase shift mask, 245
phase transformation, 204
phonon thermalization time, 169
phonon–phonon coupling, 215
photo-acoustic deflection, 239
photochemical ablation, 233
photochemical CVD, 261
photochemical model, 236
photochemistry, 177
photochemistry, surface enhanced, 118
photocoagulation, 267
photoelectron spectroscopy, 48
photolysis, 72
photomechanical ablation, 273
photon bunching, 7
photon statistics, 7
photothermal ablation model, 221, 230

INDEX

physisorption, 24
PI, 231
picosecond RHEED, 183
piezoelectric contraction, 131
PLA, 215
plasma pressure, 209
plasmon frequency, 188
plasmon lifetime, 165
PLD, 263
plume absorption, 238
PMMA, 220, 231
Poisson distribution, 8
polarization grating, 158
polarization, macroscopic, 120
polyimide, 220
PRK, 269
projection photolithography, 243
PTFE, 220
pulse generation, 5
pulsed nozzle, 44
pump/probe experiment, 154

Q-switch, 15
quantum well oscillations, 137

radiation amplification, 10
radiation damping, 106
radiation hardening, 242
Raman scattering, 114
rate equation, 8
Rayleigh criterion, 243
Rayleigh range, 3
Rayleigh scattering, 114
rearside ablation, 246
reconstruction, 21
reflectance, 185
reflectron, 62
refractive power, 268
relaxation, 21
REMPI, 92
resonant heating, 67
RETPI, 71
ripple formation, 256
rotational eigenenergies, 82
rotational temperature, 57, 226

SAM, 34
SAP, 178
saturable absorber, 19

SAW, 252
scanning electron microscope, 46
scanning tunneling microscope, 47
Schönflies notation, 131
selective desorption, 254
selective reflection spectroscopy, 96
self-assembled molecules, 34
self-focusing, 209
self-organization, 34
self-phase modulation, 20
selvedge region, 111
semiconductor treatment, 176
SERS, 115
SEXAFS, 49
SFG, 139, 172
SH adsorbate orientation, 133
SH coverage dependence, 126
SH microscopy, 126
SH resonance enhancement, 137
SH signal intensity, 124
SH spectroscopy, 137
SH surface symmetry, 131
SH, anisotropic, 131
SH, magnetic field induced, 134
shock wave, 273
Shockley state, 22
silicon reconstruction, 131
size parameter A, 166
SMOKE, 134
SNOM, 47
SODs, 32
Sommerfeld's radiation theory, 102
spallation, 247, 273
spatial dispersion, 123
spectral energy density, 9
spin dynamics, 137
spontaneous emission, 9
SPS, 218
sputtering, 30
state-selective scattering, 179
Stefan–Boltzmann law, 191
stereoreactivity, 180
sticking coefficient, 29
stimulated absorption, 9
stimulated emission, 9
stimulated Wood's anomaly, 254
Stokes line, 115
Stranski–Krastanoff growth, 30
stress confinement, 276
stretching modes, 84

subsurface superheating, 265
surface alloying, 208
surface CARS, 146
surface charge, 110
surface circular dichroism, 136
surface decoating, 250
surface dipoles, 253
surface enhanced photochemistry, 118
surface magnetization, 134
surface melting, 183, 216
surface nitriding, 208
surface phonon, 25
surface plasmon frequency, 190
surface plasmon mirror, 29
surface plasmon polariton, 105
surface plasmon resonance, 161
surface plasmon, 26, 110, 128
surface recombination, 161
surface roughening, 247
surface roughness, 106, 115, 256
surface shock wave, 252
surface state, 21, 138
surface structuring, 208
surface susceptibility, 122
surface temperature, 200
surface tension, 31
susceptibility, 120
symmetry classes, 131

T_1, 171
T_2, 87, 171
T-matrix theory, 163
Tamm state, 22
thermal conductivity, 192
thermal diffusion length, 194
thermal diffusion, 78, 220
thermal diffusivity, 194
thermal velocity, 226
thermochemical CVD, 261
THG, 142
thin film switch, 158
thin metallic films, 155
tilt angle, 88
time-resolved TPPE, 159
titanium sapphire laser, 20
total internal reflection, 95
TPLIF, 90
tracer diffusion coefficient, 78
transient grating scattering, 143

transient infrared spectra, 171
transient reflectivity, 155
transient ripples, 258
transient thermal grating, 78
transverse correlation length, 116
tunneling rate, 64
two-level scheme, 9
two-photon absorption, 188
two-photon photoemission, 23, 100
two-photon transition, 91
two-temperature model, 156

ultrashort pulses, 151
UPS, 48

vacuum energy, 23
van der Waals coefficient, 251
van der Waals potential, 102
vibrational deactivation, 179
vibrational eigenenergies, 81
vibrational predesorption, 67
vibrational relaxation rate, 171
vibrational temperature, 226
Vickers hardness, 205
Voigt profile, 87
Volmer–Weber growth, 30

waveguide Fresnel coefficient, 112
waveguide, 147
Wiedemann–Franz law, 192
Wigner–Seitz radius, 123
work function, 23

X-ray lithography, 213
XPS, 48